# Advances in Intelligent Systems and Computing

## Volume 1052

The series "Advances in Intelligent Systems and Computing" contains publications on theory, applications, and design methods of Intelligent Systems and Intelligent Computing. Virtually all disciplines such as engineering, natural sciences, computer and information science, ICT, economics, business, e-commerce, environment, healthcare, life science are covered. The list of topics spans all the areas of modern intelligent systems and computing such as: computational intelligence, soft computing including neural networks, fuzzy systems, evolutionary computing and the fusion of these paradigms, social intelligence, ambient intelligence, computational neuroscience, artificial life, virtual worlds and society, cognitive science and systems, Perception and Vision, DNA and immune based systems, self-organizing and adaptive systems, e-Learning and teaching, human-centered and human-centric computing, recommender systems, intelligent control, robotics and mechatronics including human-machine teaming, knowledge-based paradigms, learning paradigms, machine ethics, intelligent data analysis, knowledge management, intelligent agents, intelligent decision making and support, intelligent network security, trust management, interactive entertainment, Web intelligence and multimedia.

The publications within "Advances in Intelligent Systems and Computing" are primarily proceedings of important conferences, symposia and congresses. They cover significant recent developments in the field, both of a foundational and applicable character. An important characteristic feature of the series is the short publication time and world-wide distribution. This permits a rapid and broad dissemination of research results.

**\*\* Indexing: The books of this series are submitted to ISI Proceedings, EI-Compendex, DBLP, SCOPUS, Google Scholar and Springerlink \*\***

More information about this series at http://www.springer.com/series/11156

Zofia Wilimowska · Leszek Borzemski ·
Jerzy Świątek
Editors

# Information Systems Architecture and Technology: Proceedings of 40th Anniversary International Conference on Information Systems Architecture and Technology – ISAT 2019

Part III

Springer

*Editors*
Zofia Wilimowska
University of Applied Sciences in Nysa
Nysa, Poland

Jerzy Świątek
Faculty of Computer Science
and Management
Wrocław University of Science
and Technology
Wrocław, Poland

Leszek Borzemski
Faculty of Computer Science
and Management
Wrocław University of Science
and Technology
Wrocław, Poland

ISSN 2194-5357          ISSN 2194-5365  (electronic)
Advances in Intelligent Systems and Computing
ISBN 978-3-030-30442-3        ISBN 978-3-030-30443-0  (eBook)
https://doi.org/10.1007/978-3-030-30443-0

This Springer imprint is published by the registered company Springer Nature Switzerland AG
The registered company address is: Gewerbestrasse 11, 6330 Cham, Switzerland

# Preface

We are pleased to present before you the proceedings of the 2019 40th Anniversary International Conference Information Systems Architecture and Technology (ISAT), or ISAT 2019 for short, held on September 15–17, 2019 in Wrocław, Poland. The conference was organized by the Department of Computer Science, Faculty of Computer Science and Management, Wrocław University of Science and Technology, Poland, and the University of Applied Sciences in Nysa, Poland.

The International Conference on Information Systems Architecture and Technology has been organized by the Wrocław University of Science and Technology from the eighties of the last century. Most of the events took place in Szklarska Poręba and Karpacz—charming small towns in the Karkonosze Mountains, Lower Silesia in the southwestern part of Poland. This year 2019, we celebrate the 40th anniversary of the conference in Wrocław—the capital of Lower Silesia, a city with a thousand-year history. A beautiful and modern city that is developing dynamically and is a meeting point for people from all over the world. It is worth noting that Wrocław is currently one of the most important centers for the development of modern software and information systems in Poland.

The past four decades have also been a period of dynamic development of computer science, which we can recall when reviewing conference materials from these years—their shape and content were always created with current achievements of national and international IT.

The purpose of the ISAT is to discuss a state-of-art of information systems concepts and applications as well as architectures and technologies supporting contemporary information systems. The aim is also to consider an impact of knowledge, information, computing and communication technologies on managing of the organization scope of functionality as well as on enterprise information systems design, implementation, and maintenance processes taking into account various methodological, technological, and technical aspects. It is also devoted to information systems concepts and applications supporting the exchange of goods and services by using different business models and exploiting opportunities offered by Internet-based electronic business and commerce solutions.

ISAT is a forum for specific disciplinary research, as well as on multi-disciplinary studies to present original contributions and to discuss different subjects of today's information systems planning, designing, development, and implementation.

The event is addressed to the scientific community, people involved in a variety of topics related to information, management, computer and communication systems, and people involved in the development of business information systems and business computer applications. ISAT is also devoted as a forum for the presentation of scientific contributions prepared by MSc. and Ph.D. students. Business, Commercial, and Industry participants are welcome.

This year, we received 141 papers from 20 countries. The papers included in the three proceedings volumes have been subject to a thoroughgoing review process by highly qualified peer reviewers. The final acceptance rate was 60%. Program Chairs selected 85 best papers for oral presentation and publication in the 40th International Conference Information Systems Architecture and Technology 2019 proceedings.

The papers have been clustered into three volumes:

**Part I**—discoursing about essential topics of information technology including, but not limited to, Computer Systems Security, Computer Network Architectures, Distributed Computer Systems, Quality of Service, Cloud Computing and High-Performance Computing, Human–Computer Interface, Multimedia Systems, Big Data, Knowledge Discovery and Data Mining, Software Engineering, E-Business Systems, Web Design, Optimization and Performance, Internet of Things, Mobile Systems, and Applications.

**Part II**—addressing topics including, but not limited to, Pattern Recognition and Image Processing Algorithms, Production Planning and Management Systems, Big Data Analysis, Knowledge Discovery, and Knowledge-Based Decision Support and Artificial Intelligence Methods and Algorithms.

**Part III**—is gain to address very hot topics in the field of today's various computer-based applications—is devoted to information systems concepts and applications supporting the managerial decisions by using different business models and exploiting opportunities offered by IT systems. It is dealing with topics including, but not limited to, Knowledge-Based Management, Modeling of Financial and Investment Decisions, Modeling of Managerial Decisions, Production and Organization Management, Project Management, Risk Management, Small Business Management, Software Tools for Production, Theories, and Models of Innovation.

We would like to thank the Program Committee Members and Reviewers, essential for reviewing the papers to ensure a high standard of the ISAT 2019 conference, and the proceedings. We thank the authors, presenters, and participants of ISAT 2019 without them the conference could not have taken place. Finally, we thank the organizing team for the efforts this and previous years in bringing the conference to a successful conclusion.

We hope that ISAT conference is a good scientific contribution to the development of information technology not only in the region but also internationally. It happens, among others, thanks to cooperation with Springer Publishing House, where the AISC series is issued from 2015. We want to thank Springer's people who deal directly with the publishing process, from publishing contracts to the delivering of printed books. Thank you for your cooperation.

September 2019                                                      Leszek Borzemski
                                                                   Jerzy Świątek
                                                                   Zofia Wilimowska

# Organization

## ISAT 2019 Conference Organization

### General Chair

Leszek Borzemski, Poland

### Program Co-chairs

Leszek Borzemski, Poland
Jerzy Świątek, Poland
Zofia Wilimowska, Poland

### Local Organizing Committee

Leszek Borzemski (Chair)
Zofia Wilimowska (Co-chair)
Jerzy Świątek (Co-chair)
Mariusz Fraś (Conference Secretary, Website Support)
Arkadiusz Górski (Technical Editor)
Anna Kamińska (Technical Secretary)
Ziemowit Nowak (Technical Support)
Kamil Nowak (Website Coordinator)
Danuta Seretna-Sałamaj (Technical Secretary)

## International Program Committee

Leszek Borzemski (Chair), Poland
Jerzy Świątek (Co-chair), Poland
Zofia Wilimowska (Co-chair), Poland
Witold Abramowicz, Poland
Dhiya Al-Jumeily, UK
Iosif Androulidakis, Greece
Patricia Anthony, New Zealand

Zbigniew Banaszak, Poland
Elena N. Benderskaya, Russia
Janos Botzheim, Japan
Djallel E. Boubiche, Algeria
Patrice Boursier, France
Anna Burduk, Poland
Andrii Buriachenko, Ukraine

## ISAT 2019 Reviewers

Hamid Al-Asadi, Iraq
S. Balakrishnan, India
Zbigniew Banaszak, Poland
Agnieszka Bieńkowska, Poland
Grzegorz Bocewicz, Poland
Leszek Borzemski, Poland
Janos Botzheim, Hungary
Krzysztof Brzostowski, Poland
Anna Burduk, Poland
Wojciech Cellary, Poland
Haruna Chiroma, Malaysia
Grzegorz Chodak, Poland
Piotr Chwastyk, Poland
Gloria Cerasela Crisan, Romania
Anna Czarnecka, Poland
Mariusz Czekała, Poland
Yousef Daradkeh, Saudi Arabia
Grzegorz Debita, Poland
Anna Jolanta Dobrowolska, Poland
Jarosław Drapała, Poland
Maciej Drwal, Poland
Tadeusz Dudycz, Poland
Grzegorz Filcek, Poland
Mariusz Fraś, Poland
Piotr Gawkowski, Poland
Dariusz Gąsior, Poland
Arkadiusz Górski, Poland
Jerzy Grobelny, Poland
Krzysztof Grochla, Poland
Houda Hakim Guermazi, Tunisia
Biju Issac, UK
Jerzy Józefczyk, Poland
Ireneusz Jóźwiak, Poland
Krzysztof Juszczyszyn, Poland
Tetiana Viktorivna Kalashnikova,
    Ukraine
Jan Kałuski, Poland
Anna Maria Kamińska, Poland
Radosław Katarzyniak, Poland
Agata Klaus-Rosińska, Poland
Grzegorz Kołaczek, Poland
Iryna Koshkalda, Ukraine
Zdzisław Kowalczuk, Poland

Dorota Kuchta, Poland
Binod Kumar, India
Jan Kwiatkowski, Poland
Wojciech Lorkiewicz, Poland
Marek Lubicz, Poland
Zbigniew Malara, Poland
Mariusz Mazurkiewicz, Poland
Vojtěch Merunka, Czech Republic
Rafał Michalski, Poland
Bożena Mielczarek, Poland
Peter Nielsen, Denmark
Ziemowit Nowak, Poland
Donat Orski, Poland
Michele Pagano, Italy
Jonghyun Park, Korea
Agnieszka Parkitna, Poland
Marek Pawlak, Poland
Dolores Rexachs, Spain
Paweł Rola, Poland
Stefano Rovetta, Italy
Abdel-Badeeh Salem, Egypt
Joanna Santiago, Portugal
Danuta Seretna-Sałamaj, Poland
Anna Sikora, Spain
Marcin Sikorski, Poland
Jan Skonieczny, Poland
Malgorzata Sterna, Poland
Janusz Stokłosa, Poland
Grażyna Suchacka, Poland
Joanna Szczepańska, Poland
Edward Szczerbicki, Australia
Jerzy Świątek, Poland
Kamila Urbańska, Poland
Jan Werewka, Poland
Zofia Wilimowska, Poland
Marek Wilimowski, Poland
Bernd Wolfinger, Germany
Józef Woźniak, Poland
Krzysztof Zatwarnicki, Poland
Jaroslav Zendulka, Czech Republic
Chunbiao Zhu,
    People's Republic of China

## ISAT 2019 Keynote Speaker

Professor Cecilia Zanni-Merk, Normandie Université, INSA Rouen, LITIS, Saint-Etienne-du-Rouvray, France

Topic: **On the Need of an Explainable Artificial Intelligence**

# Contents

### Models of Production Management

### Models of Organization Management

# Models of Financial and Investment Decisions

# The Influence of Strong Changes of Chosen Macroeconomic Factors on Some Parameters of the State of Organization

Tadeusz Gospodarek[(✉)] [ORCID] and Sławomir Pizoń [ORCID]

University of High Banking School, Fabryczna 29-31, 53-609 Wrocław, Poland
tadgospo@gmail.com

**Abstract.** This paper describes the problem of relations between some macroeconomic indexes and microeconomic financial indicators of an organization during rapid and disastrously changes in the surroundings. The analysis of these relationships was carried out for ten Polish transport companies operating on international market, covered the period of the financial crisis 2008–2011, when catastrophic changes of macroeconomic indicators have taken place. Particularly interesting is impact of some changes of the GDP, PMI and IFO indexes on the economic state of organizations from logistic and transport sector. They are the most flexible, continuously adapting themselves to conditions arisen temporarily in the business environment. Changes of their economic state are mainly characterized by the ROE, ROS, ROA and the income ratio values. We propose a relational model of macro and micro economic indicators enabling optimization of organization's behaviour under conditions of critical changes in macro environment. We conclude that there exists a rational forecast of determining the economic state of an organization resulting from the analysis of changes in some macroeconomic indexes.

**Keywords:** Macro-micro relation · Economic state · Financial factors · Prediction model · Quantitative approach · Relations force measure

## 1 Introduction

It is an interesting problem of relations of some global macroeconomic factors and microeconomic indicators related to a financial position of organizations which must be monitored continuously. The research question in this paper is: how some strong changes of macroeconomic indicators and global indexes as GDP, PMI and IFO affect changes of some micro indicators as ROA, ROE and ROS along time. The changes of macroeconomic parameters may arise during economic crisis or critical changes of economic situation derived from natural or social disasters and some political decisions. In all cases, observations of these changes allow to build relations to microeconomic states defined by some local variables. These local parameters are mainly related to key performance indicators of an organization (Parmenter 2010; Gospodarek 2015). It is possible to obtain some macro-micro relations synchronized in time and find those where the reaction of micro indicators on macro changes will give useful information from the decision-making point of view on micro level. Of course, it is

Z. Wilimowska et al. (Eds.): ISAT 2019, AISC 1052, pp. 3–14, 2020.
https://doi.org/10.1007/978-3-030-30443-0_1

necessary to estimate the term "strong changes" if anyone would like to build an adaptive model of suggested behaviours in micro scale. Estimations may be based on scenarios derived from what if (Rizzi 2009, Gospodarek 2018) or sensitivity analysis (Pannell 1997).

The choice of macro indicators for prediction purposes should be based on availability of source data for comparison and their influence on national economies. Therefore, the GDP is one of the most important, because it is a basic measure of the overall size of a country's economy. Suitable data have been collected by International Monetary Fund for 189 countries for a couple of years (Monetary 2018).

The next chosen index in our research is the purchasing managers' index PMI related to evaluation of current situation on the market and presented as a survey of evaluations of 400 purchasing managers in the manufacturing sector on seven different fields: production level, new orders from customers, speed of supplier deliveries, inventories, order backlogs and employment level. PMI data are collected and offered by Markit Group (Markit 2019). These data offer the base for prediction of changes of some microeconomic indexes.

In this paper IFO means Business Climate Index (IFO 2019), leading indicator for economic activity in Germany prepared by the IFO Institute for Economic Research in Munich. Three data series are compiled from raw scores and grouping in three fields: business climate, current business situation and business outlook.

There are some other very useful and interesting business data surveys, as: The Global Competitiveness' Report (GCI) of the World Economic Forum (WEF 2018), Global Innovation Index (GII) of the Cornell University, INSEAD and WIPO (WIPO 2018), Doing Business of the World Bank (World Bank 2019). But they have too complex structure and be too much synthetic for models based on financial data of organizations. Looking for some relations of macro-micro indicators, the rational reduction of multi-pillar-based data (e.g. GCI Index) and too much processed and interpreted data (e.g. Doing Business Report) is a crucial aspect of scientific methodology, because further generalization may affect crude relations of source data.

Quantitative analysis of some global data along a finite period when minimum one catastrophic change has been observed was the occasion to verify the influence of some macro on some micro economic indicators. It also allows to explain a dynamic of changes before the moment of crisis or catastrophe, and after, during the recovery processes. Some changes of macro values during 2006–2014 period are presented on the Fig. 1. One can see catastrophic changes of the GDP for the Eurozone during 2008–2010 period. The amplitude of decreasing is near 2% of absolute measure (60–100% relatively), what had to influence on all economies in the EC and therefore on the related macro indexes e.g. intracommunity trade, household available income, freight transport. As the logical consequence, some changes in micro state indicators determining the financial position of organizations should take place. It was observed that very sensitive aspects of European economy during the crisis time had been domestic trade and transport. Also, a disposable income has been changed along the GDP, but in a moderate scale (see the Fig. 1). These facts suggest that observations of financial indicators of firms from logistic and transport sector would be the rational choice for experimental data analysis for hypotheses about existing measurable relations between macro and micro economic factors. Based on available macro and micro data, we have observed that since 2004 the logistic sector reacts very quick and sharp on some economic changes in the surroundings.

The aim of this paper is to create the model of quantitative changes of some financial micro indicators of the organization's state related to changes of the selected, well elaborated and commonly accessed economic macro indexes. As the result, suitable model of usable relations macro-micro will be presented and discussed for a set of transport organizations from Poland offering services on the European market. The authors try to answer two scientific questions.

1. Which relations of micro and macroeconomic indicators are particularly useful for building short-term scenarios of keeping the economic stability of organizations?

2. How to simulate the economic position of an organization when strong changes of some macro indicators derive?

Both questions are important for adaptative managing (Gospodarek 2018) and strongly depend on the changes in the surroundings. Especially transport firms are vulnerable because of strong competitiveness on the market, small margin base in their pricing strategies and serious engagement of resources for leading business.

## 2   Characteristic of the Financial Micro State

There are three basic financial statements of an organization: income statement, balance sheet and cash flow statement (Deloitte 2018, IFRS 2019).

The income statement indicates the organization's profits over a defined period (monthly, quarterly or annually) represented by EBT or EBITDA value for the whole organization or its structural part. It answers the question "how profitable the analysed business is" and reports on five areas: sales (revenue), costs of creating and delivering goods or services, operating expenses, financing costs (interest expense), tax payments. For each area, it is possible to define synthetic indicators as a measure of effectiveness or accuracy.

Balance sheet represents equality between total assets (what the organization owns) and a sum of outstanding debt and the owner's equity. It indicates an organization's financial position in terms of the assets owned and how these assets have been financed. The time-based difference between balance sheets is equal to the income statement in the analysed period. Together with the value of revenue, it is the base of dividing companies on categories: micro, small and large. From balance sheet, it follows a lot of consequences for controlling parameters and indicators related to the ratio of income to engaged capital and resources.

At last the cash-flow statement answer the fundamental question "where did the cash come from and where did the cash go". It is very important aspect of management, because from the experience it follows, that cash flow problems are a major reason for organizations failing especially small ones—even at times when the business seems to be profitable. In accounting systems, there are a difference between the income statement and the cash-flow one. The balance sheet is based on accrual accounting whereas the cash-flow one is realized on cash-basis account. Therefore, profits and cash flows usually are not equal. More details about "how to" in relation to the financial statement calculations are available from International Accounting Standards (Deloitte 2018) and International Financial Reporting Standards (Deloitte 2018, IFRS 2019).

From the managing point of view, the set of precisely tailored key performance indicators (Parmenter 2010; Gospodarek 2018) is the most important for making tactic and strategic decisions under uncertainty, and for realizing continuous adaptation of an organization to different changes in the surroundings. In practice, management uses IT support for measurement the feed-back aspects for optimization of the economic state of the organization. As the nowadays standard the ERP class integrated IT systems support monitoring of the defined set of KPI's according to strategic model and goals definition. Among the set of KPIs these related to financial report described by Dupont's model of controlling are crucial (Gospodarek 2015 and 2018).

All the above areas on micro level depend on the situation in the business surroundings, characterized by macro parameters, and they can't be separated as isolated entities for quantitative analyses (as it is possible in physics). It is the main idea of this research, and abduction inferring for the hypothesis, that it is possible to predict the behaviour of some financial indicators of the firm, when significant changes in the related indexes macro are detected. And that some relations may be useful for decision making in advance.

## 3   Source Data and Analyses

We have carefully selected ten logistic organization from Poland operating on international market, mainly EC countries. All of them had been stable organizations of average and big class since 2004 (date of Polish membership in EC) and obligated to prepare annual financial reports according to international accounting standards (Deloitte 2018). These reports are available from court files. For these firms, we have retrieved financial reports for the period 2006–2012, where big changes in macroeconomic data had been observed between 2008–2010.

We have also collected suitable information from Polish Governmental Statistic Office GUS and Eurostat as source data of macro level. We've also used reports offered by the World Bank, Markit, WIPO, etc. All the selected data (micro and macro) were synchronized in time, elaborated according to the assumed analytical model, compared and statistically verified as derived from the same population or not and in the case of relational macro-micro aspects, if they are important or not based on statistical criteria.

For confirmation of importance of the assumed changes between annual macro data and the related different micro factors derived from the source data, the Wilcoxon-Mann-Whitney rank-sum test was applied (Ruland 2018; Gospodarek 2018). It is non-parametric test which allows to verify the hypotheses regardless the type of statistic data distribution. The null hypothesis that it is equally likely that a randomly selected value from one sample will be less than or greater than a randomly selected value from a second sample. The sample of 10 items is statistically representative for such kind of comparisons (Gospodarek 2018). It is simple and effective tool nearly as efficient as the t-test for normal distributions.

Hypotheses that the variances from the sample based on the selected organizations are equal on significance level p = 0,05 were confirmed with Snedecor's F-test of equality of variances (Taboga 2017). The null hypothesis that two normal populations have the same variance was assumed. It was useful in comparative analyses of some micro data between time periods.

Due to behaviour of micro and macro indicators along time, analyses were performed in three ranges: 2006–2008 (period just before the crisis); 2008–2010 (crisis period); 2010–2011 (period of recovery from the crisis).

**Table 1.** Analysed macro indicators in the period 2006–2011

| Macro indicator | 2006 | 2007 | 2008 | 2009 | 2010 | 2011 |
|---|---|---|---|---|---|---|
| GDP (Euro zone) | 3,2% | 3,0% | 0,4% | −4,4% | 2,1% | 1,6% |
| GDP Poland | 6,2% | 7,2% | 3,9% | 2,6% | 3,7% | 5,0% |
| CPI | 1,0% | 2,5% | 4,2% | 3,5% | 2,6% | 4,3% |
| Reference interest rates National Bank of Poland | 4,0% | 5,0% | 5,0% | 3,5% | 3,5% | 4,5% |
| PMI Industrial Poland | 53,3% | 50,7% | 38,3% | 52,4% | 56,3% | 48,8% |
| IFO EC (Euro zone) | 15,1% | 3,3% | −52,0% | −26,4% | −2,6% | 31,5% |

Illustration of macro indicators changes from Table 1 is presented on Fig. 1.

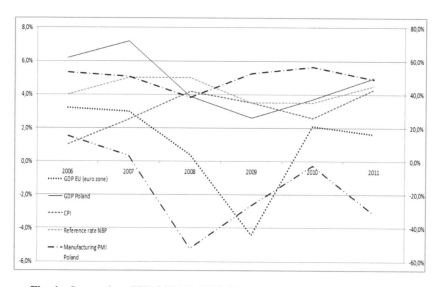

**Fig. 1.** Source data: EUROSTAT, GUS, NBP, Markit Group and IFO Institute

As one can see, some interesting relation of IFO and GDP were observed in the period 2008–2009 when the changes of IFO had been ahead of the GDP minimum (the deepest crisis) almost on a year. It may suggest that IFO and GDP should be valuable indexes for building the relations macro-micro. PMI was also relatively decreased on

1,5% to its minimum value in 2008, but it had been not so sensitive in reaction on crisis situation as IFO.

## 4  Relations Macro - Micro

For quantitative evaluations of the strength of micro-macro relation presented in this paper, the Coefficient of Dependence's Relative Force (CDF) was introduced. It may be interpreted as a ratio of relative changes of a macro variable and a micro one, according to Formula 1.

$$CDF = \frac{\frac{x_{ti}-x_{ti-1}}{|x_{ti-1}|}}{\frac{M_{tj}-M_{tj-1}}{|M_{tj-1}|}} \qquad (1)$$

*where, $x_{ti}$ – represents a given micro parameter in i-th period,*
*$M_{tj}$ represents a given macro index in j-th period respectively.*
The CDF may be positive (when increasing of a micro variable is correlated with increasing the related macro index) and it may be negative (when positive changes of micro are correlated with negative changes in macro and reverse). It is the useful measure but not very sensitive. Rational inferring needs supporting analyses of the source differences (xti – xti-1 and Mtj – Mtj-1). It is also necessary for correct interpretation of a CDF's sign.

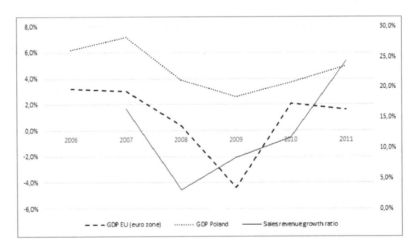

**Fig. 2.** The pace of changes in sales revenues compared to Poland's GDP and GDP for the Euro zone [source data: Eurostat]

Changes of income rate (IR) of the analysed organizations in respect to GDP of Poland and EC are presented on Fig. 2: The IR changes in relation to the GDP of Poland and EC countries were derived quicker with similar intensity as changes of

GDP. It can be noticed that in the period 2007–2008 there was a significant decrease of the rate of changes in sales revenues and GDP ratios. At the same time, GDP indicators continued to decline in 2009, while the rate of change in sales revenues began to increase. It results from the capacity of transport and logistic companies to adapt to the surroundings changes and favourable external conditions.

The values of CDF for the relations presented on the Fig. 3 are presented in Table 2.

**Table 2.** The CDF values for the micro macro relation income/GDP

| CDF of ΔIR/GDP | Period of analysis | | |
|---|---|---|---|
| | 2007–2008 | 2009–2009 | 2010–2011 |
| Poland | 1,77 | −5,16 | 2,05 |
| EC | 0,94 | −0,14 | 1,39 |

In the analysed periods, the significant positive relationship between GDP changes and the rate of change in sales revenues occurred in 2007–2008, in the initial phase of the crisis. Faster rebound of the rate of change in sales revenues than the GDP causes negative values of coefficients of the relative dependence in 2008–2009. However, in the later period 2010–2011 (beginning of recovery), the dependence is again strong and takes a positive direction. Therefore, only the relationship existing in the initial phase of the crisis may be considered for further analyses. The decline in GDP causes a drop in the rate of sale. GDP growth until the end of the analysed period (2009–2011) is related to increase the rate of sale.

Similarly, to the above, the following relations of the income to different macro indicators have been considered. The results are presented in Table 3, and Fig. 3.

**Table 3** Considered relations of income changes (micro) with some macro indicates. The differences between the average values in pointed periods were confirmed statistically as important with Snedecor's test of variance equality at 95% level.

| Relation of macro indicators and the ratio of the ROA changes | Compared periods | CDF value |
|---|---|---|
| GDP Euro zone | 2007 and 2008 | 0,94 |
| GDP Euro zone | 2009 and 2011 | 1,39 |
| GDP Poland | 2007 and 2008 | 1,77 |
| GDP Poland | 2009 and 2011 | 2,05 |
| CPI | 2007 and 2008 | −1,19 |
| Interest rate (National Bank of Poland) | 2006–2007 and 2007–2008 | −3,25 |
| PMI industrial Poland | 2006–2007 and 2007–2008 | 16,63 |
| IFO (Euro zone) | 2006–2007 and 2007–2008 | 1,04 |

From the Fig. 3 some conclusions may derive. Big changes of GDP in Euro zone and Poland between 2007 and 2008 also 2009 and 2011 are correlated with decreasing of the ratio of income changes. The level of significance measured by the CDF value was 0,94 – 2,05. The influence of CPI index on the ratio of income changes between 2007 and 2008 shown negative tendency represented by the CDF value. The most important changes of the CDF value are observed for PMI changes. It is interesting, that the IFO index had no such spectacular correlation estimated by the CDF value.

As an example of interesting relation macro and micro the dependence of the ROA on the GDP may be presented as on the Fig. 4. It means that if the ROA for diagnostic organizations will decrease, then after some time one can observe decreasing of the GDP. Therefore, the GDP is not appropriate indicator for prediction the ROA changes. It may be simply explained. The increasing/decreasing of the GDP is a consequence of the ROA changes.

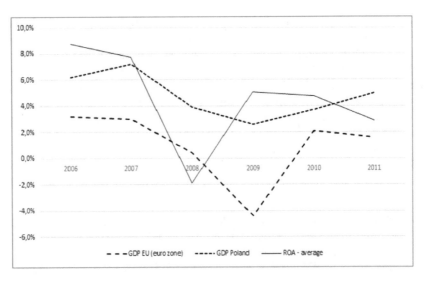

**Fig. 3.** Changes of ROA in relation to changes of GDP Poland and Euro Zone.

In case of PMI or IFO the relations macro-micro will be useful for prediction of the micro behaviour based on macro changes (Fig. 4).

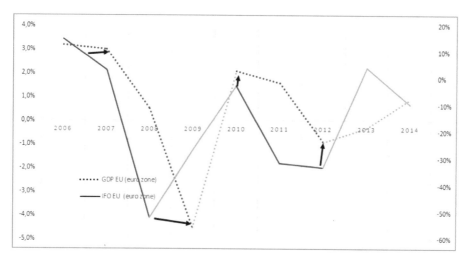

**Fig. 4.** Changes of GDP EU in relation to changes of IFO.

The first symptoms of prosperity decreasing were observed in 2007 according to decreasing the PMI industrial for Poland and the IFO for Euro Zone. Significant decreasing of the ROS for transport firms was observed in 2008, where adaptative processes in micro had been insufficient in relation to the changes in macro indicators. It is interesting how the values of the Coefficient of Dependence's Relative Force (CDF) were changed during the 2008–2011 period. This data is presented in Table 4.

**Table 4** Considered relations of the ROS (micro) with some macro indicates where differences between the average values in pointed periods were confirmed statistically as important with Snedecor's test of variance equality at 95% level.

| Relation of macro indicators and the ratio of income changes | Compared periods | CDF value |
|---|---|---|
| PMI industrial Poland/ROS - av | 2008–2010 | 4,68 |
| PMI industrial Poland/ROS - av | 2008–2011 | 6,41 |
| PMI industrial Poland/ROS - av(t + 1) | 2006–2007/2007–2008 | 28,60 |
| IFO EU (Euro Zone)/ROS - av | 2008–2010 | 2,31 |
| IFO EU (Euro Zone)/ROS - av(t + 1) | 2006–2007/2007–2008 | 1,79 |
| IFO EU (Euro Zone)/ROS - av | 2008–2011 | 4,46 |

## 5   Simulations of Some Macro-Micro Dependences

Based on the presented model, some simulations of changes of micro indexes along the macro parameters were performed. These procedures allow to estimate the sensitivity of the CDF model for decision making purposes.

As the most interesting relations, the ROA, ROS and ROE changes depending on the income ratio changes are presented on the Fig. 5. Similar results were calculated for the dependencies of the ROS, ROA, ROE and Income Ratio on PMI.

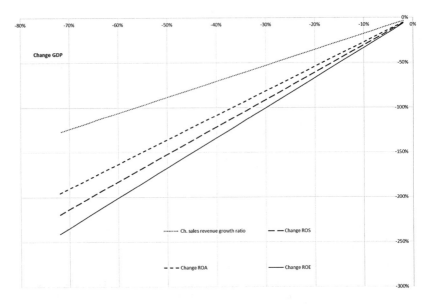

**Fig. 5.** Simulation of ROS, ROA, ROE and income ratio changes along the changes of the GDP

It may be concluded, that PMI is more sensitive index than GDP, but round 20% change in GDP ratio significantly influences the changes in micro indexes such as ROA, ROE and ROS. Assuming that standard GDP increasing of rational economy is about 3% per year, this 20% of changes is round 0,6% of the GDP increasing. In the case of PMI changes, the 3% decrease causes almost 100% changes in ROE/ROS. So, the relation is far much more valuable for decision supporting. Assuming IFO as the macro index the relations macro-micro are not so sensitive, about 30% of decreasing IFO causes 50% of the described micro indexes. But it is sufficient to support the decisions in micro.

## 6   Conclusions

The research shows that, just before the crisis starts (decrease of macro indexes), the leading indicators: PMI and IFO and interest rates are changing. In the acute phase of the crisis there is a decline in GDP and an increase in interest rates. The response of microeconomic indicators to changes in macroeconomic indicators takes place in two stages. In the initial, acute phase of the crisis yields are falling and the pace of changes in sales revenues, and a year later the receivables turnover period is extended, and the current assets debt increases. The highest sensitivity of microeconomic indicators occurs in the case of relations with PMI and interest rates, and the lowest in relations with IFO and GDP. In the case of relations characterized by lower sensitivity of microeconomic indicators to the volatility of macroeconomic indicators, it is necessary to observe a much larger change in the macroeconomic indicator, so that the volatility of microeconomic indicators can be predicted. It allows to avoid premature activities based on random changes in macroeconomic indicators. The existence of relations in the model of relationships characterized by lower sensitivity of microeconomic indicators to changes in macroeconomic indicators increases the stability of the decision-making model.

From the presented results it may be concluded, that there exist the model supporting decision making in micro (organization) level based on the changes of some macroeconomic indexes. Quantitative approach suggests the proposed CDF value defined in this paper as satisfactory measure of the relationship force between some macro and micro indexes. Based on the CDF evaluations it is possible to state the sensitivity of each relation of macro-micro indexes and make a choice suitable pairs for decision purposes. The most valuable are the relationships between the PMI and ROA, ROE and ROS indexes as well as the Income Ratio which is closely related to ROS.

It may be also concluded that based on the presented relationships it is possible to state the rational forecast of economic position of an organization using the PMI and IFO data. From the presented simulations it follows that the sensitivity of CDF model is enough to support the tactic decisions regarding financial state of the organization.

The presented heuristic and structure of quantitative approach based on CDF model, described in this paper offer the method of optimization of the micro state on acceptable level of rationality. Optimum solution is related to dynamics of some macroeconomic factors as the forecast for the micro state estimation for not too far future (tactic level of management).

# References

Deloitte: Conceptual Framework for Financial Reporting 2018 (2018). https://www.iasplus.com/en/standards/other/framework. Accessed 02 Mar 2019

Gospodarek, T.: ERP Systems. Modeling Design and Implementation (PL), Gliwice, Helion, ISBN 978-83-283-1417-7 (2015)

Gospodarek, T.: White Book of Management. (PL). Warszawa, Difin (2018)

IFO: Calculating the ifo Business Climate (2019). http://www.cesifo-group.de/ifoHome/facts/Survey-Results/Business-Climate/Calculating-the-Ifo-Business-Climate.html. Accessed 02 Mar 2019

IFRS: IFRS® Standards and IFRIC® Interpretations (2019). https://www.ifrs.org/issued-standards/. Accessed 02 Mar 2019

International Monetary Fund: World Economic and Financial Surveys, World Economic Outlook Database (2018). https://www.imf.org/external/pubs/ft/weo/2018/02/weodata/index.aspx. Accessed 02 Mar 2019

Markit Group: Purchasing Managers' Index® (2019). http://www.markit.com/product/pmi. Accessed 01 Mar 2019

Parmenter, D.: Key Performance Indicators: Developing, Implementing, and Using Winning KPIs, 2nd edn. Wiley, Hoboken (2010)

Pannell, D.J.: Sensitivity analysis of normative economic models: theoretical framework and practical strategies. Agric. Econ. **16**(2), 139–152 (1997)

Rizzi, S.: What-If analysis. In: Liu, L., Özsu, M.T. (eds.) Encyclopedia of Database Systems. Springer, Boston (2009)

Ruland, F.: The Wilcoxon-Mann-Whitney Test – An Introduction to Nonparametrics With Comments on the R Program wilcox.test. Independently published ISBN 978-1728873251 (2018)

Taboga, M.: Lectures on Probability Theory and Mathematical Statistics, 3rd ed. CreateSpace Independent Publishing Platform, ISBN 9781981369195 (2017)

WIPO: Global Innovation Index (2018). https://www.globalinnovationindex.org/gii-2018-report. Accessed 02 Mar 2019

World Economic Forum: The Global Competitiveness Report 2018 WEF (2018) https://www.weforum.org/reports/the-global-competitveness-report-2018. Accessed 02 Mar 2019

World Bank: Doing Business. Measuring Business Regulations (2019). http://www.doingbusiness.org/en/reports/global-reports/doing-business-2019. Accessed 02 Mar 2019

# Branch and Bound Method in Feature Selection Process for Models of Financial Condition Evaluation

Zofia Wilimowska[1($\boxtimes$)], Marek Wilimowski[1],
and Tetiana V. Kalashnikova[2]

[1] University of Applied Sciences in Nysa, Nysa, Poland
{zofia.wilimowska,marek.wilimowski}@pwsz.nysa.pl
[2] Simon Kuznets Kharkiv National University of Economics,
Kharkiv, Ukraine
kalashnikova.tv@gmail.com

**Abstract.** Uncertainty and risk which are associated with company's activity require a special instrument supporting process of the managers' decision making. The objective of the research is to determine the number features characterizing the financial risk – financial condition of companies. A quality of financial condition evaluation depends on the selection of variables (features) and criteria of the assessment. The choice of financial ratios in the study of financial standing of companies is crucial. The article presents the proposal to apply branch and bound method to choose sub-optimal subsets of financial ratios that best describe the subject of the research, which is the company, under the assumption that classification algorithms are used for evaluation function creation. The aim of this study is to present a solution that allows the selection of financial ratios with a very high cognitive value, enabling the building of integrated measures assess the financial condition of the company.

**Keywords:** Selection of information · Financial ratios · Optimization · Discriminatory models · Branch and bound method

## 1 Introduction

In the rapidly changing market economies continuous assessment of financial phenomena occurring in businesses, in particular continuous evaluation of their financial condition is expected. Proper evaluation of the processes occurring in the enterprise enables prediction of the financial situation of the company and taking pre-emptive action which could protect the company from bankruptcy [1, 7, 10, 11], it means enables risk reduction of the activity. A primary source of risk in human activities is a feeling of uncertainty connected with future unknown events, due to the fact that decisions are made today and the effects of the decision will be known in the future.

There have been carrying out calculations of financial ratios of public and private companies since the 19th century. As the years passed the development of statistical methods, that were used to predict business failure, were followed. The sixties of the twentieth century were a turning point in the study of the early diagnosis of the symptoms of risk of bankruptcy.

© Springer Nature Switzerland AG 2020
Z. Wilimowska et al. (Eds.): ISAT 2019, AISC 1052, pp. 15–25, 2020.
https://doi.org/10.1007/978-3-030-30443-0_2

High volatility of the business environment and high risk of management result in a large number of bankruptcies. The tools of economic analysis allow for the rapid assessment of financial condition of companies and their financial risk [1, 10, 11]. For the past twenty five years many models have been constructed to examine the financial condition of companies and to classify them as "ones with good condition" as the ones with "bad condition" which mean high risk of bankruptcy.

Enterprises can be described by certain characteristics, features that can be financial and non-financial indicators, ratios. The use of synthetic indicators in the assessment process allows the assessment of a company financial standing, this is integrated assessment. Of course, it is clear that not every financial indicator (feature) is equally important in the evaluation of companies, therefore is crucial in this respect to choose (select) financial indicators most valuable, useful and crucial from the point of view of the assessing enterprise.

Multicriterial methods for company condition evaluation – discriminant methods, taxonomic methods, classifications (discrete risk assessments) – require the definition of a size vector (vector of features) that is the basis for assessments [5, 6, 8, 11].

Why some indicators are more often used than others? Various aspects effect the frequency of their use. One of them is the availability of data, for example not all companies are listed on the stock exchange, what means that mostly the market ratios of companies are not known, and therefore should be removed from the set of financial ratios.

In the "bankruptcy" models, Polish and foreign authors, there is a lot of talk about the quality of their assessment of enterprises [11, 12], but not much about the selection of indicators in these models. Dozens of attempts to use models are carried out, estimating their diagnostic quality, but not much about the diagnostic quality of selected indicators. Of course, expert knowledge should be appreciated, but attention should also be paid to the possibility of using the methods already used, e.g. the methods of selecting information and selecting the features for example branch and bound method.

In this paper we propose well known method, branch and bound method, for selecting features for the construction of the synthetic index of company financial condition evaluation using classification methods.

## 2   Feature Selection for Financial Condition Evaluation

The literature suggests several methods of selection features (indicators) to build discriminatory models [2, 4–6, 9]. Very often correlation matrix is used for features selection, but keep in mind that a strong correlation dependence between $x_1$ and $x_2$ does not exclude a weak relationship between $x_1$ and $x_3$, as well as between $x_2$ and $x_3$.

The second technique is to set yourself up as an expert in the selection of appropriate indicators. Currently, the authors are inspired by these indicators, which are often used to assess the insolvency of companies, something discussed in a number of publications.

A company has specific characteristics (in the assessment of the financial condition it can be financial ratios) that describe the object. These characteristics are expressed by a sequence $s$ of $N$ variables $x_1, x_2, \ldots, x_N$. The larger the $N$, e.g. the number of features, more difficult to choose of financial indicators that can be used to build the synthetic indicator, which is more difficult to make a selection.

Usually information about the assessed object is collected in excess. Therefore, there are many unnecessary features. Features may be superfluous because

- they do not introduce any differentiating information risk levels,
- sometimes it is enough to use one feature, if others are very strongly dependent on the former, or
- they have no relation to the purpose of classification – company condition evaluation.

## 2.1 Feature Selection

Before undertaking a discreet risk assessment of an enterprise (classification) [4–6, 9], competent persons and experts are given a set of characteristics of objects to be classified on the basis of these features. However, the number of elements of such a set is usually very large. Some of these elements do not provide any or very little information about the class of the object. They are, from the point of view of recognition of company financial condition, not very useful and even complicate the algorithm. Therefore, it is necessary to select from a complete set a certain subset of features on the basis of which appropriate algorithms are built [2, 3, 9].

The set of features describing a given phenomenon should be constructed in such a way that it fully describes them. If this condition is met, then it will be possible to talk about the accuracy of assessments and analyzes, predictions as well as the accuracy of decisions made by the user on their basis. The reduction in the number of features is carried out on the basis of certain selection criteria.

One of the methods of reducing the number of features is the reduction of the set of features. The reduction of the set of features is understood as a procedure of reducing the set of features on the basis of which the classification algorithm will be implemented, most often by non-linear transformation of the $n$-dimensional space into $m$-dimensional, $n > m$.

The particular case of reduction is selection. Selection is understood as a method of selecting a subset of features from a larger set, ensuring minimization or maximization of the appropriate criterion. In other words, the selection task is to select a subset of features that bring as much amount information as possible (the amount of information is understood here as the value of the appropriate criterion). One of the basic selection criteria is the probability of incorrect classification, that is, the classification of an object into a class $i$ while it belongs to the class $j$.

Feature selection methods try to find a minimal subset of the features that meet the following criteria:

- the classification quality will not be significantly reduced,
- class distribution obtained using only selected features is as close as possible to the original distribution of these classes.

Depending on the a priori information about the recognized object, there are different selection algorithms. The algorithms proposed in the literature are very similar and differ only in the form of the selection criterion [9]. Typically, the criteria proposed by the authors of the algorithms are different estimates of incorrect classification, so the algorithms are suboptimal to incorrect classification.

## 2.2   Combinatorial Method of Selecting Features in Classification Based on an Increase in Risk of Incorrect Classification

Analytical methods of selecting features [9] do not provide absolute certainty as to the correctness of the choice of features – they are only the best, in the sense of a given selection criterion, approximations of the optimal selection. If the goal of the selection is to reduce the n-element set, to (n-m)-element set, it is only an overview of all possible combinations of features, i.e. analysis of all subsets $\binom{n}{n-m}$ can give such a guarantee. To reduce the number of reviewed solutions, you can use the branch and bound method. The algorithm explores branches of a tree representing the problem solving space, which represent subsets of the solution set.

For this case, two basic tasks have been formulated [3]:

1. rejection $m$ features from $n$, $m < n$, such that the increase in risk is minimal,
2. rejection of as many features as possible, so that the increase in risk does not exceed the preset number $\varepsilon > 0$.

The proposed selection algorithms are mainly based on the following three methods:

1. Calculation for each feature $x_i, i = 1, 2, \ldots, n$ appropriate estimates and rejection of those features for which these estimates reach extreme values.
2. Designation for every possible solution of an appropriate estimation and selection of the one for which it assumes the smallest value, in other words a direct review of all solutions.
3. Subsequent rejection of traits, that is, the rejection of one feature, the choice of the best solution, then the rejection of two features, the previously chosen and the next and again the choice of the best solution, etc.

Let us assume that the Bayes algorithm is used to assess the financial condition of the company.

The measure of incorrect classification for the Bayes algorithm is expressed as follows [3]:

$$R = \bar{P} = \bar{P}(\psi = j) = 1 - \int\limits_{X} \max_{j=1,\ldots,L} p_j f(x \mid j) dx$$

where:

$L$ – number of classes
$p_j$ – a'priori probability of occurrence of the class $j$
$f(x/j)$ – a'priori probability density functions of the distribution of the random vector $x$ for each class $j = 1, 2, \ldots, L$

$j$ – correct classification of the object $x = (x_1, x_2, \ldots, x_n)$, described with $n$ features, which is the realization of a multi-dimensional random variable $x$ (marking with the same vector random variable and its realization should not lead to misunderstandings)

$n$ – number of features – dimension of vector $x$

$X$ – space of random variable $x$

$\psi$ – classification determined by the algorithm;

overbar means complement, for example $\bar{P}(some\ event) = 1 - P(some\ event)$

Let

$X_i$ – means the one-dimensional space for the $x_i$ feature, $i = 1, 2, \ldots, n$ feature, whereas

$X^i$ – means $(n-1)$-dimensional space, after elimination the feature $x_i$, $i = 1, 2, \ldots, n$ then by definition

$$R^{(i)} \overset{\text{def}}{=} 1 - \int\limits_{X^i} \max_{j=1,2,\ldots,L} \left[ p_j \int\limits_{X_i} f(x/j) dx_i \right] dx^i$$

$R^{(i)}$ – means the measure of incorrect classification, which will be obtained by carrying out (recognizing) the classification based on $(n-1)$ the characteristics, after rejecting the feature $i$.

Analogously, let $k_1, k_2, \ldots, k_s \in \{1, 2, \ldots, n\}$ means the numbers of removed features from the vector $x$;

$x^{k_1, k_2, \ldots, k_s}$ – the $(n-s)$ dimensional feature vector after eliminating (rejecting) the features $k_1, k_2, \ldots, k_s$;

$X^{k_1, k_2, \ldots, k_s}$ – means the $(n-s)$ dimensional space of objects obtained from space $X$ by rejection of index $k_1, k_2, \ldots, k_s$ features, so

$$R^{(k_1, k_2, \ldots, k_s)} \overset{\text{def}}{=} 1 - \int\limits_{X^{k_1, k_2, \ldots, k_s}} \max_{j=1,2,\ldots,L} \left[ p_j \int\limits_{X_{k_1}} \cdots \int\limits_{X_{k_s}} f(x/j) dx_{k_s} \ldots dx_{k_1} \right] dx^{k_1, k_2, \ldots, k_s}$$

$R^{(k_1, k_2, \ldots, k_s)}$ – means the measure of incorrect classification that will be obtained as a result of a feature-based on $(n-s)$ features after elimination $x_{k_1}, x_{k_2}, \ldots, x_{k_s}$.

The basic selection tasks can be formulated in the following way.

**Task 1.** Let there be a given $n$-element feature vector $x = (x_1, x_2, \ldots, x_n)$ and a number of features that should be rejected $m$.

The task is to find such $(n-m)$-element combination $G(k_1^*, k_2^*, \ldots, k_{n-m}^*)$, $k_i^* \in \{1, 2, \ldots, n\}$ indexes of the set of features, that

$$R\left(X^{k_1^*, k_2^*, \ldots, k_{n-m}^*}\right) = \min_{k_1, k_2, \ldots, k_{n-m}} R\left(X^{k_1, k_2, \ldots, k_{n-m}}\right), k_1, k_2, \ldots, k_{n-m} \in \{1, 2, \ldots, n\}$$

Different combinations of the selection of $(n\text{-}m)$ elements among $n$ elements are of course $\binom{n}{n-m} = \binom{n}{m}$.

**Task 2.** Let there be a given $n$-element feature vector $x$ and the number $\varepsilon > 0$ by which the risk may increase

$$\varepsilon = R_1 - R$$

where
$R_1$ – means the measure of incorrect classification after selection.

The task is to find the maximum number of features that can be rejected on the assumption that the risk will not increase more than $\varepsilon$, i.e. it should be found $t_0, k_1^*, k_2^*, \ldots, k_{t_0}^*$ that

$$\left.\begin{aligned} t_0 = \min_{t<n} t : R\left(X^{k_1^*, k_2^*, \ldots, k_{t_0}^*}\right) &= \min_{k_1, k_2, \ldots, k_{t_0}} R\left(X^{k_1, k_2, \ldots, k_{t_0}}\right), \\ R\left(X^{k_1^*, k_2^*, \ldots, k_{t_0}^*}\right) - R &\leq \varepsilon \end{aligned}\right\}$$

The feature selection tasks defined in this way can be treated as combinatorial tasks.

Let denote $\bar{w} = (w_1, w_2, \ldots, w_n,)$ binary vector which corresponds to the set of indexes. The values of the elements of the vector $\bar{w}$ are defined as follows:

$$w_i = \begin{cases} 0, & \text{when the feature } i \text{ was rejected} \\ 1, & \text{otherwise} \end{cases} \quad i = 1, 2, \ldots, n.$$

The problem of the optimal selection of features in relation to the measure of incorrect classification consists in finding such a set $\bar{w}$, for which the goal function, in this case the measure of incorrect classification, achieves the smallest value. So it is a combinatorial task. The goal function, which should be minimized when selection tasks is defined as combinatorial tasks, cannot be presented in a general way. It should be defined for each set $\bar{w}$ separately.

When considering **task 1**, i.e. the task of rejecting from among $n$ features, $m$ features, such that the increase in risk is minimal, the function of the goal can be formulated as follows.

It is assumed that only $m$ elements of the set $\bar{w} = (w_1, w_2, \ldots, w_n,)$ can simultaneously accept zero values. If $w_{k_1} = w_{k_2} = \ldots = w_{k_m} = 0, k_1, k_2, \ldots, k_m \in \{1, 2, \ldots, n\}, \underset{i \neq j}{\forall} k_i \neq k_j, m < n$, the other elements of this set are equal to 1, then the goal function for such a set has the form

$$g(\bar{w}) = R(\bar{w}) = R^{k_1, k_2, \dots, k_m}(\bar{w})$$

The constraint in this case is in the form

$$\sum_{i=1}^{n} w_i = n - m$$

The task is to find such a set $\bar{w}$ that satisfies the this constraint so that objective function $g(\bar{w})$ will be the smallest.

**Task 2** can be formulated in the following way. It is assumed that $t$ elements of the set $\bar{w}$ can take zero values, $t = 1, 2, \dots, n - 1$.

The goal function to be minimized is in the form

$$g(\bar{w}) = \sum_{i=1}^{n} w_i$$

with constraint $R(\bar{w}) - R \leq \varepsilon$, and $R(\bar{w}) = R^{k_1, k_2, \dots, k_t}(\bar{w})$, if $w_{k_1} = w_{k_2} = \dots = w_{k_t} = 0, k_1, k_2, \dots, k_t \in \{1, 2, \dots, n\}, \; \forall_{\substack{i \neq j}} k_i \neq k_j, t < n$, the other elements of this set are equal to 1.

From the formulation of the optimal selection task in the above-mentioned way, it follows that only some of the discrete programming methods can be used in these tasks. Both tasks have nonlinearities. In **task 1** the objective function is non-linear, while in **task 2** there are restrictions. It is possible to reject all methods of cutting planes, because it was shown that these tasks should be treated as a combinatorial type task and look for a suitable method among the methods that solve this type of task.

One of the most suitable methods for solving selection tasks is the branch and bound method, and this method has continued to be adapted to the selection tasks mentioned. The branch and bound method is aimed at finding in a large set of solutions characterized by a specific feature, an element with an extreme value of this trait. This is done by appropriately sequentially dividing the set of all solutions into smaller and smaller subsets to receive a one-piece collection, which is a sought-after solution to the task. Finding the shortest way to get a solution requires the adoption of an appropriate selection strategy. For this purpose, depending on the formulation of the objective function, subsets are subdivided according to the minimum or maximum bounding values of the features present in particular subsets.

The ideas of this method can be presented in the following way.

Minimize the function

$$g(Z), \text{ under the condition } Z = (z_{k_1}, z_{k_2}, \dots, z_{k_m}) \subset \mathbb{Z},$$

where $\mathbb{Z}$ – it is some set of selected features.

At the basis of this method are the following activities, which in many cases allow to reduce the area of the review:

1. Calculation of the lower bound (estimation) $\xi(Z)$,

$$g(\mathbf{Z}) \geq \xi(\mathbf{Z}).$$

2. Division into subsets of the set of solutions $\mathbf{Z}$.
3. Conversion of bounds (estimates).

## 2.3   Selection Algorithms

It is assumed that the purpose of the relevant algorithms is to determine the features that should be rejected. The following points will be adopted as the basis for the adaptation of the division method and constraints:

1. The lower bound (estimation) of the objective function on the set $G$ of features is $R$

$$\xi(G) = R(G)$$

2. The division of the set $G$ into subsets will be made according to the following rule $G \equiv G^0 = G_1^1 \cup G_2^1 \cup \ldots \cup G_n^1$, in $G_i^1$ removed the feature $i$ and recursively split them further (branching), removing further features

$$G_k^p = G_{k_1}^{p+1} \cup G_{k_2}^{p+1} \cup \ldots \cup G_{k_p}^{p+1},$$

where:

$p$ – level of branching
$G_{k_i}^p$ – a set of these features contained in the "superior" solution (level $p-1$), in which the $k_i^{th}$ feature is additionally rejected

3. The lower bound of the objective function (estimation) for each set $G_{k_i}^p$ will be $R\left(G_{k_i}^p\right)$ – the risk relevant to features that remained in the set $G_{k_i}^p$.

Optimization of algorithms for both tasks results from the optimality of the branch and bound method [3].

In order to clearly present the idea of applying the branch and bound method to the selection of features, the selection task marked with the number 1 was analyzed. The algorithm is based on the branch and bound method according to the strategy, in which two stages can be distinguished. In the first stage, the solutions are reviewed along one branch and the optimal solution is found in it, in the second stage, on the basis of the solution obtained in the first stage, the remaining branches are reviewed [3].

The algorithm is showed in [3].

Figures 1 and 2 show an illustration of stage 1 and stage 2 of the branch and bound method on the example of the selection of three features of five; $n = 5$, $m = 3$. Of course, the example is demonstrative only and its task is to show the mechanism of the algorithm – the way to choose the next solutions.

The solution $\mathbf{Z}$ obtained in the first stage is: $\mathbf{Z} = \{3, 2, 4\}; G_4^3 = \{1, 5\}$

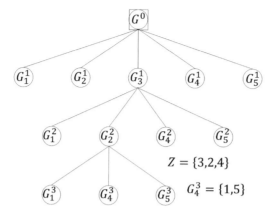

$$Z = \{3,2,4\}$$

$$G_4^3 = \{1,5\}$$

**Fig. 1.** Graphical representation of stage 1

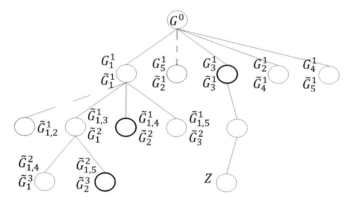

**Fig. 2.** Graphical representation of stage 2

If for a given branch tree the lower bound is greater than the already known solution, then this branch can be omitted.

Before proceeding to stage 2, it is necessary to "renumber" elements of the feature vector $x = (x_1, x_2, \ldots, x_n)$. Let the vector of features have a form

$$\tilde{x} = (\tilde{x}_1, \tilde{x}_2, \ldots, \tilde{x}_n),$$

where:

$$\tilde{x}_{n-i} = x_{k_{i+1}} i, \quad i = 0, 1, \ldots, n - m$$

and other elements $\tilde{x}_1, \tilde{x}_2, \ldots, \tilde{x}_{n-m}$ correspond to the elements $x_i, i = 1, 2, \ldots, n, i \neq k_1, k_2, \ldots, k_m$ in anyway.

In the example $\tilde{x}_1 = x_1, \tilde{x}_2 = x_2, \tilde{x}_3 = x_3, \tilde{x}_4 = x_4, \tilde{x}_5 = x_5$.

The circumflex accent indicates a solution for which the bound value is greater than the bound value for the solution $Z$. In this case, it was assumed here that the next "best" sets of solutions are sets:

$\widetilde{G}_3^1$ – in step 1, $\widetilde{G}_2^2$ – in step 2 and $\widetilde{G}_2^3$ – in step 3. The optimal solution should be sought among the sets $\widetilde{Z}, \widetilde{G}_1^3, \widetilde{G}_2^3$

## 3 Conclusion

Presenting the task of selecting traits as combinatorial tasks allowed for the construction of an algorithm for optimal risk selection of incorrect classification. The branch and bound method is the optimal method i.e. it allows finding the best, in the sense of a given criterion, solution. This is particularly important for the solution of task 2, where the number of all solutions in the extreme case can be equal $2^n - 2$, if the best solution is to be looked at by reviewing all possible acceptable solutions. an exhaustive search.

For **task 1**, the reduction in the number of solutions reviewed is less visible, but has the advantage that in some cases the solution is faster, and besides, realizing the algorithm at computer you can interrupt calculations after a while to obtain a suboptimal solution.

The authors realize that the algorithm is complicated and requires many calculations. But they take the position that it is better to try to find even a suboptimal set of features from the point of view implemented algorithm of company financial condition evaluation than to select features in a random way.

The reduction in the number of viewed files in the proposed algorithm can be achieved by using, in place of appropriate estimates, measures of incorrect classification that would satisfy the property of multiplicative or additivity to features. However, finding such an estimate is very difficult. Only for some special cases of recognition it is possible to meet these requirements.

## References

1. Karol, T.: Nowe podejście do analizy finansowej, wyd. Oficyna a Kluwer Business, Warszawa (2013)
2. Sobczak, W., Malina, W.: Metody selekcji i redukcji informacji. NT, Warszawa (1985)
3. Walesiak, M.: Uogólniona miara odległości w statystycznej analizie wielowymiarowej, Wyd. AE we Wrocławiu, Wrocław (2002)
4. Wilimowska, Z.: Kombinatoryczna metoda selekcji cech w rozpoznawaniu obrazów na podstawie wzrostu ryzyka. Archiwum Automatyki i Telemechaniki. **25**(3), 405–415 (1980)
5. Wilimowska, Z.: Względna dyskretna ocena ryzyka w szacowaniu wartości firmy. In: Information Systems Applications and Technology ISAT 2002 Seminar. Modele zarządzania, koncepcje, narzędzia i zastosowania. Materiały międzynarodowego seminarium, Karpacz, 11–13 grudnia, Wrocław (2002)

6. Wilimowska Z.: Dyskretny pomiar ryzyka w badaniu kondycji finansowej przedsiębiorstwa. W: Naukovi Praci Kirovograds'kogo Nacional'nogo Technicnogo Universitetu. Ekonomicni Nauki. vyp. 7 (2005)
7. Wilimowska, Z., Wilimowski, M.: Bayesian model of the firm's risk of bankruptcy diagnose. In: Information Systems Architecture and Technology: System Analysis in Decision Aided Problems, Oficyna Wydawnicza Politechniki Wrocławskiej, pp. 69–82 (2009)
8. Wilimowska, Z.: Models of the firm's financial diagnose. In: Information Systems Applications and Technology ISAT 2003 Seminar. Proceedings of the 24th International Scientific School, Szklarska Poreba, 25–26 September 2003, Wroclaw (2003)
9. Wilimowska, Z., Wilimowski, M., Koszalka, J.: Integrated models of the firm's financial diagnose. In: Information Systems Architecture and Technology: Advances in Web-Age Information Systems, PWr, pp. 303–314 (2009)
10. Wilimowska, Z., Tomczak, S., Wilimowski, M.: Selected methods of risk modelling in management process In: Wilimowska, Z., Borzemski, L., Grzech, A., Świątek, J. (eds.) Information Systems Architecture and Technology: Proceedings of 37th International Conference on Information Systems Architecture and Technology, ISAT 2016. Pt. 4, cop. 2017. s. 3–19. Advances in Intelligent Systems and Computing, ISSN 2194-5357, vol. 524. Springer, Cham (2016)
11. Winakor, A., Smith, R.F.: Changes in financial structure of unsuccessful industrial companies. Bureau of Business Research, Bulletin No. 51 (1935)
12. Wyrobek, J.: Efficiency of gradient boosting decision trees technique in Polish companies' bankruptcy prediction. In: Wilimowska, Z., Borzemski, L., Grzech, A., Świątek, J. (eds.) Information Systems Architecture and Technology: Proceedings of 38th International Conference on Information Systems Architecture and Technology, ISAT 2017, Pt. 4, cop. 2018. Advances in Intelligent Systems and Computing. Springer, Cham (2018)

# Influence Threshold Margin on Value Migration Processes

M. J. Kowalski[✉] and Ł. Biliński

Wroclaw University of Science and Technology, Wroclaw, Poland
Michal.Kowalski@pwr.edu.pl

**Abstract.** The paper presents results of empirical research on the relationship between reaching threshold margin by the companies and value migration processes. The analyzes were conducted based on companies listed on the Polish capital market, data from 2001–2014 were used. The dependence between the surplus of the company's profitability and threshold margin determined according to the Rappaport formula and the assignment companies to the value migration phase according to Siudak was examined. The research has shown that achieving ROS over the minimum margin is not the only factor conditioning value creation, other non-financial factors can be decisive. In the group of companies qualified for the value building phase, there were companies that did not reach the threshold margin. Moreover, on average in the entire research sample, the operating profitability was lower than the threshold margin by over 3% points, although the median in the group of companies included in the value building phases is positive. It was also noted that companies which are qualified for value building phase showed a lower negative difference between profitability and threshold margin than companies qualified for the stabilization phase and value destroying phase. The statistical significance of these results was confirmed.

**Keywords:** Value migration processes · Threshold margin · Enterprise value

## 1 Present State of Knowledge and Research Aims

### 1.1 Threshold Margin Concepts

In the literature subject there are numerous examples of mainly model research which purpose is to analyze the relationship between the value drivers and their impact on the enterprise value. These studies focus primarily on the threshold margin concept initiated by Rappaport (1995). Rappaport introduces the concept of threshold margin, directly related with value drivers and additional threshold margin (Rappaport 1995). He defines the threshold margin as: "the minimum operating profit margin that company must achieve in any period in order to preserve shareholder value during this period (Rappaport 1995)." The development of this definition is the statement that the operating profit margin is the minimum rate of return achieved by the company. Rappaport presents two approaches of using threshold margin. The first one refers to additional sales units (additional threshold margin) (Rappaport 1995). The second is focused on total sales. Threshold margin is suitable both for evaluation the company's

© Springer Nature Switzerland AG 2020
Z. Wilimowska et al. (Eds.): ISAT 2019, AISC 1052, pp. 26–35, 2020.
https://doi.org/10.1007/978-3-030-30443-0_3

current financial results as well as future ones. The additional threshold margin is an elementary component for determining the total margin and constitutes a "break-event-point" from the point of view of the company's ability to build value in relation to additional sales (Dudycz 2005). In order to collapse shareholder value, the company must obtain a minimum value. This means that additional sales are treated as an investment aimed at providing an additional margin. This margin must be sufficient for allocating cash to working capital and fixed assets that are necessary to obtain new sales.

As a development of the Rappaport's concept Dudycz proposes a correction of Rappaport's formulas: "anticipating the investment in relation to the cash flows received from it" (Dudycz 2005) and assumes two conditions: (1) the occurrence of investment at the beginning of the period, which means that cash flows will appear at the end of this period, (2) the occurrence of investment in fixed assets at the beginning of the period and in working capital at the moment of growing sales, parallel to its growth, which means investing at the end of the period.

The Balachandran is also presenting his theory related with threshold margin. Similarly, to Rappaport and Dudycz, Balachandran states that in order to build company's value it is necessary to achieve a minimum profit margin. He considers three market situation cases. The first model concerns the overall development and growth of the enterprise value, the second takes into account the phenomenon of inflation, and the third concerns price fluctuations that affect the entire process of creating the company's value (Balachandran et al. 1986). He also prepared a proposal for calculating the marginal value for each model, the aim of which is to balance the funds invested in working capital with the demand for investment.

The modification of Rappaport, Dudycz and Balachandran's considerations about threshold margin is introduced by M.J. Kowalski. Kowalski introduces the concept of the minimum profitability of value creation and denotes it as the MIN VC RentEBIT symbol (Kowalski and Świderski 2005).

The threshold margin theory is an analytical tool enabling to choose strategies that ensure a possible company's value increase and eliminate strategies that cause a value drop. Threshold margin becomes the central theory in the analysis of value-shaping factor and the business portfolio building tool. The concept of threshold margin allows for effective decision making regarding the selection of appropriate business units for the investment portfolio and selection development strategies for these companies to maximize the value of the entire investment and the entire portfolio.

Studies on threshold margin indicate that, if company doesn't achieve a minimum margin it's not able to build its value by increasing sales. Based on preliminary models, we showed the dependence in our studies, illustrating the impact of the value drivers on the threshold margin and consequently on the company's value based on classic DCF model. We showed that there is an interdependence between the value drivers, threshold margin and company's value (Kowalski and Biliński 2016a, b). However, these dependencies weren't verified empirically. We haven't seen the research which presents relations between reaching by the company a minimum margin and its value measures, in particular market measures.

## 1.2 Value Migration Processes - Theory and Concepts

The value migration processes have been the subject of research since the 1990s. The situation when company's value decreases Slywotzky, Morisson and Andelman called value migration process (Slywotzky et al. 1997). It can be defined as a situation when risk and resulting cost of capital are higher than company's financial results. It means that business unit becomes unprofitable for investors. Moreover, the company's value decreases. Capital which was invested is relocated to other business units which are more attractive, and which give the chance to create stronger market positions. According to Slywotzky, value migration process shows directions for creating shareholder value (Slywotzky 1996).

According to Brabazon value migration process is flow of opportunity to growth, make a profit and increase shareholder value added from one company or sector to other (Brabazon 1999). Based on it, it's hard to say, using financial terms, how should we measure "flow of opportunity". Philips defined value migration as a shareholder asset return on profits, which moves between business units of low attractiveness for investors (value outflow) to companies with the highest growth potential (value inflow) and with the highest possible return (Philips et al. 2012). Siudak pointed on effectiveness allocation as the result of value migration and understood value migration as process based on outflow the amount of value from one business to another as the result in searching effectiveness possibilities capital allocation (Siudak 2000). According Siudak value migration means flow of the value between companies as the result of searching effectives business models which can create shareholder value, including the highest returns and the lowest risk (Siudak 2013). Currently available literature underlines the importance and meaning of the value migration Billington (1997), Donol (1997), Griffiths (1997), Slywotzky and Linthicum (1997), Baptista (1999), Brabazon (1999), Siudak (2000), Strikwerda et al. (2000), Campbell (2001), Moster and Moukanas (2001), Herman and Szablewski (1999), Sharma et al. (2001), Klincewicz (2005), Owen and Griffiths (2006), Slywotzky et al. (2006), Wiatr (2006), Woodard (2006), Szablewski (2008), Szablewski (2009), Woźniak-Sobczak (2011), Jabłoński (2013), Siudak (2013), Skowron (2014). In the mentioned papers authors only discuss about the value migration topic as a theoretical consideration. Some of them attempt to define this phenomenon.

Slywotzky proposed besides a three-phase model to measure and analyze the value migration process, which main idea is to classify an enterprise into one of three phases: outflow value, inflow value and value stability (Slywotzki 1996). His concept was developed by Siudak and his linear ordering method based on calculating synthetic variable, Synthetic Index of Valuation (Siudak 2013), which enables quantitative classification sectors and companies for the phases specified. Proposed measure based on measuring market value added MVA.

The construction of index proposed by Siudak is based on three diagnostic variables: share in the migration balance of the economy, share in the sector's migration balance, and change of market value added to market capitalization (Siudak 2013). In the context of factors determining the value of enterprises and sectors as well as the value migration, they can be identified with value drivers.

One of the characteristic features of the migration phenomenon is its instability, which perfectly illustrates the search for the most effective investment opportunities. Our sectoral analysis in the years 2010–2014 for the Polish Capital Market has shown that the value migration to the Main Market and New Connect demonstrates a strong fluctuation of enterprises and sectors between the outflow, stabilization and inflow (Kowalski and Biliński 2016a, b). Our research has also shown that the process of value migration is unstable in 2010–2014, thus confirming the thesis put forward by Siudak who covered in his research the period in years 2003–2006.

## 1.3 Research Aims

In our research, we analyze the interdependence between value drivers and value migration process. In our study from 2017, we proved on the example of the Polish Capital Market, that companies located in different value migration phases have different value levels of value drivers and that value migration processes are accompanied by changes in the company's value factors. Company's transition from build-to-destroy or destroy-to-build value migration phases are accompanied by a significant change in value drivers, such as sales dynamics, profitability and fixed assets profitability (Kowalski and Biliński 2018). The purpose of this paper is to supplement previous results by the threshold margin results. The aim of this paper is to empirically analyze the dependence between achieving minimum margin and value migration process.

Based on our previous paper (Kowalski and Biliński 2016a, b) it can be said that there is a similar interdependence between phases of value migration processes and threshold margin, what will be the subject of our research, as we proved between threshold margin, value drivers and company's value.

In particular, we want to show whether achieving profitability of sales above the threshold margin determines the building of value and if the process of value inflow and value outflow depends on the threshold margin achievement by the company.

# 2 Empirical and Research Findings

## 2.1 Database Structure

Empirical research was conducted on companies listed on the Warsaw Stock Exchange. The research sample includes both companies listed on Main Market, as well as on the New Connect. We used the company's financial statements (singular and consolidated) published at the end of the year for the years 2001–2014 as the initial research material.

First, we have separated companies with complete and standardized financial statements. Then based on the data available on the Warsaw Stock Exchange and the data published on the Damodaran's science website, we have classified the companies into 29 separate economic sectors.

To determine the market capitalization, we used the average market valuation of the shares in a given year, based on daily quotes.

Originally, the material included 2656 observations understood as company data in a given year. Because the calculation of threshold margin is based on the annual changes in the working capital and fixed assets, the number of observations decreased to 1910. In our research procedure, we rejected companies that have negative sales profitability as well as extreme observations, which finally allowed the research sample of 1147 observations.

## 2.2   Research Procedure

In our study, we analyzed migration phases based on model of (Siudak 2013). To investigate the process of value migration on the Polish Capital Market as well as the allocation and companies to one of the three migration phases, we used the linear ordering method proposed by Siudak. Additionally, using proposed modification, involving the use of consolidated data, not individual data (Kowalski and Biliński 2016a, b).

First, we calculated synthetic index of value migration for each company in a sample and assigned companies into three migration phases: destroy (D), stability (S) and build (B). We classified companies separately for each year using Siudak approach. The Siudaks method is based on market value added, and therefore we analyzed also the following measurers: MVA, dMVA, MVA/IC.

To classify companies and sectors to value migration phases, we have identified three diagnostic variables that have been normalized and then used to calculate the synthetic variable. We have extracted the input data to calculate threshold margin.

We used the classic Rappaport model to calculate threshold margin (Rappaport 1981):

$$p'_{min} = \frac{(f+w)k}{(1-T)(1+k)} \tag{1}$$

where:

p' - ΔEBIT/Δsales, i.e., incremental operating margin on incremental sales,
f - capital expenditures minus deprecation per money unit of sales increase,
w - cash required for net working capital per money unit of sales increase,
k - weighted average cost of capital,
T - income tax rate.

In the WACC calculation, we used the classic CAPM model to determine the cost of equity. Data for WACC estimation were adopted in the following way (Table 1):

Finally, we analyzed the difference between the operating profitability of sales (ROS), where ROS was set as EBIT/Revenue, the threshold margin (TH).

## 2.3   Research Results and Discussion

Table 3 presents obtained results of analysis for adopted variable. Descriptive statistics are presented. Data were presented in the relation to the entire research sample and companies assigned to the three phases of value migration based on modified Siudak's model (Table 2).

**Table 1.** WACC calculation

| Variable | Definition |
|---|---|
| Risk free | Defined as an interest on 10-year treasury bonds in a given year of analysis |
| Beta | Determined on the basis of company's D/E ratio and Damodaran's formula |
| Unlevered beta | Defined as the sector beta, to which was assigned company specified on the basis of Damodaran Beta data, unlevered beta and other risk measures |
| Risk premium | Determined on the basis of Country Default Spreads and Risk Premiums by Damodaran |
| Debt cost | Defined as risk free increased by the risk premium on the basis of Country Default Spreads and Risk Premiums by Damodaran |

**Table 2.** Descriptive statistics for ROS-TH

| ROS - Th | N | Avg | Q1 | Med | Q3 | Var | St. Dev |
|---|---|---|---|---|---|---|---|
| All | 1 147 | −0.0302 | −0.0710 | −0.0040 | 0.0333 | 0.0123 | 0.1110 |
| B | 333 | −0.0077 | −0.0488 | 0.0120 | 0.0557 | 0.0134 | 0.1159 |
| S | 556 | −0.0342 | −0.0707 | −0.0066 | 0.0231 | 0.0101 | 0.1003 |
| D | 258 | −0.0508 | −0.1073 | −0.0134 | 0.0274 | 0.0148 | 0.1215 |

Table presents descriptive statistics for the difference between the operating profitability of sales and threshold margin (ROS-TH). The data was presented for whole analyzed sample and separately for companies classified to destroy (D), stability (S) and build (B) migration phases. N – indicate numbers of observations, AVG – average, Q1, Q3 – first and third quartile respectively, Med – median, Var – variance, St. Dev. – standard deviation.

**Table 3.** Results of range tests

| Panel A: Statistics | |
|---|---|
| F | 25.44*** |
| K-W | 32.80*** |
| Panel B: Pairs comparison | |
| D vs B | ***/*** |
| D vs S | / |
| S vs B | ***/*** |

The one-way ANOVA and Kruskal-Wallis tests (K-W test) were used to compare results. (*), (**) and (***) indicate that the differences are significant at 10, 5 and 1% levels of significance, respectively. Panel A presents the results of range tests, panel B presents results of pairs comparison. Destroy (D), stability (S), and build (B) identify migration phases. The results of post-hoc Tukey's test (the first indicator before the slash) and non-parametric Kruskal-Wallis test (the second indicator after the slash) are presented. Statistical significance is identified by (*), (**) and (***) in the same way

The obtained results can be considered surprising. On average, in the entire research sample analyzed, companies achieve profitability below the threshold margin calculated based on Rappaport's model. On average, the analyzed companies achieve the operating profitability of sales lower than the threshold margin by over 3% points. We also carried out our analyzes by setting a threshold margin on longer 2- and 3-year trends in change in sales and working capital as well as fixed assets. Then, the difference between the operating profitability of sales and threshold margin was smaller, while it remained negative[1].

Based on the obtained results, it can be stated that empirical data doesn't confirm that the financial value drivers clearly determine the potential of companies to value creation. It should be assumed that the potential of companies to value creation is further shaped by other factors, including probably non-financial which is widely mentioned in the literature. Marcinkowska identified 13 non-financial and off-balance sheet value drivers that determine company's value (Marcinkowska 2000).

Additionally, obtained results indicate one more distinct trend. The analyzed difference between ROS and TH is the largest among in the companies classified to destroy value phase and amounts to −5.1% point and the smallest in the group of companies assigned to value building phase is 0.77% point. For companies in stabilization phase the analyzed parameter reaches the value −3.4% point. It should be also noted that the median difference between ROS and TH for companies in the value building phase is positive and amounts to +1.2% point. Detailed analysis shows that over 60% of companies assigned to the value building phase show an operating margin of sales higher than the Rappaport's threshold margin.

The purpose of further analysis was to verify whether the observed differences between the results could be considered statistically significant. For this purpose, parametric and non-parametric statistical test were carried out.

The obtained results confirm the initial assumptions. Companies in the phase of building value achieve lower differences between the operating profitability of sales and threshold margin than companies classified to the stabilization and value destruction phases and these differences are statistically significant. We also noted that companies in the phase of value destruction achieved greater difference between the operating profitability of sales and threshold margin then companies in the stabilization phase. However, statistical tests don't show that these differences can be considered statistically significant. The obtained results indicate that achieving profitability adequate to the capital demand related to new sales is one of the factors determining company's value creation.

## 3   Conclusion

The research has shown that achieving ROS over the threshold margin is not the only factor determining value creation, and other non-financial factors can be decisive. In the group of companies classified to the value building phase, there were companies that didn't reach the threshold margin, moreover, on average in the research sample the

---

[1] We didn't decide to use this data for analysis because this approach significantly reduced the research sample size.

operating profitability was lower than threshold margin by over 3% points, though the median in the group of companies included in the value building phase is positive. It was also noted that companies classified to the value building phase showed a lower negative difference between profitability and threshold margin than companies classified to stabilization phase and the value destroy phase. The statistical significance of these results was confirmed.

The obtain results lead to the formulation of further research directions related with value migration processes. First of all, in our study, we analyzed the short-term impact of the dependence between the operating profitability of sales and threshold margin on the value building phase. This influence may be more unambiguous in the long-term dimension. While other than financial factors may have an impact on the value inflow to companies that don't reach the threshold margin in the short-term associated with investors discounting the belief that this trend will reverse in the future, this long-term failure to reach the threshold margin should clearly indicate the value outflow and allocate the company in the phase of value destruction. Our results undoubtedly strengthen the existing voices in the literature about the need to search for models describing the value that would take into account non-financial value drivers. It seems to be particularly important in the short-term perspective. In our results, we applied the basic model of Rappaport's threshold margin, the use of another extended concept could bring other results, however it requires the collocation of a large amount of additional data and acceptance of additional assumption.

# References

Balachandran, B.V., Nagaraja, N.J., Rappaport, A.: Threshold margins for creating economic value. Financial Management, Spring 1986

Baptista, J.P.A.: The new profit zones: growing shareholders value in the communications industry. Commentary, Mercer Management Consulting (1999)

Brabazon, T.: Value migration. Where is the value in your industry going? Accountancy 31(3) (1999)

Billington, J.: Understanding value migration and putting it to work. Harvard Management Update 2(6) (1997)

Campbell, M.K.: The ascent of the customer. IEEE Potentials 20(2) (2001)

Donol, J.P.: The great value migration. Chief Exec. 123, 64–73 (1997)

Dudycz, T.: Zarządzanie wartością przedsiębiorstwa. Polskie Wydawnictwo Ekonomiczne, Warszawa (2005)

Griffiths, E.C.: Where does your profitability reside? Equip. Leasing Today 9(5) (1997)

Herman, A., Szablewski, A.: Orientacja na wzrost wartości współczesnego przedsiębiorstwa. In: Herman, A., Szablewski, A. (eds.) Zarządzanie wartością firmy, wydawnictwo Poltext, Warszawa, pp. 13–56 (1999)

Jabłoński, A., Jabłoński, M.: Cykl życia wartości przedsiębiorstw wobec kondycji modelu biznesu. Kwartalnik Nauk o Przedsiębiorstwie, Nr. 4 (2013)

Klincewicz, K.: Przemiany w branży high-tech i zjawisko migracji wartości. Studia i materiały Nr 2, Wydział Zarządzania Uniwersytetu Warszawskiego, Warszawa (2005)

Kowalski, M.J., Świderski, K.: Wpływ dynamiki sprzedaży na wycenę przedsiębiorstwa. Prace Naukowe Uniwersytetu Ekonomicznego we Wrocławiu, nr 252, Wrocław, pp. 248–262 (2005)

Kowalski, M.J., Biliński, Ł.: A value-oriented quantitative model for strategy formulation regarding strategic business units. The essence and measurement of organizational efficiency. In: Springer Proceeding in Business and Economics, pp. 97–121. Springer, Switzerland (2016a)

Kowalski, M.J., Biliński, Ł.: Migracja wartości-wyniki badań empirycznych dla polskiego rynku kapitałowego. Zeszyty Naukowe Uniwersytetu Szczecińskiego, Finanse Rynki Finansowe, Ubezpieczenia, Szczecin, nr 4/2016 (82) cz.2, pp. 179–190 (2016b)

Kowalski, M.J, Biliński, Ł.: Interdependence between value driver and value migration processes. Evidence from Warsaw stock exchange efficiency in business and economics. In: Springer Proceedings in Business and Economics, pp. 99–118. Springer, Switzerland (2018)

Marcinkowska, M.: Kształtowanie wartości firmy. Wydawnictwo Naukowe PWN, Warszawa (2000)

Moser, T., Moukanas, H.: Finding the right drivers of value growth. With so many options, which initiatives really matter? The value growth agenda. Mercer Manag. J. **13** (2001)

Owen, D., Griffiths, R.: Mapping the Markets. A Guide to Stock Market Analysis. Bloomberg Press, New York (2006)

Philips, P.: Time to revisit value migration. Business Corner, "Strategies & Analysis". Rodman Publishing, July 2012

Sharma, A., Krishnan, R., Grewal, D.: Value creation in markets. A critical area of focus for business-to-business markets. Ind. Mark. Manag. **20** (2001)

Rappaport, A.: Selecting Strategies that Create Shareholder Value, pp. 139–149. Harvard Business Review, Brighton (1981)

Rappaport, A.: Creating Shareholder Value: A Guide for Managers and Investors. The Free Press, A Division of Simon & Schuster Inc., New York (1995)

Siudak, M.: Analiza migracji wartości w Polsce. In: Knosala, R. (ed.) Materiały konferencyjne: Komputerowo zintegrowane zarzadzanie, WNT, Warszawa, pp. 170–179 (2000)

Siudak, D.: Pomiar procesów migracji wartości przedsiębiorstw na polskim rynku kapitałowym, wyd. C.H. Beck, Warszawa (2013)

Skowron, S.: Migracja wartości w warunkach sieci organizacyjnych. In: Gregorczyk, S., Sopińska, A. (eds.) Granice strukturalnej złożoności organizacji, pp. 163–172. Szkoła Główna Handlowa, Warszawa (2014)

Slywotzky, A.J.: Value Migration. How to Think Several Moves Ahead of the Competition. Haravard Business School Press, Boston (1996)

Slywotzky, A.J., Morrison, D.J., Andelman, B.: The Profit Zone: How Strategic Business Design Will Lead You to Tomorrow's Profits. Random House, New York (1997)

Slywotzky, A.J., Baumgartner, P., Alberts, L., Moukanas, H.: Are you enjoying globalization yet? The surprising implications for business. J. Bus. Strategy **27**(4), 23–32 (2006)

Slywotzky, A.J., Linthicum, F.: Capturing value in five moves or less: the new game of business. Strategy Leadersh. **25**(1) (1997)

Strikwerda, J., Amerongen, M., Wijck, M.: How do firms adapt their organization to the new or informational economy? Research Memorandum #5, Nolan Norton Institute (2000)

Szablewski, A.: Migracje kapitału i wartości w warunkach globalnej niestabilności. Kwartalnik Nauk o Przedsiębiorstwie **3**(8) (2008)

Szablewski, A. (ed.): Migracja kapitału w globalnej gospodarce. Difin, Warszawa (2009)

Wiatr, M.: Migracja wartości a procesy przedsiębiorczości. In: Strużycki, M. (ed.) Przedsiębiorczość w teorii i praktyce. Szkoła Główna Handlowa w Warszawie, pp. 27–47 (2006)

Woodard, J.C.: Architectural control and value migration in platform industries. In: Academy of Management 2006 Annual Meeting, Atlanta, August (2006)

Woźniak-Sobczak, B.: Warunki współczesnego tworzenia wartości przedsiębiorstwa. Wyższa Szkoła Ekonomiczno-Społeczna w Ostrołęce, Zeszyty Naukowe Nr 8 (2011)

# Extrapolation of Maxima with Application in Chi Square Test

Mariusz Czekała[1]([⊠]), Agnieszka Bukietyńska[1], Marek Gurak[2],
Jacek Jagodziński[3], and Jarosław Kłosowski[1]

[1] Wroclaw School of Banking, Finance and Management Department,
Wroclaw, Poland
{mariusz.czekala,
agnieszka.bukietynska}@wsb.wroclaw.pl,
j.klosowski@mongos.pl
[2] MG4-LTD Co., Wroclaw, Poland
m.gurak@mg4.eu
[3] Faculty of Electronics, Wroclaw University of Science and Technology,
Wroclaw, Poland
jacek.jagodzinski@pwr.edu.pl

**Abstract.** The paper presents the method of estimating the parameters of extreme statistics distribution by the maximum likelihood method. Knowing the distribution of extreme statistics enables the prediction of the maximum value for periods outside the analysed sample. Such forecasting has many applications. From the analysis of weather phenomena, through floods to financial data. An important part of the work are examples related to the error assessment when making decisions. The examples relate to the selection of the optimal profile when assessing the demand for clothing goods. The method of regression (or classification) trees used in this problem requires at each step the process of selecting the profiles by means of the chi square statistics. For a small number of profiles, one can view all the options and select the highest value for the chi square statistic. However, if the number of possible features increases to several dozen, finding the optimal solution is not possible in a reasonable time. The issue of choosing the optimal profile is equivalent to the problem of set partition.

**Keywords:** Recursive partitioning · Decision trees · Extreme statistics · Bootstrap · Clothing industry

## 1 Introduction

The work concerns the application of order statistics in the method of recursive partition [1]. The equivalence of the analysed problem with the problem of the partition of the set is presented. The next part presents data on the decision problem in the clothing industry. A brief overview of the selected facts from the order statistics theory along

This is a part of the Project implemented as part of the Program Operacyjny Inteligentny Rozwój 2014–2020 of Narodowe Centrum Badań i Rozwoju. This is a part of the Project title: "Demand forecasting system controlled by dynamic fashion trends for clothing, textile and lingerie enterprises". Number of application: POIR.01.01.01-00-0886/17-00.

Z. Wilimowska et al. (Eds.): ISAT 2019, AISC 1052, pp. 36–48, 2020.
https://doi.org/10.1007/978-3-030-30443-0_4

with the maximum likelihood method was presented. A method based on drawing partitions was proposed. The last part of the article is devoted to the analysis of the error resulting from the replacement of parameters from the population with the parameters from the sample.

## 2   The Problem of Partitioning the Set into Classes

The problem of partitioning a set into classes is a problem known in combinatorics for a long time. It consists in finding the number of various ordered partitions of the k-element set for r-classes. It can be proved that the number of these partitions is [7]

$$\frac{1}{r!}\sum_{s=1}^{r}(-1)^{s}\binom{r}{s}(r-s)^{k}$$

If $S(k, r)$ is the number of different partitions of the k-element set into r classes, the following recursive formulas take place [7]

$$S(k, k) = 1, \quad S(k+1, r+1) = S(k, r) + (r+1)S(k, r+1)$$

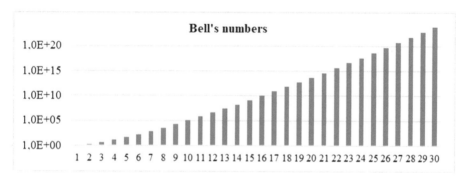

**Fig. 1.** Bell's numbers – logarithmic scale

The numbers $S(k, r)$ are called Stirling numbers, and the numbers that are their sums are called Bell's numbers. One can express them using the formula

$$B_n = \sum_{k=0}^{n} S(n, k)$$

In this work, Bell's numbers $B_{10}$ and $B_{20}$ will be of particular interest as numbers related to the analysed decision problem. By using the formulas above, one can calculate that

$$B_{10} = 115\,975 \text{ and } B_{20} = 51\,724\,158\,235\,372$$

## 3   Contingency Tables - A Decision Problem

The chi square test is one of the most important tests for studying the relationship between attributes on a nominal scale. In the recursive partition method [2, 15], the chi-square statistic is the basis for making decisions. In practice, this depends on the selection of such a set of profiles for which this statistic reaches its maximum. It turns out that the computational complexity of this problem is increasing exponentially, as evidenced by the data from Fig. 1.

As can be seen in Fig. 1, the problem of determining all set sets of partitions is the most time-consuming element of the tested algorithm. In fact almost every available math package provides functions to solve this problem. For Matlab there is available the `SetPartition()` function [13], in the R software [20] is known as `setparts ()` [9], and it can be found also in Mathematica [19] under the name `SetParti- tions()` [17]. For research purposes, the generation times of profiles sets were estimated. The Wolfram Mathematica was used as well as the Intel Core i7-8550U 1,80 GHz computer with 16 GB RAM. Based on the calculation times for n < 15, it was estimated that in the case of n = 20, the computation time will reach several years, and for $n = 30$ it is about $10^7$ years.

Table 1 presents the contingency table for a specific problem concerning the optimal selection (with the criterion chi square) of the clothing industry characteristics. Five variants of colours and two variants of the type of product are considered here. In the case under consideration, there are therefore ten product variants, which means that the selection of the optimal partition of the product set is equivalent to the analysis of the partition of a 10-element set into r classes, with the number of classes not known. This problem was considered in the work [4]. The method proposed there, consisting of drawing 100 partitions, gave a suboptimal solution. It also turned out that in this case, using the Mathematica program [19], it was also possible to find an accurate solution to review all 115 975 cases. To avoid having to consider three-dimensional tables in Table 1, the value of an explanatory variable was placed next to the value of one of the features. In this case, it is a demand-related variable (1-satisfying demand, 0-unsatisfactory). The basic goal is to get the best fit model explaining the demand for clothing items. In Table 1, the MX0 symbol means boxer briefs with a satisfactory

**Table 1.**  Contingency table

|     | C1 | C2 | C3 | C4 | C5 | SUM |
| --- | --- | --- | --- | --- | --- | --- |
| MX0 | 38 | 78 | 748 | 90 | 87 | 1041 |
| MX1 | 44 | 109 | 2314 | 83 | 97 | 2647 |
| MP0 | 68 | 528 | 52 | 1583 | 1302 | 3533 |
| MP1 | 74 | 582 | 86 | 1559 | 1462 | 3763 |

Source: Data obtained from MG4 Company

**Table 2.** Colour and type

| Features | C1 | C2 | C3 | C4 | C5 |
|---|---|---|---|---|---|
| MX | 82 | 187 | 3602 | 173 | 184 |
| MP | 142 | 1110 | 138 | 3142 | 2764 |
| Sum | 224 | 1297 | 3200 | 3315 | 2948 |

Source: Table obtained from Table 1.

margin size (subjectively determined threshold). Similarly, in the case of panties briefs, where the MP symbol was used. The symbols C1 to C5 are colour groups.

The partition problem concerns data from Table 2, and one of the goodness of fit criteria is the chi square value. A number of other measures of this type are given in Gatnar [8].

An example of a subdivision into 4 separate subsets could look like this: $MX \times (C_1 \cup C_2 \cup C_3 \cup C_4)$ and $MX \times C_5$ and $MP \times (C_2 \cup C_3)$ and $MP \times (C_1 \cup C_4 \cup C_5)$.

From a practical point of view, the solution proposed as a result of drawing 100 divisions turned out to be satisfactory because it was not much different from the exact solution. In this work, however, we intend to provide another universal method for assessing the error made as a result of replacing the population with a sample.

In fact, the practical problem is usually more complex. In addition to the features listed in Table 1, size, pattern, price, design and other features should be considered. The method proposed in the work gives an approximate solution together with the necessary error assessment in this case. The solution will use the order statistics theory.

## 4 Order Statistics

The order statistics theory provides many tools to assess the probability of extreme events. In the initial period mainly dealing with sequences of independent random variables were dealt with. After [12] introduced $D(u_n)$ and $D'(u_n)$ conditions, many claims of this theory were extended to stationary processes, and in some cases [10], non-stationary processes. Often, the conditions required in the aforementioned claims are impossible or difficult to check, so practitioners try to match existing models with trial and error.

Let $M_n = \max(X_1, X_2, \ldots, X_n)$ be a random variable whose distribution is considered. Where the variables $X_1, X_2, \ldots, X_n$ are i.i.d, it can be shown that the random variable $M_n$ has either a Gumbel distribution, either a Frechet distribution or a Weibull distribution [12]. It turns out that a similar thesis can be obtained for a wide class of stationary processes, and even for some cases of non-stationary processes. This justifies the attempt to describe the phenomena analysed by means of the maximum distribution even in the case of inability to check some assumptions. This article will present a part of the research resulting in the presentation of a model proposal, thanks to which it is possible to obtain an assessment of the error resulting from the replacement of the population with the sample. We will attempt to describe the error assessment method using the Gumbel distribution. For this distribution, the distribution function is:

$$F(x) = \exp(-e^{-x}), x \in R$$

After taking into account the location parameter and the shape parameter of the distribution function, distribution function of Gumbel takes the form:

$$F_{\mu,\sigma}(x) = \exp\left(-e^{-(x-\mu)/\sigma}\right), x \in R$$

The $\mu$ and $\sigma$ parameters in our case are not known. There are only known observations regarding the values of chi square statistics from a random sample of partitions. The size of this sample is 100. One can calculate the (one) maximum based on the sample one have taken. Of course, this is not enough to estimate any distribution parameters. At work [6], it is postulated that the original data should be:

$$\left.\begin{aligned}
X^{(1)} &= \left(X_1^{(1)}, X_2^{(1)}, \ldots, X_s^{(1)}\right) \\
X^{(2)} &= \left(X_1^{(2)}, X_2^{(2)}, \ldots, X_s^{(2)}\right) \\
&\phantom{=}\cdots \\
X^{(n)} &= \left(X_1^{(n)}, X_2^{(n)}, \ldots, X_s^{(n)}\right)
\end{aligned}\right\} \tag{1}$$

Based on this data, an $n$-element sample of maxima can be obtained and unknown parameters can be estimated on this basis. Such a procedure is admissible in hydrological research, where observations were available anyway, which leads to some contradiction in the issue under consideration. Estimating parameters with this method means the availability of $n$ times $s$ observations. If so many observations would be available then one could certainly guess the maximum with more precision than on the basis of one series. However, if such an increase in the sample would be possible, we would suggest using the method presented here for the increased sample. Of course, the increase in the sample is also possible in the analysed case, but the method is aimed at limiting calculations due to computational complexity. Therefore, the work will show how to get the parameter estimation based on a single series. For this purpose, the bootstrap [5] method will be used. The essence of this method is the re-randomization of the realization of the chi-square values from the drawn set of values. After receiving the bootstrap samples, the maximum likelihood method can be used to estimate the parameters. In the case of the distribution function, the likelihood function is in the form of:

$$L(\mu, \sigma, X) = \prod_{i=1}^{n} \exp\left(-e^{-\frac{x_i-\mu}{\sigma}} - (x_i - \mu)/\sigma\right)\frac{1}{\sigma}$$

The estimator of the maximum likelihood is expressed by the formula:

$$(\hat{\mu}, \hat{\sigma}) = argmaxL(\mu, \sigma, X)$$

The logarithm of the likelihood function is more convenient in the analysis [6]:

$$\text{Ln } L(\mu, \sigma, X) = -\sum_{i=1}^{n} \exp\left(-\frac{x_i - \mu}{\sigma}\right) + \sum_{i=1}^{n} \left(-\frac{x_i - \mu}{\sigma}\right) - n \ln \sigma \quad (2)$$

After differentiation with respect to $\mu$ and $\sigma$, a system of two non-linear equations is obtained

$$\frac{\partial lnL(\mu, \sigma, X)}{\partial \mu} = \frac{1}{\sigma} \sum_{i=1}^{n} \exp\left(-\frac{x_i - \mu}{\sigma}\right) - \frac{n}{\sigma} = 0$$

and

$$\frac{\partial lnL(\mu, \sigma, X)}{\partial \sigma} = -\sum_{i=1}^{n} \left(\exp\left(-\frac{x_i - \mu}{\sigma}\right)\right)\left(\frac{x_i - \mu}{\sigma^2}\right) + \sum_{i=1}^{n} \left(\frac{x_i - \mu}{\sigma^2}\right) - \frac{n}{\sigma} = 0.$$

After simple algebra one can get a system of two equations to solve.

$$\sum_{i=1}^{n} \exp\left(-\frac{x_i - \mu}{\sigma}\right) - n = 0, \quad (3)$$

$$-\sum_{i=1}^{n} \left(\exp\left(-\frac{x_i - \mu}{\sigma}\right)\right)\left(\frac{x_i - \mu}{\sigma}\right) + \sum_{i=1}^{n} \left(\frac{x_i - \mu}{\sigma}\right) - n = 0. \quad (4)$$

By solving this system, one will get the evaluation of the desired parameters.

## 5  Applications in Clothing Industry

In 2022 fashion products market in Poland will reach the value of 43 billion PLN. On the one hand, we have a growing group of conscious customers looking for clothing, lingerie, footwear and fashion gadgets on the market. The interest in fashion and image is not only the domain of young women, it is estimated that 68% of society is interested in and responds to fashion trends. On the other hand, the Polish clothing, footwear and lingerie industry after the period of stagnation and crisis at the end of the last century is experiencing a significant boom. Several Polish clothing and footwear companies are listed on the Warsaw Stock Exchange. The annual revenues of the largest of them are counted in billions of PLN, and the net profit - in hundreds of millions. Polish companies such as LPP, CCC, or companies from the Białystok Lingerie Cluster are recognizable and present on the European market and in the case of women's underwear also in Asia. By increasing revenues by continually increasing production, they still bear the risk of a "fashion caprice". All previously produced collections can suddenly become unsellable (dead stock). Due to the speed of turnover, the shorter seasons of fashion trends [18], globalization and the influence of trend-setter, social media portals, aggressive advertising campaigns, placement products, producers of fashion products are exposed to an unacceptable product by an increasingly capricious and demanding market. The phenomenon of fast fashing, disposable collections, the use of new (usually cheaper) materials and methods of dyeing puts the manufacturer in fear of incurring losses as a result of proposing an unsuccessful collection. All this testifies to the growing role of the clothing industry. Therefore, any errors in the selection of the assortment may prove to be costly both from an economic, social and environmental point of view. This work is to help find solutions close to optimal.

The next step is to draw partitions. This step was made using the "sample" function available in the [20]. In the first example, the case of 10 cells will be considered. As is known in this case, the number of possible partitions of a 10-element set is 115 975.

**Example 1.** A case of 10 cells.

The basis here is the contingency Table 1. On the basis of drawn partitions, the chi-square statistics and the corresponding p-value are calculated. The chi square values are shown in Fig. 1.

In this case, the maximum value calculated on the basis of the sample is 533,004. Using the Mathematica program, the calculation of all possible values of the chi square statistics was possible. The maximum value obtained from all 115 975 cells is 536,094 [4]. By treating the chi square as a measure of the strength of dependence, one can notice a small difference between the value obtained from the sample of 100 and the population of 115 975 items (Fig. 2).

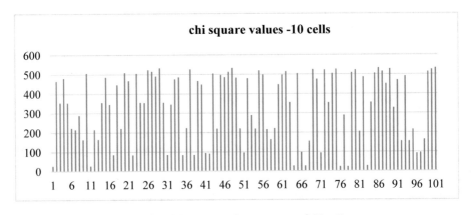

**Fig. 2.** Chi squares values - a case of 10 cells

In the first example, an empirical verification of the error resulting from the replacement of the maximum value in the population with the appropriate value in the sample proved to be possible. In the case where the number of cells is 20, the computational complexity makes the number of possible compositions reach 52 trillion. In this case, empirical verification is not possible. The second example presents data on a case of 20 cells. This is a generalization of the data contained in the contingency Table 1. The second example includes an additional feature - collection.

**Example 2.** 20 cells.

The collections were divided into two groups – base collection and non-base [3] collection (other collections). This partition results in the contingency Table 2 increasing its size to 20 cells. Each of the pair of attributes can occur in two variants when it comes to the collection.

In this case, the maximum value calculated on the basis of the sample is 701,28 [3].

Unlike the first example, empirical verification is not possible this time. Estimated computation time on generally available hardware would most likely exceed one year.

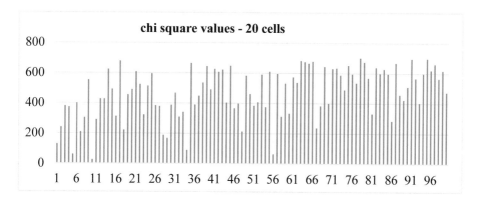

**Fig. 3.** Chi squares values - a case of 20 cells

The second example indicates that even very efficient numerical programs are not enough to assess the error resulting from replacing the maximum in the population with its counterpart in the sample.

The main proposition of this work is the application of order statistics theory as the basis for building a model useful in predicting the maximum value in order to evaluate the error. The proposed solution is general, though it will be illustrated by a practical example of the selection of attributes in the issue of the clothing industry. The presented method is one of the steps of the classification (regression) trees method [8, 16]. In the further part of the work, it is assumed that the data constitute the implementation of the stationary process. The possibilities of verification of this assumption are very limited because the sample (also the order of observations) is of random nature. The popular stationarity - Dickey Fuller test [14], assumes a natural order of observation. In this case, it should be taken into account that parameter t of the stochastic process does not mean time. It is used to number successive elements of the sample.

Otherwise, it would not make sense to study the first order autocorrelation found in the test. But even in the non-stationary case, the possible areas of research will be presented in the final part of the paper. Assuming stationarity, it remains to consider a particular form of the distribution of the analysed variable. There are a number of theorems thanks to which, knowing the distribution, one can easily find the maximum distribution [12]. Without ruling out the possibility of matching the appropriate distribution to the observations in Examples 1 and 2, only one distribution will be tested in this work. It will be for a normal distribution.

Four tests available in the Gretl program [11] were used to study normality. The results are presented in Tables 3 for data from Example 1.

**Table 3.** Test for normality of Example 1.

| Doornik-Hansen | test = 292,188, with p-value 3,56681e−064 |
|---|---|
| Shapiro-Wilk | W = 0,544228, with p-value 4,21341e−016 |
| Lilliefors | test = 0,377498, with p-value $\cong$ 0 |
| Jarque-Bera | test = 171,199, with p-value 6,67625e−038 |

Source: own work

The results for the second example are shown in Table 4.

**Table 4.** Test for normality of Example 2.

| Doornik-Hansen | test = 481,6, with p-value 2,64198e−105 |
|---|---|
| Shapiro-Wilk | W = 0,474654, with p-value 3,05738e−017 |
| Lilliefors | test = 0,370257, with p-value $\cong$ 0 |
| Jarque-Bera | test = 1114,75, with p-value 8,60289e−243 |

Source: own work using Gretl programme

As can be seen in both examples, for all four tests used, the normality hypothesis must be rejected at all the levels of significance commonly used.

In the absence of reliable hypotheses regarding the distribution of the observed sequence of observations, there remains a direct adjustment of the distribution of extremal statistics. As indicated above, however, this requires data on the form (1), which requires a sample extension. Realizing that in many issues such an extension is not possible (in the case of time series it is impossible to supplement the sample with unobserved data from the past), and where it leads to a significant increase in computational complexity, the bootstrap method is proposed in this work. This method consists in re-sampling the results obtained on the basis of one series. The re-sampling was done in the Excel system. In each case, 100 bootstrap attempts were made. In each of 100 cases, maximum values were calculated. These values are to constitute the sample postulated in (1). The results are shown in Figs. 3 and 4.

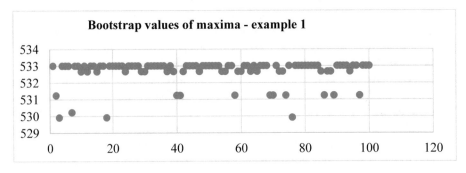

**Fig. 4.** Maxima - the case of 10 cells

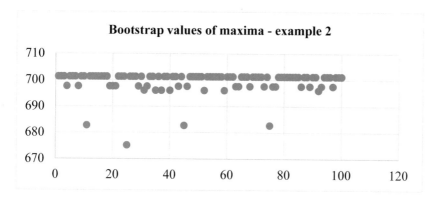

**Fig. 5.** Maxima - the case of 20 cells

In the second case, the majority of the replicated maximum values exceeds 700.

Having observations in the form (1), one can proceed to the estimation of parameters by the maximum likelihood method. For this purpose, the Mathematica program was used. With it, a system composed of Eqs. (3) and (4) was solved. Figure 5 shows a fragment of the log plot from the likelihood function.

Figure 6 shows a three-dimensional graph of the left sides of both equations for data from the first examples. The solution is on the level $z = 0$ (Fig. 7).

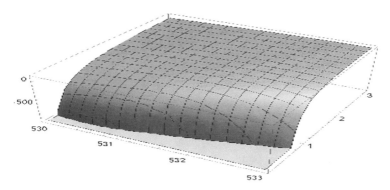

**Fig. 6.** Likelihood function – Example 1 Source: own work using Mathematica

The following printout from the Mathematica program contains the results of the estimation for the data from the first example.

$$\{mi \rightarrow 532{,}189, si \rightarrow 1{,}06688\}$$
$$1{,}19371 * 10^{-12}$$
$$6{,}75016 * 10^{-13}$$
$$-148{,}009$$

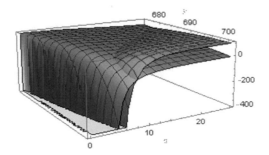

**Fig. 7.** System of normal equations – Example 2 Source: own work using Mathematica

The parameter estimates are given, i.e. $\hat{\mu} = 532{,}189$ and $\hat{\sigma} = 1{,}06688$. These estimates are supplemented with information on maximum numerical estimation errors. In addition, the value of the natural logarithm from the likelihood function is given.

For the data from Example 2, the estimates obtained in the Mathematica program are:

$$\{\mathbf{mi} \rightarrow \mathbf{696{,}699}, \mathbf{si} \rightarrow \mathbf{7{,}47285}\}$$
$$\mathbf{5{,}96856 * 10^{-13}}$$
$$\mathbf{3{,}76588 * 10^{-13}}$$
$$\mathbf{-337{,}122}$$

## 6  Prediction

In the case of the Gumbel distribution analysed here, based on the data for the sample size of 100, the estimated cumulative distribution obtained allows for extrapolation of the maximum. Here, the approximate independence of the maxima for disjoint samples is assumed. Assuming $T = 100$ for a sample increased to size $nT$, the following approximate equality is obtained [12].

$$P(Max(nT) \le x) \approx P^n(Max(T) \le x)$$
$$= \exp\left(-ne^{-(x-m_T)/\sigma_T}\right) \tag{5}$$
$$= \exp\left(-e^{-\frac{(x-m_T-\sigma_T \ln n)}{\sigma_T}}\right)$$

The above equation allows forecasting the maximum value after increasing the sample. A point forecast will be presented which is the median of the found distribution and a one-sided confidence interval at the confidence level of 0,95.

**Example 3.** Continuation of the first example.

For the data from the first example, $nT = 115\,975$. Thus, $n \approx 1160$ and $\ln n = 7{,}056$. The median value of the maximum distribution is $540{,}104$. In addition, the following equality defines a one-tailed confidence interval.

$$P(Max(115\,975) \leq 542{,}872) = 0{,}95$$

In this case, it is possible to verify the result. Since the true maximum value was $536{,}09$, and the value calculated on the basis of the sample from the first example was $533{,}004$, it is possible to calculate the relative estimation error. It is $1{,}31\%$ with respect to the median, $0{,}58\%$ with respect to the true maximum value and $1{,}82\%$ with respect to the upper limit of the confidence interval. In most applications these errors are acceptable.

**Example 4.** Continuation of the second example.

For the data from the second example, $nT = 51\,724\,158\,235\,372$. Thus, $n \approx 517\,241\,582\,353$ and $\ln n = 26{,}972$. The median value of the maximum distribution is $901{,}033$. In addition, the following equality defines a one-tailed confidence interval.

$$P(Max(nT) \leq 920{,}491) = 0{,}95$$

The relative error in relation to the median is $22{,}17\%$, and in relation to the extreme value of the confidence interval $23{,}81\%$.

## 7  Conclusions

The examples presented in the article allow us to suppose that the decision process based on the selection of appropriate variable compositions can be significantly improved. It was important to use the bootstrap method, thanks to which it was possible to obtain the distribution values of the maximum distribution parameters. The use of extreme statistics makes it possible to assess the error resulting from replacing the population parameter with the maximum value from the sample. Thanks to Eq. (5) it is possible to calculate the required sample size depending on the requirements for precision of estimation. In future studies, the assumption of stationarity does not have to be maintained. The method presented by Horowitz [10] may be useful here.

## References

1. Athey, S., Imbens, G.: Recursive partitioning for heterogeneous causal effects. PNAS **113** (27), 7353–7360 (2016)
2. Breiman, L.: Bagging predictors. Mach. Learn. **24**, 123–140 (1996)
3. Czekała, M., Bukietyńska, A., Gurak, M., Jagodziński, J., Kłosowski, J.: Condition analysis and forecasting in the fashion industry. Int. J. Econ. Financ. Manag. Sci. (2019). http://www.sciencepublishinggroup.com

4. Czekała, M., Bukietyńska, A., Gurak, M., Jagodziński, J., Kłosowski, J.: Fashion, expectations and reality. Methodology for researching stakeholders preferences. Article submitted to the Conference "Społeczna odpowiedzialność biznesu – perspektywa zarządzania i ekonomii". Szczecin (2019)

5. Efron, B.: The Jackknife, the Bootstrap and Other Resampling Plans. CBMS-NSF Regional Conference Series in Applied Mathematics, Stanford University (1982)

6. Embrechts, P., Klueppenberg, C., Mikosch, T.: Modelling Extremal Events: for Insurance and Finance. Springer (1997)

7. Flachsmeyer, J.: Kombinatorik. VEB Berlin (1969)

8. Gatnar, E.: Nieparametryczna metoda dyskryminacji i regresji. PWN, Warszawa (2001)

9. Hankin, R.K.S., West, L.J.: Set partitions in R. J. Stat. Softw. **23**, Code Snippet 2 (2007)

10. Horowitz, J.: Extreme values from a nonstationary stochastic process: an application to air quality analysis. Technometrics **22**, 469–478 (1980)

11. Kufel, T.: Ekonometria. Rozwiązywanie problemów z wykorzystaniem programu Gretl. PWN, Warszawa (2011)

12. Leadbetter, M.R., Lindgren, G., Rootzen, H.: Extremes and Related Properties of Random Sequences and Processes. Springer, Heidelberg (1986)

13. Luong, B.: MATLAB script for computing all partitions of a set of $n$ elements, May 2019. http://www.mathworks.cn/matlabcentral/fileexchange/24133-set-partition/content/SetPartFolder/SetPartition.m

14. Maddala, G.S.: Ekonometria. PWN, Warszawa (2008)

15. Mugridge, J., Wang, Y.: Applying decision tree in food industry – a case study. In: Wang, K., Wang, Y., Strandhagen, J., Yu, T. (eds.) Advanced Manufacturing and Automation VIII, IWAMA. LNEE, vol. 484. Springer, Singapore (2018)

16. Walesiak, M., Gatnar, E.: Statystyczna Analiza Danych. PWN, Warszawa (2012)

17. Weisstein, E.: Set Partition. From MathWorld – A Wolfram Web Resource, May 2019. http://mathworld.wolfram.com/SetPartition.html

18. Wen, X., Choi, T.M., Chung, S.H.: Fashion retail supply chain management: a review of operational models. Int. J. Prod. Econ. **207**, 34–55 (2019)

19. Wolfram Research, Inc.: Mathematica, Version 11.3 (2019)

20. R Core Team: R: a language and environment for statistical computing. R Foundation for Statistical Computing, Vienna, Austria (2019)

# GDPR Implementation in Local Government Administration in Poland and Republic of Lithuania

Dominika Lisiak-Felicka[1](✉) ⓘ, Maciej Szmit[2] ⓘ, Anna Szmit[3] ⓘ,
and Jolanta Vaičiūnienė[4] ⓘ

[1] Department of Computer Science in Economics, University of Lodz,
Lodz, Poland
dominika.lisiak@uni.lodz.pl
[2] Department of Computer Science, University of Lodz, Lodz, Poland
maciej.szmit@uni.lodz.pl
[3] Department of Management, Lodz University of Technology, Lodz, Poland
anna.szmit@p.lodz.pl
[4] Faculty of Social Sciences, Arts and Humanities,
Kaunas University of Technology, Kaunas, Lithuania
jolanta.vaiciuniene@ktu.lt

**Abstract.** The year 2018 was a breakthrough year for all European Union countries due to the need to adapt the organization to the management of personal data in accordance with the General Data Protection Requirements (GDPR). The Regulation specified some important obligations that should be implemented in all entities processing personal data in all sectors, including public administration offices. The aim of the article was an analysis and evaluation the degree of preparation for the implementation of this Regulation in the offices of local government administration in Poland and Republic of Lithuania. As part of the work on the article, a research was conducted using the Computer Aided Web Interview (CAWI) method and data from 472 offices of self-government administration was collected. The results of the research were subjected to statistical analysis using the Mann-Whitney U Test. On the basis of the data the process of implementing changes resulting from the GDPR was assessed and some significant problems in the implementation of these changes were identified.

**Keywords:** Personal data protection ·
General Data Protection Regulation (GDPR) · Local government administration

## 1 Introduction and Literature Review

The Regulation (EU) 2016/679 of the European Parliament and of the Council of 27 April 2016 on the protection of natural persons with regard to the processing of personal data and on the free movement of such data, and repealing Directive 95/46/EC (General Data Protection Regulation, GDPR) has been in force since the 25th of May, 2018 [26, 29, 35]. The Regulation has changed the approach to the personal data

© Springer Nature Switzerland AG 2020
Z. Wilimowska et al. (Eds.): ISAT 2019, AISC 1052, pp. 49–60, 2020.
https://doi.org/10.1007/978-3-030-30443-0_5

protection system, introduced a number of significant changes and unified the rules on the protection of personal data in all EU countries [7, 10, 16, 19]. The GDPR specified some important obligations that should be implemented in all entities processing personal data. The process of adapting the organization to changes resulting from the Regulation certainly required appropriate preparation and enough time to implement these changes [30].

The implementation of GDPR in various European countries has become the subject of a number of studies (e.g. [4, 6, 14, 21, 25]), especially comparative law research [20]. A number of recommendations and - more or less formalized - good practices regarding GDPR awareness, readiness and preparation for implementation in various organizations, in particular commercial companies, were also developed (see e.g. [1, 12, 13, 25, 35]). The issues of the GDPR's maturity are also being considered, including different commercial GDPR maturity models (e.g. KPMG GDPR Discovery and Maturity Assessment - see [15], GDPR Maturity Framework - see [27], The Forrester Privacy And GDPR Maturity Model - see [11] etc.). The concept of maturity is understood in the same way as process maturity in software engineering or management sciences (see e.g. [2, 22, 28]).

## 1.1 Local Government Administration in Poland

The local government administration in Poland has been based on three levels of subdivision since 1999 [31]. The area of Poland is divided into voivodeships (województwa in Polish), which are further divided into districts (powiaty in Polish), which in turn are divided into municipalities (gminy in Polish). Major cities have the status of both the municipality and the district.

There are currently 16 voivodships in Poland, 380 districts (including 66 cities with district rights) and 2,478 municipalities in Poland [9]. The Constitution of the Republic of Poland specifies in art. 169 s. 1 that units of local self-government shall perform their duties through constitutive and executive organs. In municipalities there are an elected council as well as a directly elected mayor (known as "prezydent" in large towns, "burmistrz" in most urban and urban-rural municipalities, and "wójt" in rural municipalities). In districts - an elected council ("rada powiatu"), which elects an executive board ("zarząd powiatu") headed by the "starosta". In voivodship - an elected assembly called the "sejmik", and an executive board ("zarząd województwa") with the leader that is called "marszałek". Organizational units whose purpose is to help the heads in the tasks defined by the law of the state are as follows: municipal offices, districts offices and marshal's offices [18, 32–34].

## 1.2 Local Government Administration in the Republic of Lithuania

Lithuania is a unitary state with two levels of government – the central government and local governments. Lithuania has one tier of local self-government composed of 60 municipalities (savivaldybė in Lithuanian), which have the right to self-rule exercised through their municipal councils and mayors.

Members of municipal councils and mayors (since 1st March 2015) are elected by direct universal suffrage for four years term. As the municipal council and mayor are representative institutions, the director of a municipal administration plays an executive role. Since the director is appointed by the municipal council upon the mayor's proposal, this position has the status of a civil servant of political (personal) confidence.

The sixty municipalities vary in size from 3,224 to 547,484 citizens, with an average of 47,000 and in area from 40 $km^2$ to 2,218 $km^2$, with an average of 1,088,3 $km^2$ [24].

The country used to have another administrative tier between the central government and local authorities – county (apskritis in Lithuanian) administrations. In 2010, all ten county administrations were abolished. Currently, counties serve as territorial and statistical units. The territory of the Republic of Lithuania currently comprises 10 counties: Alytus, Kaunas, Klaipeda, Marijampole, Panevezys, Saulai, Taurage, Telsai, Utena, Vilnius [5, 23, 24].

## 2   Research Questions, the Aim and Method

Prior to the entry into force of the GDPR regulations, doubts were raised as to whether all EU countries would be able to prepare national legislation and implementing acts on time (see e.g. [8, 17]). We decided to investigate the situation in local government administration offices because it can be expected that such specific type of organizations are commitment to complying with the law, so their readiness for GDPR should be relatively high and, to some extent, reflect the situation in the whole country.

The research questions for this study were:

1. What are the differences in preparation for the implementation of GDPR between Polish and the Lithuanian offices of local government administration?
2. How the maturity of processes of the implementation of GDPR were assessed by the employees of particular offices responsible for it in both countries?
3. What were the main issues and weaknesses identified in the process in both countries?

The aim of the article was an analysis and evaluation the degree of preparation for the implementation of this Regulation in the offices of local government administration in Poland and Republic of Lithuania. As indicators of this degree, self-assessment declared by officials and formal procedures/strategies and metrics known from various maturity models were adopted.

The survey was conducted using Computer Aided Web Interview (CAWI) method between March and April 2018 in Poland and June and July in Republic of Lithuania. The survey invitation was sent by email to all local government administration offices in both countries. It was explained that the obtained data would be used in an aggregated form only for the statistical summaries and analyses in scientific publications. The questionnaire was anonymous. The survey questionnaire in Polish and Lithuanian has been available for several weeks on the webankieta.pl website [17].

# 3  Results

Obtained 462 responses from offices of local government administration in Poland: 6 offices at the voivodeship level (marshal offices), 66 at the districts level (district offices) and 390 at the municipalities level (municipal offices). Only 10 responses from local government administration of Republic of Lithuania were obtained (see Table 1). Table 2 presents the numbers of employees in the surveyed offices.

**Table 1.**  Types and sources of support. Source: own survey, [9] and [24].

|  | Population | Sample |
|---|---|---|
| *Poland – total offices* | **2874** | **462 (16%)** |
| Voivodeships | 16 | 6 (38%) |
| Districts | 380 | 66 (17%) |
| Municipalities | 2478 | 390 (16%) |
| *Republic of Lithuania - municipalities* | **60** | **10 (17%)** |

**Table 2.**  Numbers of employees in the offices. Source: own survey.

| Numbers of employees | Number of responses | |
|---|---|---|
| | Poland | Republic of Lithuania |
| Up to 50 people | 279 | 0 |
| 51 to 100 people | 104 | 0 |
| 101 to 500 people | 59 | 10 |
| 501 to 1,000 people | 7 | 0 |
| 1,001 to 2,000 people | 7 | 0 |
| 2,001 to 3,000 people | 4 | 0 |
| Over 3,000 people | 2 | 0 |

Among 462 offices of local government administration in Poland, only 96 formally defined implementation strategy for the GDPR (objectives, deadlines, responsible persons, procedures), 12 of them provided a link or attached a file that includes strategy specification, while among 10 offices of local government administration in the Republic of Lithuania only two offices had the subject strategy developed and one of the respondents attached a file with the ordinance (see Fig. 1a).

For the above data, the Mann-Whitney U Test [3] value was calculated. The asymptotic significance value was 0.952, which means that there are no statistically significant differences between the two groups (Polish offices and Lithuanian offices) in assessing the readiness to introduce changes resulting from the GDPR.

Only 32 offices in Poland (less than 7%) defined the indicators of readiness/ maturity of the GDPR implementation. Two of the surveyed Lithuanian offices defined such measures, but did not specify the types of measures (see Fig. 1b).

a)                                             b)

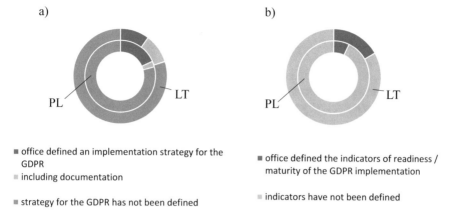

- office defined an implementation strategy for the
  GDPR
- including documentation

- strategy for the GDPR has not been defined

- office defined the indicators of readiness /
  maturity of the GDPR implementation

- indicators have not been defined

**Fig. 1.** (a) Implementation strategies for the GDPR, (b) Indicators of readiness/maturity of the GDPR. Source: own survey.

Respondents were also asked if the evaluation process of the GDPR implementation is conducted at the office? In Polish offices, there were 139 affirmative answers, including 92 in self-evaluation (in addition, 32 offices indicated that they had evaluation tools). In other units, the assessment is conducted by external companies. In the case of offices from Lithuania, the evaluation is conducted in 4 units (in one office evaluation is conducted independently, but the office does not have any evaluation tools, and in three cases through an external company), (see Fig. 2a).

a)                                             b)

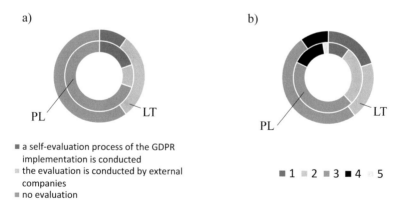

- a self-evaluation process of the GDPR
  implementation is conducted
- the evaluation is conducted by external
  companies
- no evaluation

- 1  - 2  - 3  - 4   5

**Fig. 2.** (a) Evaluation process of the GDPR implementation, (b) Degrees of office readiness for implementing changes resulting from the GDPR. Source: own survey.

Similarly to the previous question, there are no statistically significant differences in the assessment of the readiness to introduce changes resulting from the GDPR (the significance value of the asymptotic the Mann-Whitney U Test was 0.5).

Next question concerned the assessment degree of office readiness for implementing changes resulting from the GDPR? (on a scale of 1 to 5, where 1 – no readiness, 5 – all GDPR requirements have been already implemented). Results are shown on Fig. 2b.

The significance value of the asymptotic the Mann-Whitney U Test in this case was 0.548, which means that also in this case there are no statistically significant differences between the two surveyed groups.

In the next question respondents were asked to specify the level of difficulty in implementing changes resulting from the GDPR to the office on a scale of 1 to 5 (1 – very easy, 5 – very difficult):

– Extended rights of the data subject,
– Data Protection Officer position,
– Information obligation and consent to data processing,
– Notification of a personal data breach obligation,
– Records of processing activities,
– Data protection by design and by default,
– Data protection impact assessment,
– Limitations on profiling.

Results for offices from Poland and Republic of Lithuania are shown on Figs. 3a–h.

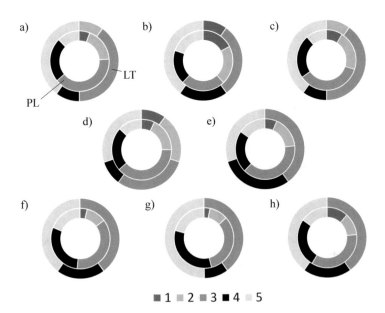

**Fig. 3.** The assessment of difficulty in implementing changes resulting from the GDPR: (a) Extended rights of the data subject, (b) Data Protection Officer position, (c) Information obligation and consent to data processing, (d) Notification of a personal data breach obligation, (e) Records of processing activities, (f) Data protection by design and by default, (g) Data protection impact assessment, (h) Limitations on profiling. Source: own survey.

On the other hand, the use of external companies' services was significantly different: among Polish offices, only 47% declared using such support, and 80% from Lithuanian ones (see Fig. 4a). The significance value of the asymptotic the Mann-Whitney U Test in this case was 0.037, which means that there are statistically significant differences between the two surveyed groups.

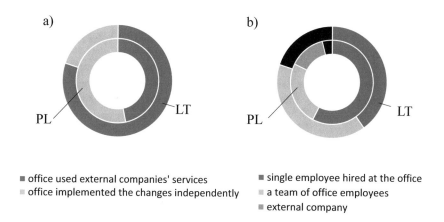

**Fig. 4.** (a) The way of implementing changes, (b) Responsibility for ensuring the security of personal data. Source: own survey.

In more than half of the surveyed offices in Poland, one person is responsible for ensuring the security of personal data, a special team of employees is set up in approximately 25% of offices, about 15% – external data security. In the Lithuanian responses from offices were spread evenly between the "single employee" and "a team of office employees" options. In two cases, the respondents indicated that no one is responsible for such duties (see Fig. 4b).

In Polish offices, only 5% declared that in the last year (before the implementation of the GDPR) there were cases of personal data breaches, while in 21 cases the number of instances of such cases ranged from 1 to 5, and in 3 offices from 6–20. On the other hand, in almost half of the surveyed offices in Lithuania there were breaches of personal data protection. In all cases, 1–5 incidents were reported (see Fig. 5a). Also in this respect, the difference measured by the Mann-Whitney U Test between the two groups was statistically significant ($p < 0.05$).

The next questions concerned information about the support received from higher organizations in implementing the GDPR (e.g. training, information materials). In Poland, about 1/3 of the surveyed offices received such assistance, mainly in the form of training, information materials, materials from websites, brochures, support in interpreting regulations, participation in conferences, webinars, using publications and consultations. Half of the surveyed offices in Lithuania received such support in the form of: trainings, seminars and workshops, as well as in financial form (see Fig. 5b).

The next question concerned training for employees in the scope of changes resulting from the GDPR. In the Polish offices: 16% of offices have trained all employees, 44% have conducted such training, but not all employees have been trained yet and 40% of offices did not conduct any training on this subject. In Lithuania, the majority of offices indicated the option that "not all employees were trained yet". One third of the offices did not conduct any training (see Fig. 5c).

Respondents were asked to indicate the biggest concern (in their opinions) in preparing the office for the GDPR implementation. They mentioned the following problems (grouped in legal, financial, organizational, essential and human aspects), see Table 3.

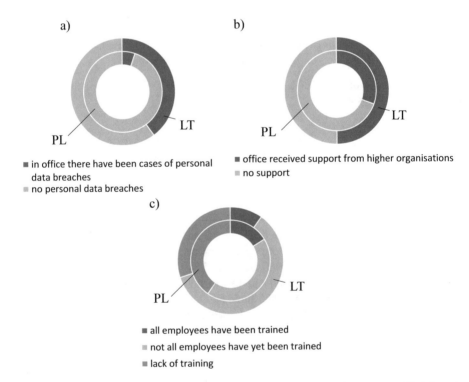

a)

b)

LT
PL

■ in office there have been cases of personal
  data breaches
▨ no personal data breaches

■ office received support from higher organisations
▨ no support

c)

LT
PL

■ all employees have been trained
▨ not all employees have yet been trained
■ lack of training

**Fig. 5.** (a) Personal data breaches, (b) Support from higher organizations, (c) Training for employees. Source: own survey.

**Table 3.** Types of problems indicated by offices. Source: own survey.

| Types of problems | Poland | Republic of Lithuania |
|---|---|---|
| Legal | The Polish Personal Data Protection Act has not been adopted yet (during the research), the rules are unclear, the absence of specific legal acts, implementing regulations and specific guidelines | Many ambiguities in the regulations |
| Financial | The lack of sufficient financial resources for the GDPR implementation | The lack of sufficient financial resources |
| Organizational | The lack of time, excess of duties, lack of support from superior authorities, problem in finding qualified staff, very few trainings | Lack of qualified trainers who could lead trainings, lack of an advisory institution, failure of the management to pay attention to serious processes and serious infringements, insufficient management commitment, lack of interest on the part of the Government |
| Essential | Problems with the procedures implementation, preparation of documentation, risk analysis and assessment, implementation of tasks resulting from changes introduced in the GDPR, technical barriers, lack of tested solutions | Inappropriate assessment of GDPR |
| Human | Resistance and reluctance against changes, lack of awareness of employees and management staff, lack of sufficient knowledge about GDPR | Dissatisfaction with additional work, lack of understanding of the importance of personal data security |

## 4   Discussion and Conclusion

The survey covered slightly more than 16% of offices both in Poland and Republic of Lithuania. Since the offices agreed to participate voluntarily, due to the fact that the method of representation was not met, the results cannot be generalized to the entire population, so as to statistically estimate the amount of the error. Accordingly, the Mann-Whitney U Test values calculated above should also be regarded as indicative. On the other hand, the relatively high percentage of the sample combined with the relatively high repeatability of the results obtained allows for the formulation of some general conclusions.

First of all, in both countries the way of implementing changes resulting from the GDPR seems to be – from the point of view of management sciences – not enough mature. A vast minority of offices in both countries did not conduct this process based on a formalized strategy and did not define measures to determine the degree of readiness for the GDPR implementation.

The self-assessment of the degree of preparation of offices to introduce changes resulting from the GDPR was also similar in both countries: on the five-point Likert scale, the central value was chosen the most, the average answer for Poland was 2.73 and for Lithuania – 2.5 which can be interpreted as very moderate optimism.

Similar problems were also found in the implementation of the GDPR identified by respondents in both countries. The difference was evident in the answers in two questions: for support from external companies (significantly higher in Lithuania) and the number of incidents (significantly fewer incidents in Poland). Because it is difficult to expect fewer attacks against information security in offices in Poland (a large part of the attacks on the Internet is non-directional), it can be expected that this phenomenon is largely due to poorer identification or classification of events related to information security in Poland, not actually a lower number of attacks.

The research indicated that both among the surveyed offices in Lithuania and in Poland, the readiness to implement changes resulting from the GDPR on the eve of its introduction was low, and the manner of its introduction was not enough mature. Employees of offices in both countries similarly identified the reasons for this state and problems occurring during implementation.

# References

1. Addis, M.C., Kutar, M.: The General Data Protection Regulation (GDPR), emerging technologies and UK organisations: awareness, implementation and readiness. UK Academy for Information Systems Conference Proceedings 29 (2018). https://aisel.aisnet.org/ukais2018/2
2. Becker, J., Niehaves, B., Poeppelbuss, J., Simons, A.: Maturity models in IS research. In: 18th European Conference on Information Systems, ECIS (2010)
3. Corder, G.W.: Nonparametric Statistics for Non-statisticians: A Step-by-Step Approach. Wiley, New York (2009)
4. Deloitte Legal: The GDPR: six months after implementation practitioner perspectives. https://www2.deloitte.com/bg/en/pages/legal/articles/gdpr-six-months-after-implementation-2018.html#. Accessed 08 July 2019
5. European Committee of the Regions, Division of Powers. https://portal.cor.europa.eu/divisionpowers/Pages/Lithuania-Introduction.aspx. Accessed 15 Apr 2019
6. Finocchiaro, G.: GDPR implementation series • Italy: the legislative procedure for national harmonisation with the GDPR. Eur. Data Prot. Law Rev. 4(4), 496–499 (2018). https://doi.org/10.21552/edpl/2018/4/12
7. Gellert, R.: Understanding the notion of risk in the General Data Protection Regulation. Comput. Law Secur. Rev. 34(2), 279–288 (2018)
8. Giurgiu, A.: GDPR implementation series • Luxembourg: reshaping the national context to adjust to the GDPR. Eur. Data Prot. Law Rev. 3, 372–375 (2017). https://doi.org/10.21552/edpl/2017/3/11
9. Główny Urząd Statystyczny, Liczba jednostek podziału terytorialnego kraju (eTERYT). http://eteryt.stat.gov.pl/eteryt/raporty/WebRaportZestawienie.aspx. Accessed 15 Apr 2019

10. Hoofnagle, C.J., Sloot, B., Borgesius, F.Z.: The European Union General Data Protection Regulation: what it is and what it means. Inf. Commun. Technol. Law **28**(1), 65–98 (2019)
11. Iannopollo, E.: The five milestones to GDPR success, Forrester, 25 April 2017. https://interwork.com/wp-content/uploads/2018/03/The-Five-Milestones-To-GDPR-Success.pdf. Accessed 08 July 2019
12. IBM Corporation: IBM's Journey to GDPR Readiness. https://www.ibm.com/downloads/cas/V2B2VBXY. Accessed 08 July 2019
13. IBM: IBM Pathways for GDPR Readiness. https://www.ibm.com/downloads/cas/QE2NL4GP. Accessed 08 July 2019
14. Korpisaari, P.: GDPR implementation series • Finland: a brief overview of the GDPR implementation. Eur. Data Prot. Law Rev. **5**(2), 232–237 (2019). https://doi.org/10.21552/edpl/2019/2/13
15. KPMG, GDPR Discovery and Maturity Assessment. https://assets.kpmg/content/dam/kpmg/xx/pdf/2018/03/gdpr-discovery-maturity.pdf. Accessed 08 July 2019
16. Krystlik, J.: With GDPR, preparation is everything. Comput. Fraud Secur. **2017**(6), 5–8 (2017)
17. Lisiak-Felicka, D., Szmit, M., Szmit, A.: The assessment of GDPR readiness for local government administration in Poland. In: Wilimowska, Z., Borzemski, L., Świątek, J. (eds.) Information Systems Architecture and Technology: Proceedings of 39th International Conference on Information Systems Architecture and Technology – ISAT 2018. Advances in Intelligent Systems and Computing, vol. 854, pp. 417–426. Springer, Cham (2018)
18. Lisiak-Felicka, D., Szmit, M.: Cyberbezpieczeństwo administracji publicznej w Polsce. Wybrane zagadnienia, European Association for Security, Kraków 2016, pp. 36–37 (2016)
19. Lopes, I.M., Guarda, T., Oliveira, P.: EU General Data Protection Regulation implementation: an institutional theory view. In: Rocha, Á., Adeli, H., Reis, L., Costanzo, S. (eds.) New Knowledge in Information Systems and Technologies. WorldCIST 2019. Advances in Intelligent Systems and Computing, vol. 930, pp. 383–393. Springer, Cham (2019)
20. Mc Cullagh, K., Tambou, O., Bourton, S.: National adaptations of the GDPR. https://blogdroiteuropeen.files.wordpress.com/2019/02/national-adaptations-of-the-gdpr-final-version-27-february-1.pdf. Accessed 08 July 2019
21. McSweeney, A.: GDPR - context, principles, implementation, operation, data governance, data ethics and impact on outsourcing. In: Warwick Legal Network Conference (2019)
22. Mettler, T.: Maturity assessment models: a design science research approach. Int. J. Soc. Syst. Sci. **3**(1/2) (2011)
23. Ministerstwo Spraw Zagranicznych, Informator ekonomiczny. https://informatorekonomiczny.msz.gov.pl/pl/europa/litwa/. Accessed 15 Apr 2019
24. Official Statistics Portal, Administrative Territorial Division. https://osp.stat.gov.lt/regionine-statistika-pagal-statistikos-sritis. Accessed 15 Apr 2019
25. Pandit, H.J., O'Sullivan, D., Lewis, D.: Queryable provenance metadata for GDPR compliance. In: Fensel, A., de Boer, V., Pellegrini, T., Kiesling, E., Haslhofer, B., Hollink, L., Schindler, A.: Proceedings of the 14th International Conference on Semantic Systems, Vienna, Austria, 10–13 September 2018, vol. 137, pp. 262–268 (2018). https://doi.org/10.1016/j.procs.2018.09.026
26. Polkowski, Z.: The method of implementing the General Data Protection Regulation in business and administration. In: Proceedings of the 10th International Conference on Electronics, Computers and Artificial Intelligence, ECAI 2018 (2018)
27. Privacy Culture Ltd.: The GDPR maturity framework. https://iapp.org/media/pdf/resource_center/PrivacyCulture_GDPR_Maturity_Framework.pdf. Accessed 08 July 2019

28. Proença, D.: Methods and techniques for maturity assessment. In: 11th Iberian Conference on Information Systems and Technologies (CISTI), pp. 1–4 (2016). https://doi.org/10.1109/cisti.2016.7521483
29. Regulation (EU) 2016/679 of the European Parliament and of the Council of 27 April 2016 on the protection of natural persons with regard to the processing of personal data and on the free movement of such data, and repealing Directive 95/46/EC (General Data Protection Regulation)
30. Tikkinen-Piri, C., Rohunen, A., Markkula, J.: EU General Data Protection Regulation: changes and implications for personal data collecting companies. Comput. Law Secur. Rev. **34**(1), 134–153 (2018)
31. Ustawa z 24 lipca 1998 r. o wprowadzeniu zasadniczego trójstopniowego podziału terytorialnego państwa (Dz. U. z 1998 r. Nr 96, poz. 603)
32. Ustawa z 5 czerwca 1998 r. o samorządzie powiatowym (Dz. U. z 2001 Nr 142, poz. 1592, z późn. zm.)
33. Ustawa z 5 czerwca 1998 r. o samorządzie województwa (Dz. U. z 2001 r. Nr 142, poz. 1590 z późn. zm.)
34. Ustawa z 8 marca 1990 r. o samorządzie gminnym, (Dz. U. z 2001, nr 142, poz. 1591, z późn. zm.)
35. Voight, P., von dem Bussche, A.: The EU General Data Protection Regulation (GDPR). A Practical Guide. Springer (2017). https://doi.org/10.1007/978-3-319-57959-7

# Models of Management of Public Organization

# Cognitive Modeling and Cognitive Map Applying to the Knowledge Management in the Higher Education System

Nataliya Kobets[1]([⊠]) [iD], Tetiana Kovaliuk[2] [iD],
and Daryna Mozoliova[2]

[1] Borys Grinchenko Kyiv University,
18/2 Bulvarno-Kudriavska Street, Kyiv 04053, Ukraine
nmkobets@gmail.com
[2] National Technical University of Ukraine "Igor Sikorsky
Kyiv Polytechnic Institute", 37 Prospekt Peremohy, Kyiv 03056, Ukraine
tetyana.kovalyuk@gmail.com, srslyreader@gmail.com

**Abstract.** The article analyzes and summarizes the problems of higher education in the era of the knowledge economy. The authors use the methodology of cognitive modeling of weakly structured systems to build cognitive maps of the educational process. The author's goal is to show the possibility of using cognitive modeling and cognitive maps in the university's knowledge management system. To build cognitive maps, the authors identified target risk factors, basic risk factors and risk factors that affect the quality of the educational process in the higher education system. Cognitive maps are used as a tool to analyze the higher education quality and risk factors. The cognitive analysis of factors affecting the higher education quality and risks showed the need to change the role the universities play in the knowledge economy.

**Keywords:** Knowledge management · Cognitive map · Higher education ·
Risk factors · Quality factors

## 1 Introduction

Knowledge economy is an economy where knowledge is created, distributed and used to promote personal refinement and achieve better competitiveness. The features of the new knowledge-based economy include a rapid growth of knowledge-intensive goods and services with reduced lifecycle, a more pronounced intellectual content of technologies increasing the labor productivity by order of magnitude, the emergence of a large market for intelligent products and services, a brisk tempo of knowledge updates and the ever-growing need in them. A successful development strategy in the context of the "knowledge economy" hinges on high quality of human capital that generates transforms and multiplies the intangible assets of the organization.

The World Bank has elaborated a system of metering the "knowledge economy" in four main areas [1]:

© Springer Nature Switzerland AG 2020
Z. Wilimowska et al. (Eds.): ISAT 2019, AISC 1052, pp. 63–73, 2020.
https://doi.org/10.1007/978-3-030-30443-0_6

- institutional regime;
- educational level of the population;
- status of information and communication infrastructure;
- national innovation system.

One of the components of the country's labor potential qualitative attribute is the level of education and knowledge across individuals. Education and training are factors contributing to a society of well-qualified, dynamic and creative people where opportunities for good education and lifelong learning are accessible for everyone and where a favorable environment is created for a rational combination of public and private funding. Tailoring the modern methods of knowledge management to the needs of educational institutions helps accumulating knowledge in subject areas in a meaningful way and effectively transferring this knowledge to the future specialists.

## 2 Knowledge Management Issues in Higher Education System

Knowledge management in an organization is a systematic process of creating, updating and applying knowledge to maximize the organization's overall effectiveness [2, 3]. This is the process where the organization generates, accumulates and uses knowledge to gain a competitive edge [4]. Modern educational system requires knowledge management mechanisms to address the issues related to the training of most in-demand specialists and enabling them to quickly take on the challenges in the industry and elsewhere.

The intellectual capital of a higher educational institution is composed of both implicit intellectual resources (individual knowledge of the faculty members, collective experience, accumulative knowledge across departments, etc.) and explicit intellectual funds (codified knowledge, perceptive knowledge, physical description of specific knowledge, etc.) [5–8]. In an educational institution, the intellectual capital embodies the accumulated intellectual activity of its staff and structural units, on which this institution can capitalize to gain a competitive advantage.

The knowledge management of the educational process in universities is of particular relevance. Here, the gist of learning process is to make the transfer of knowledge from teachers to students as effective as possible. In practice, the universities come across the following issues of knowledge management:

- the classic process of knowledge transfer is not perfect and fails to stimulate the students;
- teachers without stimuli are reluctant to improve their skills;
- the university does not provide ample access to information required in the labor market and promoting the goals of scientific and technological development;
- the universities are lukewarm about innovations;
- the universities are underfinanced;
- the number of students is dwindling as many of them prefer studying in the European universities or drop out.

The urgency of issues faced by the knowledge management in higher education requires the analysis of quality and risk factors in the educational process.

## 3 Cognitive Modeling in Knowledge Management System

Like any social system the higher education system is characterized by the complexity of the management process, which can be described in the following way:

- processes that take place in a system are multifactorial and interconnected so much so that it is impossible to isolate and investigate individual phenomena;
- the quantitative information about the processes' dynamics is insufficient, which makes it necessary to limit the research with only qualitative analysis;
- processes vary over time.

Because of these features, the social systems are categorized as weakly structured. The cognitive approach is used to simulate systems of this type [9, 10]. Cognitive analysis and modeling are fundamentally new elements in the knowledge management and decision-making support system. The cognitive analysis of weakly structured systems and situations helps acquiring new knowledge and skills, which can be used to identify and adapt the existing environment or decide how to change in it.

Another feature of the cognitive modeling process is the lack of detailed quantitative information. The cognitive modeling is used to produce a cognitive model, which is based on expert evaluation, i.e., the knowledge of individual experts and as such is in fact a subjective reflection of the system. The subjective model of a weakly structured dynamic system or situation is called a cognitive map. The cognitive map reflects (individual or collective) subjective representations about the problems and situations that are associated with the functioning and development of the studied systems. The key elements of a cognitive map are factors and causal relationship between them [11–13].

## 4 Building a Cognitive Map to Analyze Education Process Quality and Risks

### 4.1 Definition of System Factors

The idea behind the cognitive modeling in the analysis and management problems of weakly structured systems and situations is to use a cognitive map for researching the systems' functions and evolvement. The cognitive map is a graph model of a complex situation in the form of causal relationships between factors. In other words, the cognitive map is a causal network (1).

$$G = <E, W >$$ (1)

where $E = \{e_1, e_2, \ldots, e_n\}$ – set of factors (concepts); $W$ – binary relation on set $E$, which determines a set of relations between its elements.

Let's consider an example where cognitive maps are used to describe the higher education system from the risk management perspective. The expert surveys and reference literature review [14–17] helped identify the key factor types affecting the higher education system's quality and risks:

- risk factors $C^r = \{C_i^r | i = \overline{1..I}\}$;
- base factors $C^b = \{C_j^b | j = \overline{1..J}\}$ affecting the institution's activity;
- target factors $C^t = \{C_k^t | k = \overline{1..K}\}$.

Let's consider the factors that generate risks in the higher education system. Table 1 summarizes some of the risk factors discussed in the example.

**Table 1.** Some of the risk factors in higher education system

| Notation keys | Name of the factor | Characteristic of the factor |
|---|---|---|
| $\{C_1^r\}$ | Number of successful applicants | The factor determines the total student enrolment in a given year and reflects the risks associated with demographic problems. The contingent of students began diminishing in 2008 and since 2011 almost all Ukrainian universities suffered a sharp decline [17]. In previous years, the HE system had substantially increased the staffing and institutional potential: the number of faculty with doctoral degrees grew and the technical and methodological support of the educational process improved. However, the critical dependence of the universities in terms of financial and economic stability on the number of enrolled students jeopardizes the HE system's chances for betterment |
| $\{C_2^r\}$ | Solvency of the population | The factor determines the solvency of the population, i.e., the ability of individuals and legal entities to pay for educational services and reflects trends in the public demand for higher education. The higher the cost of education the less the solvency of the population and, as a consequence, the less students and financial resources coming in the higher education system. These changes run contrary to the interests of educational organizations and are a source of risks for higher education |
| $\{C_3^r\}$ | Interaction of HE with enterprises | The factor determines the interaction between business and universities. This factor reflects the interest that the enterprises may have in scientific research of the areas they consider relevant. It is calculated by adding up the number of requests filed by business companies to the universities. Today, HE and business seem to run in parallel, each system |

*(continued)*

**Table 1.** (*continued*)

| Notation keys | Name of the factor | Characteristic of the factor |
|---|---|---|
| | | pursuing its own goals, barely interacting with each other. Business seems not to be ready for a systematic relationship with universities and tackles HR issues on its own. Meanwhile, the HE system is self-sufficient, does not feel any dependence on and keeps a distance from business |
| $\{C_4^r\}$ | Curriculum quality | This factor determines the total number of curriculum flaws discovered in the current year and is associated with the following potential risks: insufficient theoretical knowledge base; lack of practical experience; underdeveloped basic psychological qualities; inability of graduate students to achieve a decent social status |
| $\{C_5^r\}$ | Pre-university training | The pre-university training factor reflects the quality of additional educational services provided to applicants in the form of preparatory courses. The factor is calculated as the ratio of the number of applicants who have taken the preparatory courses and entered the university and the total number of applicants who have taken the preparatory courses. This factor is considered to be a source of risk due to the low solvency of the population |
| $\{C_6^r\}$ | University reputation | The university's reputation factor is defined as a rating awarded to the university through surveys conducted on educational, national and international web-portals. It is significantly affected by the reputation of the graduate students as direct consumers of educational services. The other parameters include: the individual socio-psychological traits of student behavior, educational interests, endeavor to master sciences, social status |
| $\{C_7^r\}$ | Basic knowledge level of the applicant | Factor determines the entrance level of the applicant's knowledge and is defined as the arithmetic mean of external independent assessment or internal admission tests. The lower level of an applicant's knowledge the less are his or her chances to achieve high performance in the university |
| $\{C_8^r\}$ | Loss of university's contingent of students | The factor reflects the number of students who are either expelled or transferred to other universities. The permanent loss of students in the learning process can occur for social, domestic, family, psychological and other reasons as well as at the request of a student. The students may be expelled because of falling behind in their studies and failing to finish their study program |

Let's consider some basic factors affecting the universities' well-being. Since most educational institutions in Ukraine are state-owned, we believe that one of the main factors is financing. Table 2 contains a list of basic factors.

**Table 2.** List of basic factors

| Notation keys | Name of the factor | Characteristic of the factor |
|---|---|---|
| $\{C_1^b\}$ | Amount of financing | The factor is defined as the amount of budgetary and extra budgetary funds earmarked to support the educational process of the university |
| $\{C_2^b\}$ | Number of research projects | The factor determines the total number of completed research by the end of the reporting period, broken down by funding sources |
| $\{C_3^b\}$ | Graduate's employment | The factor is defined as the percentage of university graduates successfully employed after graduation from the total number of graduates |
| $\{C_4^b\}$ | Results of state attestation | The factor reflects the quality of bachelors and masters' training as the average score based on the performance indicator for the entire period of study and the diploma thesis grades |

Let's determine the target factors as the target assets of the university (Table 3). Since the main task of the university is to prepare graduate students who are qualified to meet the industry's needs, we will consider the factors related to professional activity.

**Table 3.** The target factors of the university

| Notation keys | Name of the factor | Characteristic of the factor |
|---|---|---|
| $\{C_1^t\}$ | Graduate's career | This factor determines the total number of university graduates who have been promoted at work and, accordingly, received a salary increase |
| $\{C_2^t\}$ | Provision of the relevant sector by human resources | This factor determines the total number of open vacancies offered by the companies during a certain period |
| $\{C_3^t\}$ | Quality of professional activity of graduates | The professional competence of graduates is evaluated by experts selected from among employers who consume the graduates' professional competencies. How well the level of professional training fits the labor market requirements is discovered by sociological surveys in the region. The factor is then calculated by dividing the number of complaints filed by the employers in the previous and the following reporting periods |

## 4.2 Identification of Cause and Effect Relationships

The relation $W$ is represented as a matrix of dimension $n \times n$, where $n$ is the number of concepts in the system, which can be viewed as the adjacency matrix of this graph, and called a cognitive matrix [18]. We will consider fuzzy cognitive maps [11, 12] where the edge weights can represent values from a certain linguistic scale that characterizes the strength (degree) of the corresponding relationship or the degree of confidence in the availability of this relationship. The numerical values of the segment $[-1, 1]$ can be assigned to these linguistic variables:

$$W = \{w_{ij} \in [-1; 1], \ i,j = \overline{1..n}\} \tag{2}$$

The linguistic scale that was used to reflect the strength of relationships in the cognitive model is given in Table 4 [13].

**Table 4.** Linguistic scale to determine the relationship between concepts

| The linguistic value of a variable | Numerical values of variables |
|---|---|
| Does not affect | 0 |
| Very weak | 0,1; 0,2 (−0,1; −0,2) |
| Weak | 0,3; 0,4 (−0,3; −0,4) |
| Moderate | 0,5; 0,6 (−0,5; −06) |
| Strong | 0,7; 0,8 (−0,7; −0,8) |
| Very strong | 0,9; 1 (−0,9; −1) |

The extreme upper and lower limit values of the scale are 1 and −1, respectively. These limit values can be chosen arbitrarily; the developers can create their scale, at their own discretion. However the above concept is more commonplace and popular and quite convenient to analyze and establish relationships between concepts.

## 4.3 Graph Description Using DOT Language and Graph Construction

A cognitive map is an oriented graph. Graphviz is a package of open-source tools for drawing graphs specified in DOT language scripts. DOT draws directed graphs as hierarchies. Its features include well-tuned layout algorithms for placing nodes and edge splines, edge labels, record shapes with ports for drawing data structures; cluster layouts; and an underlying file language for stream-oriented graph tools [19].

The cognitive map is described in the DOT language for the domain "Higher Education System" and its archetype "Quality and Risks". A fragment of cognitive map's DOT description is given below.

```
digraph map{
// snippet of cognitive map description code
node [shape = "box", fontsize=15];
edge [fontsize=18];
"Number of applicants" -> "Provision of the relevant
industry personnel" [label="+0.4", color="red"];
"Number of applicants" -> "Pre-university training"
[label="-0.25", color="blue"];
"Number of applicants  -> "Number of research papers"
[label="+0.4", color="red"];
"Solvency of the population" -> "Number of applicants"
[label="+0.5", color="red"];
"Solvency of the population" -> "Pre-university training"
[label="+0.3", color="red"];
"Solvency of the population" -> "Basic knowledge level of
the applicant" [label="+0.4", color="red"];
"Solvency of the population" -> "Amount of financing"
[label="+0.3", color="red"];
"Interaction of universities with enterprises" -> "Amount
of financing" [label="+0.7", color="red"];
"Interaction of universities with enterprises" -> "Number
of research projects" [label="+0.4", color="red"];
"Interaction of universities with enterprises" ->
"Indicator of graduate's employment" [label="+0.5",
color="red"] ;
}
```

The image of the cognitive was made by using Graphviz (Fig. 1). The red color shows the positive effects some factors produce on other factors and the blue ones are used to show negative effects. Each edge shows the edge weight or the value of influence between factors. The strength of the relationships was evaluated by the experts.

At the next stage, the cognitive map is transformed in an adjacency matrix, the elements of which are the quantitative values of the factor influences. The quantitative cognitive map in which the force of influence and the value of the factors are determined on the numerical axis [− 1; 1] is used. An ordered set of linguistic values is defined for each factor. The factor influence matrix is based on a fuzzy cognitive map. The values of the matrix coefficients are found by using the method developed and presented in [18].

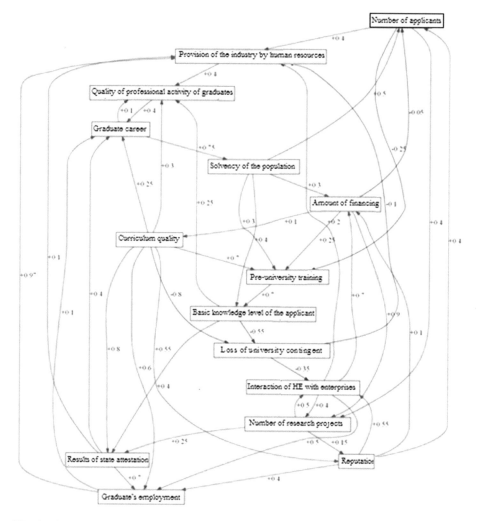

**Fig. 1.** Cognitive map of higher education system with an archetype "Higher Education Quality and Risks"

## 5 Cognitive Analysis Results

The structural characteristic analysis of the cognitive map that was created to model the HE quality and risks gave the following result:

1. All factors perform the functions of receivers. They affect and are affected by other factors.
2. Among the fifteen factors, only one has a negative effect, namely the "Loss of university contingent".

3. Five factors affect or have a negative effect. For example, the factors "Basic knowledge level of the applicant" and "Curriculum quality" affect the factor "Loss of university contingent", the factor "Number of applicants" affects the factor "Pre-university training".
4. Ten factors have and receive a positive effect. The factor "Curriculum quality" has the greatest effect on other factors. Risks and target factors depend on the "Curriculum quality".
5. The most influential for the HE quality are such factors as "Curriculum quality", "University reputation", "Interaction of universities with enterprises" and "Amount of financing".
6. The higher education risks are mostly affected by the factors "Curriculum quality" and "Graduate' employment".
7. Factors "Graduate's employment" and "Graduate's career" are the most important HE target factors from the point of view of students.

## 6  Conclusion

Education in a modern market economy can be viewed as a quite insecure area where innovative projects are exposed to greatest risks. Today, the risk management system practiced by the Ukrainian universities remains underdeveloped, which results in inadequate assessment of risks associated with the development of new educational projects and affects the universities' overall financial performance. In this article the cognitive analysis of the factors affecting the education quality was used to identify the most promising areas for the further development of higher education, in particular, the tools that allow rethinking the role of the universities in the era of the knowledge economy and the need for the knowledge management system.

## References

1. Gorji, E., Alipourian, M.: The knowledge economy & the knowledge assessment methodology. Iran. Econ. Rev. **15**(29), 43–72 (2011)
2. Wiig, K.M.: Knowledge management: an introduction and perspective. J. Knowl. Manag. **1**(1), 6–14 (1997)
3. Crhová, Z., Pavelková, D., Matošková, J.: A knowledge management literature review based on Wiig's prognosis of 1997. In: Proceedings of the 7th International Joint Conference on Knowledge Discovery, Knowledge Engineering and Knowledge Management (IC3K 2015), vol. 3, pp. 281–286. KMIS (2015)
4. Zhukova, Yu.M., Chernyaev, S.I.: Nekotoryye aspekty upravleniya znaniyami i intellektual'nym kapitalom v vuze. Fundamental'nyye issledovaniya (5), 123–130 (2016)
5. Chernoles, G.V.: Intellektualnyj kapital v strukture aktivov predprijatija, osnovannogo na novyh znanijah: sushhnost, soderzhanie i funkcionalnye roli ego sostavljajushhih. Innovacii (9), 106–127 (2008)
6. Trofimova, L.A., Trofimov, V.V.: Upravleniye znaniyami: uchebnoye posobiye – SPb: Izd-vo SPbGUEF, 77 p. (2012)

7. Panikarova, S.V., Vlasov, V.M.: Upravleniye znaniyami i intellektual'nym kapitalom: [uchebnoye posobiye]. Izd-vo Ural. universiteta, Yekaterinburg, 140 p. (2015)
8. Zaei, M.E., Kapil, P.: The role of intellectual capital in promoting knowledge management initiatives. Knowl. Manag. E-Learn. **8**(2), 317–333 (2016)
9. Gorelova, G.V.: Cognitive modelling as the instrument in the course of knowledge of large system. Int. J. Inf. Theor. Appl. **18**(2), 172–182 (2011)
10. Labunska, S., Iermachenko, I., Prokopishyna, O.: Cognitive analysis and modeling of innovation potential. J. Econ. Manag. Trade **18**(3), 2–14 (2017)
11. Axelrod, R., Papageorgiou, E., Stylios, C.D., Groumpos, P.P.: Fuzzy cognitive map learning based on nonlinear Hebbian rule. In: Australian Conference on Artificial Intelligence, pp. 256–268 (2003)
12. Lin, C.-M.: Using fuzzy cognitive map for system control. WSEAS Trans. Syst. **7**(12), 1504–1515 (2008)
13. Tristan, A.V.: Zastosuvannya kognítivnikh pídkhodív v slabostrukturovanikh sistemakh pídtrimki priynyattya ríshen. Zbírnik naukovikh prats' Kharkívs'kogo uníversitetu Povítryanikh Sil **3**(36), 133–136 (2013)
14. Tkach, Y.U.: Pobudova nechitkykh kohnityvnykh kart dlya otsinky informatsiynykh ryzykiv vyshchoho navchal'no-ho zakladu, Visnyk Chernihivs'koho derzhavnoho tekhnolohichnoho universytetu. Seriya «Tekhnichni nauky»: naukovyy zbirnyk **2**(73), 149–153. CHNTU, Chernihiv (2014)
15. Sirotkin, G.V.: Kognitivnaya model' novoy sistemy upravleniya kachestvom obrazovaniya vuza v tselom. In: Tekhnicheskiye nauki — ot teorii k praktike. Sbonik statey po materialam XXIX mezhdunarodnoy nauchno-prakticheskoy konferentsii, vol. 12, no. 25, pp. 53–68. Izd. «SibAK», Novosibirsk (2013)
16. Lytvynov, V.V., Saveliev, M.V., Skiter, I.S., Trunova, O.V.: Competence model as a tool for estimation of state of it-companies in university's business centre. Matematychni mashyny i systemy (2), 49–60 (2015)
17. Didenko, S.: Demohrafiya Ukrayiny: skorochennya naselennya i ryzyky dlya ekonomiky. https://ua.news/ua/demografiya-ukrayiny-skorochennya-naselennya-i-ryzyky-dlya-ekonomiky/. Accessed 17 Apr 2019
18. Vokuyeva, T.A.: Vychisleniye matritsy vzaimovliyaniya kognitivnoy karty. Izvestiy a Komi nauchnogo tsentra. UrO RAN, Syktyvkar. Vypusk **3**(11), 123–129 (2013)
19. Gansner, E.R., Koutsofios, E., North, S.: Drawing graphs with DOT. https://www.graphviz.org/pdf/dotguide.pdf. Accessed 25 Nov 2018

# Business Model Environment in the Segment of Language Schools

Zbigniew Malara, Janusz Kroik, and Paweł Ziembicki[✉]

Wroclaw University of Science and Technology, Wroclaw, Poland
pawel.ziembicki@pwr.edu.pl

**Abstract.** The paper presents a part of the research on business models of language schools from Lower Silesian Voivodeship. In Poland. The influence of the surrounding elements of the enterprises on the components of the business models according to the template of A. Osterwalder was assessed. The analysis covered the majority (96 – approx. 66%) of language schools in the Voivodeship. A survey consisting of 31 surrounding elements of the enterprises was prepared, and its respondents included owners or managers of these enterprises. The results allowed to isolate the most important elements according to two dimensions of the assessment of the surroundings, i.e. according to significance and favourability. In the further surroundings, such an element was constituted by, among others, seasonality of the demand on the language services market, and in the immediate surroundings, among others, the threat from the substitutes. The components of the business models in the analysed segment which should undergo modifications as a reactive form of adaptation of the models to the conditions of the environment were indicated. Concrete solutions depend on recognising the gap between the value offer expected by the clients and the value proposition of the enterprise.

**Keywords:** Research · Language schools · Small and medium enterprises · Environment · Model component

## 1 Research Premises

The paper presents a part of the results of the research conducted in the years 2018–2019 in the language schools' sector. The basic objective of the works was to diagnose the strategic gap from the perspective of prevailing business models realised by language schools in Lower Silesia in Poland. The diagnosis of this gap referred to the difference in the perception of value propositions of the offer by different segments of the clients of the schools (Ziembicki 2019). The business model canvas template (Osterwalder and Pignuer 2012) and the business model lean canvas template (Maury 2012) were used in the assessment. The authors have been interested in the language services segment since 2014 (Ziembicki 2014; Malara et al. 2015; Kroik and Ziembicki 2016). Comprehensiveness of the recent studies required assessing the impact of the isolated surrounding elements on business models of the companies from this segment. This impact can be diagnosed mono-dimensionally, through analysing the general perception of managers/owners of the enterprises, or multidimensionally, through

© Springer Nature Switzerland AG 2020
Z. Wilimowska et al. (Eds.): ISAT 2019, AISC 1052, pp. 74–85, 2020.
https://doi.org/10.1007/978-3-030-30443-0_7

several contexts of the impact. The research uses a two-dimensional approach referred to as the assessment of significance and favourability of the surrounding elements, respectively. Thanks to the obtained data on the impact of the surrounding elements (in two dimensions), it was possible to transition to the part which diagnoses the strategic gap, and the postulatory part of the research which is substantially supported by its context of environmental influences.

Business models have been a topic of deep interest for the many Polish and foreign researchers, over the last dozen or so years. Study of the definitions of these models was also performed in order to focus on the aspect of environmental influences. Own, original proposition of the description of the essence of the model was created on the basis of several other propositions and formulated in the following way: Conceptualising business model of the enterprise may be perceived as a system of inter-dependent actions which **go beyond the enterprise and cross its boundaries**. The system of actions enables the company, in consultation with its partners, to create values and to make these values available in an appropriate way (Amit and Zott 2009). The business model is considered as a reflection of the strategy of the enterprise (Baden-Fuller and Morgan 2010). Due to adopting the convention of the description of the business model according to the canvas template (Osterwalder and Pigneur 2005, 2009), this paper quotes the description proposed by its creators: a business model is a conceptual tool which that contains a set of elements and their relationships and allows expressing the business logic of a specific firm. It is a description of the value a company offers to one or several segments of customers. It is a description of the architecture of the firm and its networks of partners for creating, marketing and delivering this value and relationship capital, to generate profitable and sustainable revenue streams. When it comes to Polish authors, two names are worth noting, i.e. T. Falencikowski and M. Jabłoński. The first one asserts – in the analysis of cohesion of business models – that: the structure of a business model (components of a business model) consists of: competences, resources, organisations, supply logistics, instruments for intercepting values and values proposed to the clients (Falencikowski 2013). The second one assessed business models in creating values of the companies from the New Connect trading floor and identified the category of innovative business models (Jabłoński 2013). J. Brzóska states that: perceiving the achievement of competitiveness as a market imperative by an enterprise may be connected with the used business model (Brzóska 2009). Transformation of the industry (sector, segment) is therefore linked to the changes in prevailing business models. The condition for the transformation is using new technologies, but its catalyst are transformational business models (Kavadis et al. 2017). The above premises were the basis for developing several measuring devices - in the form of questionnaires and interviews with the use of the Likert scale (1–5) - with managers of 96 companies referred to as language schools from Lower Silesia in Poland, which belong to the category (class) of small and medium-sized enterprises (SME). They constitute nearly two thirds of all enterprises of this segment in this Region.

## 2  SME Language Services in Poland

Small and medium-sized enterprises play a key role in the economic and social functioning of Poland. Generally, SMEs are characterised by high flexibility in market activities and adaptive skills to changing conditions of the environment. SMEs, through stimulating innovative processes and modernisation of the industry structure, have a positive effect on the economic development. In 2016, non-financial business activity was conducted by 2,013 thou. enterprises in Poland. In SMEs, micro-enterprises which constitute as much as 96.2% of the whole population are predominant, whereas small and medium enterprises amount to 2.8% and 0.8% respectively[1]. On the other hand, the level of competence of persons on the labour market in terms of foreign languages has an impact on the development of many branches of economy and industry, and language schools are an organised form of raising that level. The reports analysing the language industry emphasise its dynamic and rapidity of changes. Over the years, since Poland's accession to the EU, changes in the educational services sector had an impact on the change in assessment of value of language competence in the EU. For instance, in 2012 European tendency of decrease in skills and competences in terms of knowledge of foreign languages was observed in the period from 2005. The percentage of Europeans declaring knowledge of a foreign language decreased in this period by 2% points to 54% in the EU[2]. Meanwhile, Poland due to the dynamics of integration processes showed an upward trend. On the other hand, according to the same source (Eurobarometer), the number of enterprises providing language services in Europe was growing and providing their services to an increasingly large group of clients, recording an increase in sales indicators (see footnote 2). In Poland, in the subsequent years, the tendency continued. It should be emphasised as an important context for the surroundings of language schools and their business models, that there exists jobless growth. It is the cause of an increased personalised competition on the labour market. It is an additional reason for which education and improving qualifications is becoming a prerequisite for survival for a growing number of collaborators of companies and institutions. In the segment of language services, apart from traditional language schools and certifying bodies, there are also companies which operate in the macro scale, as according to the Central Statistical Office (CSO) in 2016 there were approx. 62 companies registration in Poland which conduct business activity in the area of non-scholastic forms of education[3]. The segment of language schools constitutes an insignificant percentage of business entities among them. According to the Register of Training Institutions (Rejestr Instytucji Szkoleniowych, RIS), at the end of 2018 this

---

[1] Based on: Central Statistical Office, *Activity of non-financial enterprises in 2017*, https://stat.gov.pl/obszary-tematyczne/podmioty-gospodarcze-wyniki-finansowe/przedsiebiorstwa-niefinansowe/dzialalnosc-przedsiebiorstw-niefinansowych-w-2016-r-,2,12.html?pdf=1 (access: 19.11.2018).

[2] Special Eurobarometer 386/Wave EB 77.1, *Europeans and their languages*, Brussels, TNS Opinion & Social, 2012 http://ec.europa.eu/commfrontoffice/publicopinion/archives/ebs/ebs_386_en.pdf (access: 23.10.2017).

[3] http://stat.gov.pl/obszary-tematyczne/podmioty-gospodarcze-wyniki-finansowe/zmiany-strukturalne-grup-podmiotow/zmiany-strukturalne-grup-podmiotow-gospodarki-narodowej-w-rejestrze-regon-2015-r-,1,17.html (access: 19.11.2018).

number amounted to 1,596 entities and 146 in Lower Silesian Voivodeship in Poland. More than half of the enterprises (approx. 56%) are natural persons performing one-person economic activity. In 2015, it was possible to observe a significant growth in the number of education and training enterprises registering at the RIS. There was a new tranche of EU funding dedicated to them for the years 2014–2020.

A vast majority of enterprises (9 out of 10) has one establishment and employs approx. 21 people. Most of the employees of this sector are not employed full-time. More than 2/3 of enterprises do not have appropriate local infrastructure and their needs are fulfilled in the form of a fixed-term or periodic tenancy. It should be noted that more than 75% of enterprises has computer infrastructure with access to the Internet network[4].

The above background indicates that language companies are facing an evolutionary stage of changes which transforms the segment (in two aspects: technology and business model), whereby the range of transformations depends on the strategic gap in both aspects. The first of them, "hard," lies outside the scope of the research which according to the components of the canvas template is linked to resources, activities and cost structure (left side of the template). Of course, relations between components translate into capacities and solutions included on its right side. This side, on the other hand, is a starting point (according to the methodology of design) for modifications of the business model leading to increased cohesion between value offer proposition for the clients and their expectations. Environmental impact on these modifications is obvious, but specific surrounding elements may have more agency (for both dimensions adopted in assessing the surroundings) in adopting reactive solutions (e.g. service packages) by SME language schools.

# 3    Measurement of Impact of Surrounding Elements on Business Models of Language Schools

In order to measure the impact of the surrounding elements on the business model, a broad review of different propositions of distinguishing them (classification) was conducted. Deriving from two items – from among many in the literature – the surroundings are defined as: everything that exists outside of it, outside of its boundaries and has an impact on the enterprise, and shapes its business model (Obłój 2010) and (Koźmiński and Piotrowski 2011). Classifications focus on its two types. General (further, macroeconomic, indirect) and immediate (competitive, microeconomic, direct). Over time, a coherent picture of the classification of the elements of both types of environment has developed. In the first one, there are five areas, including the key economic area, as it encompasses all economic phenomena and processes, including manufactured, sold and purchased product and/or service markets, as well as labour markets and capital markets. In the immediate surroundings, attention is placed on suppliers, clients and competitors, although the influence of regulators may also

---

[4] http://psz.praca.gov.pl/documents/10240/1012429/RIS%20raport%202013%20PDF.pdf/b839706c-03ee-46fe-a6fc-229a560512b3?t=1417694996000 (access: 24.12.2018).

determine behaviours of enterprises and segments (and sectors). The environmental impact is continuous and may be gradual in the perception of its influence on the level of a single entity as well as market trends, events and factors targeting a segment, a sector, an industry[5]. The methods of measurement are different then. On the level of a single enterprise, the source of information may be key managers (managerial perception), and for the segment, in an integrated formula of data processing, their statistical estimations and/or modelling.

Obtaining information in the form of obtaining managerial perception was used in the paper, and the condensed data obtained on its basis give a picture of the segment in the SME category in the arrangement of the Region. Agreed segment uniformity of language schools (initial, according to the canvas, component of the construction and modification of the business model) and a large sample of analysed enterprises allows to draw conclusions on the state of the sector in Poland. Ultimately, for the research on environmental influences - based on a dozen or so literature items - 31 factors (elements $j = 1, 2 \dots 31$) representing further surroundings (21) and immediate, sectoral-competitive surroundings (10) were preselected. In the first step, a pilot study for 20 companies was conducted. Its aim was to perform sectoral valuation of the initial set, allowing to introduce additional elements. In the pilot study, the environmental impact was considered in a comprehensive perspective, that is the functioning of the enterprise. In the main study, the aspect of assessment referring to the business model in its soft part (the right side according to the canvas) was specified. The essence of measuring the two dimensions of the environment: **significance and favourability** was specified by providing appropriate interpretation of expected perception of both dimensions. Using the Likert scale, the respondents spoke simultaneously about both contexts by describing 5 states: from insignificant to very significant and from very unfavourable to very favourable. No new factors were proposed in the pilot study. As a result of the analysis of the results according to the principle that an external surrounding element remains in a set for the purposes of the study proper if it is at least once described as significant or very significant by enterprises participating in the pilot study. The **favourability** dimension was used alternatively in the situation of low significance. Such a situation applied to the factor (element) activities of trade unions. In arriving at final findings about surrounding elements of business models, a selection consisting of several steps was performed based on the data of the surveyed. They were described in greater detail in the next part of the paper.

Managers in the pilot studies drew attention to failures in the construction of the questions and incomprehensibility of a part of the research questions. Suggestions of the respondents were taken into account in the construction of the measuring device in the study proper. On this basis, the form of some questions was simplified, and a strictly uniform scale was used in order to obtain quantitative data allowing to verify research hypotheses in establishing the gap and correctly interpret the results of the research in terms of business models of language schools. A random selection of enterprises from the sample was performed with the use of the RIS database and the

---

[5] A convention for language schools that a segment is a part of the education sector was adopted in the paper. A branch of industry is equivalent to a sector.

CSO database. Concentration in the perception of environmental influence on the right side of business models allowed to assess the stability of standards developed in: segmentation of the clients, relations of cooperation with the clients, communication channels with them and price sensitivity of enterprises behind the assumed value proposition. In the next stage, it allowed to assess the pressure of the value proposition expected by the clients and the inclination of the companies towards changes in business models of language schools through an adequate modification of their components. This stage can use recommendations of the lean canvas (Maury 2012).

## 4  Sectoral Valuation of the Impact of Surrounding Elements - Result and Interpretation

Selection of the group of entities which may potentially participate in the research was based on the RIS database and the CSO database including entities from the group P (according to the Polish Classification of Activities). Verification of these databases indicated that some of the entities ceased the economic activity or suspended it. These entities were removed from the database. In order to obtain a random sample, a tool was created in MS Excel program. The questionnaire, including assessment of the surrounding elements, was handed over to 98 enterprises. The complete answers, fulfilling the methodical requirements, were received from 96 language schools. Their structure in the SME category is presented in Table 1.

**Table 1.** Structure of the enterprises according to the number of employed people assessing the impact of the surroundings on the components of their business model.

| Number of the employed | Number of companies | Percentage (%) |
|---|---|---|
| Micro-enterprise (up to 10) | 44 | 45.8 |
| Small enterprise (up to 50) | 39 | 40.6 |
| Medium-sized enterprise (up to 250) | 13 | 13.6 |

The participants of the study were people managing the companies - owners or key managers. These are enterprises from Wroclaw and surrounding areas (Lower Silesia, Poland). The structure of the managers in terms of gender and location is presented in Table 2.

**Table 2.** Structure of the enterprises in terms of gender of the respondents and location.

| Gender of the respondent | Number of people | Percentage (%) |
|---|---|---|
| Woman | 55 | 57.3 |
| Man | 41 | 42.7 |
| Place of business | Number of language schools | Percentage (%) |
| Wroclaw | 76 | 79.1 |
| Outside Wroclaw | 20 | 20.9 |

The duration of the business activity may be perceived as an important premise in expressing opinions on the environmental impact on the business model. Four classes of schools were distinguished in the research in this regard. A small, few percent group is constituted by enterprises operating for less than 2 years. It means that the perception of the sector (through own experience) may be considered sufficient. Table 3 presents the data.

**Table 3.** Duration of the companies participating in the research on the influence of the surroundings on the components of their business models.

| Number of years | Number of language schools | Percentage (%) |
|---|---|---|
| Up to 2 years | 6 | 6.3 |
| Between 2 and 5 years | 23 | 23.9 |
| Between 5 and 10 years | 27 | 28.1 |
| More than 10 years | 40 | 41.7 |

The reliability of the scale adopted in the research was analysed. The results confirmed its reliability with the use of Cronbach's Alpha statistic with reference to significance and favourability of the surroundings (62 items). Table 4 contains the assessment of reliability of the scale.

**Table 4.** The value of Cronbach's Alpha statistic in the assessment of reliability of the scale adopted for the valuation of significance and favourability of the business model environment of language schools.

| Business model environment | Cronbach's Alpha coefficient (threshold 0.700) | Number of items |
|---|---|---|
| The scale of assessments of significance and favourability of the environment | 0.902 | 62 |

Due to the adopted measuring scale (1–5) and the necessity of indicating these elements for which transfer of the influence onto the business model of an enterprise is probable, thresholds (range) of sectoral reaction were introduced. When it comes to **significance** of a given element, it was the threshold of 3.0, considering other possible values (3.5 or 4.0). It is the class Significant for the segment. In the case of the second dimension – **favourability** – situations referring to both the lower limit of the range (interpreted as threats to which an enterprise tries to react) and the upper limit of this value which may be an opportunity for the language schools are important for the diagnosis. In the first case, the threshold is 2.5 (range: 1–2.5 – favourability class: **threat**), and in the second case 3.5 (range: 3.5–5 – favourability class: **opportunities**). These findings imposed a defined method of analysis and interpretation of the results. It was performed in several steps. The final stage consisted of isolating elements in two fields: (1) significant - **threat** (respective ranges 3.0–5 and 1–2.5) and (2) significant - **opportunities** (respective ranges 3.0–5 and 3.5–5). Table 5 contains number distribution of answers for sei of the element (i = 1, 2 … 22) in the dimension of significance (Si) with the average within the range (3.0–5) and distribution of the assessment of favourability (Fi) (i = 1, 2 … 22) adequate for this selection due to significance. Average values used for the selection and threshold assessment of the impact of surrounding elements on the language schools segment ($S_{avg}$; $F_{avg}$) were also provided.

When adopting a comprehensive (on the level of a segment) overview of the impact of external surrounding elements on language schools - and a possible modification of their business models resulting from this impact - forced reactivity of the enterprises may be observed. Such a view is a result of information included in Table 5. It contains elements which stand out thanks to the considerable potential to influence business models of language schools, with **significance** exceeding the threshold and being within the range of stipulated opportunities and threats according to the dimension of favourability (**averages in bold**). The indirect influence of further surroundings does not have to have such a causative character, although several elements from this group may transform into specific reactions, including those resulting from the fact of the development of the Wroclaw agglomeration, e.g. the level of foreign investments (**4.01; 4.00**), behind which lurk new linguistic needs of the interested parties (opportunities). The prominent element of this environment is seasonality assessed as a significant threat (**4.33; 1.54**). This factor, as it seems, is known and noticeable, but also exerts pressure on changes in shaping of the revenue structure (one of the business model components). It is possible that this pressure will initiate a new perspective on the segmentation criteria. Integration with the EU (**4.11; 4.02**) and internationalisation of the economy (**4.25; 4.09**) are coupled factors which is confirmed in the perception of the respondents. Immediate surroundings (competitive-sectoral) are assessed as significant in the field of threats, which responds to the intuition of the respondents. The highest priority is given to the component of substitutes (**4.30; 1.95**). It may mean difficult to quickly react to variants of solutions in the field of supporting relations with the clients or improved channels of reaching the clients. The high bargain position of the recipients (**4.09; 1.98**) should result in identifying their needs and establishing the gap in the value proposition, as was mentioned in the beginning, in the conditions of high competitiveness (**4.23; 2.03**). The obtained results are used as hints for the projection of changes in business models in terms of the segment and individual entities. The lean canvas approach and the related approach of theory of tasks to perform may be used for that purpose. Customer personas[6], consumer panels or competition analysis may constitute the starting point for performing analysis of the tasks of the clients (Duncan et al. 2017).

---

[6] A persona is possibly the most detailed description of a model group of target recipients, characterising their profound needs and priorities. Personas are created based on consumer questionnaires and quantitative behavioural data.

**Table 5.** Distribution of the results of assessment of these surrounding elements conforming to the requirement of the threshold of significance and the average sectoral valuation of these dimensions.

| Scale | 1 | 2 | 3 | 4 | 5 | Average |
|---|---|---|---|---|---|---|
| i-th element (i = 1, 2 … 22) | Number of assessments $(S_i, F_i)$ | Number of assessments $(S_i, F_i)$ | Number of assessments $(S_i, F_i)$ | Number of assessments $(S_i, F_i)$ | Number of assessments $(S_i, F_i)$ | $(S_{avg}; F_{avg})$ |
| *Further surroundings - political-legal elements* | | | | | | |
| Social insurance system (se1) | (5.1) | (21.55) | (26.32) | (17.8) | (27.0) | (3.42; 2.49) |
| Form of running business activity (se2) | (0.10) | (11.15) | (37.15) | (17.23) | (31.33) | (3.71; 3.56) |
| Government policies towards SMEs (se3) | (4.2) | (26.20) | (31.21) | (35.24) | (0.29) | (3.01; 3.60) |
| *Further surroundings - economic elements* | | | | | | |
| Foreign investments level (se4) | (0.0) | (0.1) | (35.35) | (25.23) | (36.37) | **(4.01; 4.00)** |
| Unemployment level (se5) | (0.6) | (29.15) | (28.18) | (6.35) | (33.22) | (3.45; 3.54) |
| Economic growth (se6) | (0.8) | (0.7) | (29.7) | (22.25) | (45.49) | **(4.17; 4.04)** |
| Internationalisation of the Polish economy (se7) | (0.8) | (0.6) | (19.8) | (34.21) | (43.53) | **(4.25; 4.09)** |
| Seasonality of demand (se8) | (0.53) | (0.34) | (14.9) | (36.0) | (46.0) | **(4.33; 1.54)** |
| *Further surroundings - demographic and socio-cultural elements* | | | | | | |
| Lifestyle, ethical and moral values and norms (se9) | (3.2) | (30.14) | (10.25) | (19.25) | (34.30) | (3.53; 3.70) |
| Level of wealth of the society (se10) | (3.10) | (9.25) | (27.18) | (12.20) | (45.23) | (3.91; 3.22) |
| Need of professional development (se11) | (0.0) | (5.3) | (33.32) | (13.39) | (45.23) | (4.02; 3.83) |
| *Further surroundings - technological elements (lack of assessment according to the threshold over 3.0)* | | | | | | |
| Further surroundings - international elements | | | | | | |
| Integration with the European Union (se12) | (0.0) | (0.5) | (33.20) | (19.39) | (44.32) | **(4.11; 4.02)** |
| Globalisation processes (se13) | (14.2) | (16.10) | (15.30) | (15.38) | (36.16) | (3.45; 3.58) |
| Close surroundings - competitive-sectoral elements | | | | | | |
| Bargaining power of the suppliers (se14) | (2.1) | (24.3) | (14.30) | (21.36) | (35.26) | (3.66; 3.86) |
| Bargaining power of the recipients (se15) | (0.35) | (2.36) | (29.18) | (23.6) | (42.1) | **(4.09; 1.98)** |
| Risks posed by the substitutes (se16) | (0.36) | (1.34) | (14.24) | (36.3) | (4,0) | **(4.30; 1.95)** |
| Intensification of competition in the sector (se17) | (0.27) | (1.45) | (21.19) | (29.4) | (45.1) | **(4.23; 2.03)** |
| Entry barriers (se18) | (5.3) | (8.27) | (26.32) | (31.28) | (26.6) | (3.68; 3.07) |
| State of the local infrastructure (se19) | (0.0) | (5.26) | (32.23) | (54.40) | (5.7) | (3.61; 3.52) |

(continued)

**Table 5.** (*continued*)

| Scale | 1 | 2 | 3 | 4 | 5 | Average |
|---|---|---|---|---|---|---|
| i-th element (i = 1, 2 … 22) | Number of assessments $(S_i, F_i)$ | Number of assessments $(S_i, F_i)$ | Number of assessments $(S_i, F_i)$ | Number of assessments $(S_i, F_i)$ | Number of assessments $(S_i, F_i)$ | $(S_{avg}; F_{avg})$ |
| Absorption capacity of the market (se20) | (0.1) | (0.14) | (39.31) | (11.34) | (46.16) | (4.07; 3.52) |
| Location of the enterprise (se21) | (0.12) | (0.17) | (37.17) | (42.26) | (17.24) | (3.79; 3.34) |
| Innovativeness of the co-operators (se22) | (30.2) | (9.31) | (10.17) | (24.30) | (23.16) | (3.01; 3.28) |

## 5   Summary

Due to the nature of the presented studies which are the first step in diagnosing the strategic gap between the sectoral value offer and the offer resulting from the needs of the recipients, conclusions from the assessment of the impact of surrounding elements on business models may have more importance in the situation of structuring this gap. At this stage, it is possible to indicate the pressure of influence on the analysed components of the model – its right side. Among 22 components – from among 33

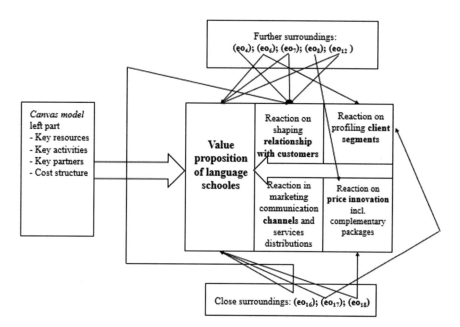

**Fig. 1.** Diagram of the most important surrounding elements influencing (narrow pointers) the components of business models of language schools in Wroclaw and Lower Silesian Voivodeship according to the canvas approach (results of the analysis of managerial perception). (Source: own elaboration)

assessed – included in the collective summary (Table 5), a certain number of them must be taken into consideration by language schools. In the above concise comment of these results, each of the components of the right side of the model canvas will be subject to the influence of surrounding elements. The strongest emphasis, according to the authors, was placed on profiling the clients. To a very large extent, from interpretation of the received results, this comment refers to relations. The necessity of remodelling the standards in terms of the component of revenue structure is also clear, whereby the area of sophisticated price tactics should be entered. These tactics are a derivative of changes in segmentation, relation and channels isolated in the business model. These finding are symbolically presented in the diagram below, factoring in the key impact of the surrounding elements with the use of symbols adopted in the analysis.

The identified impact of the surrounding elements described in Fig. 1 constitutes an important premise for the perception of the strategic gap formulated in the introduction. Its detailed structure requires transitioning to the level of analysing specific solutions in business models of language schools and the statistically significant group of recipients expressing their assessment of the expectations. A tool to measure that gap is a survey distinguishing 30 situations which model the offers of language schools and the expectations about the value proposition for 6 segments of clients, and 5 levels of price sensitivity of the offer for 11 languages (10 foreign). Statistical preparation of the results required, among others, the analysis of the measuring scales and tests: the Kolmogorov–Smirnov test and the Shapiro-Wilk test. In the second step, factor analysis of (Ziembicki 2019; Malara and Ziembicki 2019) was used.

# References

Amit, R., Zott, C.: The Business Model as the Engine of Network – Based Strategies. Wharton School Publishing, Upper Saddle River (2009)

Baden-Fuller, Ch., Morgan, M.S.: Business Models as Models. Long Range Planning, vol. 43 (2010)

Brzóska, J.: Model biznesowy – współczesna forma modelu organizacyjnego zarządzania przedsiębiorstwem. Organizacja i Zarządzanie. Kwartalnik naukowy, Wydawnictwo Politechniki Śląskiej, no. 2. Gliwice (2009)

Duncan, D., Dillon, K., Hall, T., Christensen, C.: Ustal, jakie zadania muszą wykonać klienci. Harvard Business Review, no. 168, February 2017

Falencikowski, T.: Spójność modelu biznesu. Koncepcja i pomiar. Wydawnictwo CeDeWu, Warsaw (2013)

Jabłoński, M.: Kształtowanie modeli biznesu w procesie kreacji wartości przedsiębiorstw. Difin, Warsaw (2013)

Kavadis, S., Lados, K., Loch, Ch.: Transformacyjny model biznesu. Harvard Business Review, no. 170, April 2017

Koźmiński, A., Piotrowski, W.: Zarządzanie. Teoria i praktyka. PWN, Warsaw (2011)

Kroik, J., Ziembicki, P.: Zarządzanie wiedzą źródłem innowacyjności przedsiębiorstwa – wstępna ocena sektora edukacyjnego MŚP. Zeszyty naukowe Wyższej Szkoły Ekonomiczno-społecznej w Ostrołęce 2/2016 (21). Ostrołęka (2016)

Malara, Z., Hrydziuszko, M., Ziembicki, P.: Innowacyjne modele biznesowe szansą na rozwój przedsiębiorstw. In: Dudzik-Lewicka, I., Howaniec, H., Waszkielewicz, W. (eds.) Zarządzanie wiedzą i innowacje w organizacji. Bielsko-Biała, Akademia Techniczno-Humanistyczna (2015)

Malara, Z., Ziembicki, P.: Luka w propozycji wartości modelu biznesu szkół językowych. Report Series PRE 4, Wydział Informatyki i Zarządzania Politechniki Wrocławskiej (2019)

Maury, A.: Running Lean. Iterate from Plan A to a Plan That Work. O'Reilly, Sebastopol (2012)

Obłój, K.: Pasja i dyscyplina strategii: jak z marzeń i decyzji zbudować sukces firmy. Poltext, Warsaw (2010)

Osterwalder, A., Pigneur, Y.: Clarifying business models: origins, present and future of the concept. Commun. Assoc. Inf. Sci. (CAIS) **16** (2005)

Osterwalder, A., Pigneur, Y.: Business Model. Wiley, Hoboken (2009)

Osterwalder, A., Pigneur, Y.: Tworzenie modeli biznesowych. Podręcznik wizjonera. Helion, Gliwice (2012)

Osterwalder, A. https://strategyzer.com/canvas/value-proposition-canvas. Accessed 31 Dec 2018

Ziembicki, P.: Czynniki i strategia kształtowania się cen w przedsiębiorstwach usługowych w sektorze językowym. Master's thesis, Politechnika Wrocławska (2014)

Ziembicki, P.: Modele biznesowe w segmencie usług językowych. Doctoral thesis, Faculty of Computer Science and Management, Wrocław University of Science and Technology (2019, in preparation)

https://stat.gov.pl/obszary-tematyczne/podmioty-gospodarcze-wyniki-finansowe/przedsiebiorstwa-niefinansowe/dzialalnosc-przedsiebiorstw-niefinansowych-w-2016-r-,2,12.html?pdf=1. Accessed 19 Nov 2018

http://ec.europa.eu/commfrontoffice/publicopinion/archives/ebs/ebs_386_en.pdf. Accessed 23 Oct 2017

http://stat.gov.pl/obszary-tematyczne/podmioty-gospodarcze-wyniki-finansowe/zmiany-strukturalne-grup-podmiotow/zmiany-strukturalne-grup-podmiotow-gospodarki-narodowej-w-rejestrze-regon-2015-r-,1,17.html. Accessed 19 Nov 2018

http://psz.praca.gov.pl/documents/10240/1012429/RIS%20raport%202013%20PDF.pdf/b839706c-03ee-46fe-a6fc-229a560512b3?t=1417694996000. Accessed 24 Dec 2018

# Models of Economic Efficiency of Hospitals in Poland

Agnieszka Parkitna$^{(\boxtimes)}$ ⓘ and Magdalena Gądek

Wrocław University of Science and Technology, Wrocław, Poland
agnieszka.parkitna@pwr.edu.pl,
magdalena@gadek.pwr.edu.pl

**Abstract.** The paper presents the results of research concerning the identification of differences in the models of hospital management determined by the financial result. The indication of the model structure was the result of the analysis and conclusion of the foundations contained in the source literature, and the hospital was therefore described as an organisation, and its definition became a basis for creating a superior structure reflecting the Polish reality of the health system. The main stages of empirical verification of the model were: preparation and implementation of questionnaire surveys, verification of their correctness, statistical description and analysis of the collected data, construction of the internal structure of the main model variables, separately for hospitals that generated profit and loss. The internal structure of the main model variables was ordered according to the agglomerative hierarchical clustering algorithm in a bottom-up approach. Two models of organisational determinants influencing the hospital's daily financial result, as well as organisational determinants influencing the hospital's negative financial result were compared. A mathematical comparative analysis of both models was carried out, with justification of its validity in the form of verification of research hypotheses. On this basis, model differences were determined.

**Keywords:** Efficiency · Hospital · Economic model

## 1 Problems of Economic Efficiency of Hospitals in Poland

The activity of hospitals in Poland focuses on three main goals: economic, medical and social [5] Non-financial goals of hospital activity lead to lower expenses resulting from smaller government subsidies ensuring necessary care for citizens [4]. In this situation, hospitals must be guided by cost-efficiency, as it allows them to remain in the market and provide health services. A permanent problem of the health care system in Poland is funding insufficient to meet the health needs of citizens [2]. Currently, the problem is reflected in new legislation which fails to specify their place in the system, mission in health care infrastructure, rules of carrying out non-profit health care activity in the public system [13]. The supply of resources necessary to meet growing health needs not only fails to keep up with the growing demand, but also imposes numerous limitations on the development of these resources, both financial and labour, as well as access to and use of new technologies [3].

© Springer Nature Switzerland AG 2020
Z. Wilimowska et al. (Eds.): ISAT 2019, AISC 1052, pp. 86–96, 2020.
https://doi.org/10.1007/978-3-030-30443-0_8

Hospitals in Poland operate based on a special formula – a combination of independent operation and public ownership [13]. There is also no single model of hospital supervision, thus supervisors of subordinate medical entity's activity have very limited knowledge of their situation [7]. Transforming an independent public health care facility into a commercial law company does not always bring the desired effects; [8] however, it leads to a significant increase in efficiency measured not only by the financial result, but also by the achieved health effect The efficiency of hospitals – the way they transform inputs into outputs – may be affected by environmental factors falling beyond the scope of hospital management [12]. Hospitals, in particular public, do not have much room for influencing the amount of earned revenue, placing much emphasis on cost management as a result of having the amount of points and funds for medical services from the payer limited. Such emphasis is aimed primarily at making decisions on HR management, the use of medicines, and other medical and non-medical materials, premises along with equipment and apparatus available there, etc. [6]. In this situation, the hospital's indebtedness is a measure of its success in the medical services market [9]. Hospital management influences the way the hospital is organised in terms of quality, efficiency and effectiveness. In accordance with the world literature, the main cause of ineffectiveness is hospital organisation [11]. The effectiveness of health services depends on a number of factors, in particular the shape of the health care system, as well as the organisation and management of individual facilities [15]. Medical staff are one of key resources determining the quality and effectiveness of hospital operation [1]. The modernisation strategy of health care facilities involves not only innovative, cost-effective projects, but also investment activities undertaken to ensure health services of appropriate quality [16].

Although public aid and health care restructuring programmes brought positive results, they did not solve the organisational and financial problems of hospitals. Managers in restructured hospitals often remain the same, regardless of their performance. Management staff responsible for the previous indebtedness of the hospital may, after the restructuring of the facility, remain in office and increase its indebtedness, continuing their management policy [18]. Please note, however, that the National Health Fund's (NFZ) failure to correctly value rates of health services cannot be the main reason for the indebtedness of certain hospitals. After all, some clinical hospitals discharge all their liabilities in due time [17].

Given the foregoing, the research problem concerns the identification of relationships between organisational elements of the hospital and its financial result.

## 2  Hospital Management Model Determined by the Financial Result

The term "hospital" understood as an organisation entails certain constraints in searching for its correct model. They indicate that the desired organisational scheme must emphasise the importance of capital, relationships between organisational components, the importance and role of the management and control subsystem, and must clearly define the boundaries of each element based on its functions. It can be therefore concluded that the financial result achieved by hospitals is significantly influenced by factors resulting from organisational elements, i.e. organisational determinants. We propose to illustrate the hospital as an organisation with a simple model consisting of three elements: **People, Mechanisms and Components.**

- People, being the most important element of the organisation, perceived as its intellectual capital, meaning all specific characteristics and features embodied in employees that have a certain value and are a source of future income [10],
- Mechanisms, being the second element of the model, are understood as states, processes, phenomena or activities following each other in a certain order [14],
- Components, understood as elements of a whole, and a single component – as one element of something [14].

The three elements of the conceptual organisational model referred to and defined above are assumed to be interrelated as verified empirically below. We want to prove that the financial result is influenced by organisational determinants: People, Mechanisms and Components, and thus their internal structure between hospitals with a positive and negative financial result will be different. The stage of operationalisation of research model categories involved formulating four research hypotheses regarding the adopted model:

*HP1: There is a relationship between Components and the financial result of the hospital.*

*HP2: There is a relationship between Mechanisms and the financial result of the hospital.*

*HP3: There is a relationship between the Human Factor (People) and the financial result of the hospital.*

*HP4: Model elements: Components, Mechanisms, People are interrelated.*

In order to conduct research and then verify the research hypotheses, a questionnaire for health managers in hospitals was developed. We designed it to measure the impact of organisational factors on the financial result of hospitals, containing 33 single-choice questions in the 7-point Likert scale.

## 2.1 Analysis of Experimental Data

The questionnaire surveys covered 580 public hospitals in Poland, accounting for 59.24% of the research sample. 104 complete forms were received, accounting for 17.93% of all the questionnaires sent. Analysing the financial result demonstrated in the balance sheet for the year under review, 59 hospitals generated profit, and 45 surveyed units suffered loss.

The research tool was developed for research purposes, and therefore it was necessary to assess its suitability in terms of the reliability and accuracy of the results obtained. Cronbach's statistics for the scales of the proposed research tool are as follows: Components 0.917, Mechanisms 0.972, People 0.959.

Since these are latent variables, values resulting from indicator values were applied. The hypothesis on the normality of distribution of the examined variables was verified by the Kolmogorov-Smirnov test. The test revealed that no distribution had the characteristics of a normal distribution, as the significance level of statistics for Components   p = 0.014 < 0.05,   Mechanisms   p = 0.003 < 0.05   and   People p = 0.000 < 0.05.

Variables do not have a normal distribution, so the Spearman test for the Hpn hypothesis on the lack of monotonic dependence between variables will be carried out (Table 1).

**Table 1.** Spearman's correlation coefficients together with the level of significance

| | | | Net financial result | Components | Mechanisms | People |
|---|---|---|---|---|---|---|
| Spearman's rho | **Net financial result** | Correlation coefficient | 1.00 | 0.642[**] | 0.682[**] | 0.564[**] |
| | | Significance (two-sided) | | 0.00 | 0.00 | 0.00 |
| | | N | 104 | 104 | 104 | 104 |
| | **Components** | Correlation coefficient | 0.642[**] | 1.00 | 0.943[**] | 0.808[**] |
| | | Significance (two-sided) | 0.00 | | 0.00 | 0.00 |
| | | N | 104 | 104 | 104 | 104 |
| | **Mechanisms** | Correlation coefficient | 0.682[**] | 0.943[**] | 1.00 | 0.752[**] |
| | | Significance (two-sided) | 0.00 | 0.00 | | 0.00 |
| | | N | 104 | 104 | 104 | 104 |
| | **People** | Correlation coefficient | 0.564[**] | 0.808[**] | 0.752[**] | 1.00 |
| | | Significance (two-sided) | 0.00 | 0.00 | 0.00 | |
| | | N | 104 | 104 | 104 | 104 |

\*\* Significant correlation at 0.01 (two-sided).

$$H_0 : \rho_s = 0$$

$$H_{pn} : \rho_s \neq 0$$

Based on the results obtained, the formulated research hypotheses were verified:

**HP1: There is a relationship between Components and the financial result of the hospital.**

Because: R = 0.642, p = 0.00 < 0.01, the hypothesis on the lack of correlation between Components and the financial result of the hospital should be rejected in favour of hypothesis HP1;

*HP2: There is a relationship between Mechanisms and the financial result of the hospital.*

Because: $R = 0.682$, $p = 0.00 < 0.01$, the hypothesis on the lack of correlation between Mechanisms and the financial result of the hospital should be rejected in favour of hypothesis HP2;

*HP3: There is a relationship between the Human Factor (People) and the financial result of the hospital.*

Because: $R = 0.564$, $p = 0.00 < 0.01$, the hypothesis on the lack of correlation between the Human Factor and the financial result of the hospital should be rejected in favour of hypothesis HP3;

*HP4: Model elements: Components, Mechanisms, People are interrelated.*

Since the correlation between Components and Mechanisms ($R = 0.943$, $p = 0.00 < 0.01$) is statistically significant; the correlation between Components and People ($R = 0.808$, $p = 0.00 < 0.01$) is statistically significant and the correlation between Mechanisms and People ($R = 0.752$, $p = 0.00 < 0.01$) is also statistically significant, the hypothesis on the lack of correlation between model elements should be rejected in favour of alternative hypothesis HP4.

The analysis of the significance of correlation coefficients was supplemented with the analysis of correlation direction and strength. All correlations between the analysed variables have a positive coefficient, meaning that any increase in one feature is accompanied by an increase in the value of the other (Table 2).

**Table 2.** The correlation coefficient between model variables (from the highest to the lowest value)

| Pair | Coefficient |
|---|---|
| Components – Mechanisms | 0.943 |
| Components – People | 0.808 |
| Mechanisms – People | 0.752 |
| Mechanisms – Net financial result | 0.682 |
| Components – Net financial result | 0.642 |
| People – Net financial result | 0.564 |

Having analysed the factors related to the net financial result, it can be concluded that it is the most strongly related to Mechanisms and the most weakly related to People. When examining the relationships between model components, the highest correlation coefficient is recorded between Components and Mechanisms, while the lowest one – between Mechanisms and People. The results of the analysis of significance, strength and direction of the correlation coefficient between the examined features are also confirmed by an approximation of Spearman's R correlation coefficients from the sample, using the t distribution. Together with R2 calculations, they allowed us to find out what percentage of the variance of one variable is related to the variance of the other (Table 3).

**Table 3.** Values of $R$ coefficients in the $t$ distribution and $R^2$ coefficients (%)

|  | Net financial result | Components | Mechanisms | People |
|---|---|---|---|---|
| Net financial result | - | 8.46; 41% | 9.42; 47% | 6.90; 32% |
| Components | 8.46; 41% | - | 28.62; 89% | 13.85; 65% |
| Mechanisms | 9.42; 47% | 28.62; 89% | - | 11.52; 57% |
| People | 6.90; 32% | 13.85; 65% | 11.52; 57% | - |

When interpreting the results obtained, it should be noted that the variability of the financial result may be perceived in as much as 47% of the variability of Mechanisms and 41% of the variability of Components. The variability of the financial result may be the most weakly related to the Human Factor. The analysis of the relationship between the main model components (Components – Mechanisms – People) reveals that the variability of organisational Components is 89% related to the variability of Mechanisms. There is a strong relationship (65%) between the Human Factor (People) and organisational Components. The variability of organisational Mechanisms is slightly less (57%) related to the variability of the Human Factor structure (People).

## 2.2 Differences in Hospital Management Models Determined by the Level of the Financial Result in Poland

The internal structure of the main model variables was ordered by implementing the agglomerative hierarchical clustering algorithm in the bottom-up approach. The research showed significant differences in the perception of the impact of organisational factors on the financial result between hospitals that generated profit and loss. Hospital managers, who declared that their managed entity recorded a positive financial result for the period considered, considered Components, Mechanisms and People as positively influencing, i.e. increasing, the financial result. The model for this group of respondents is presented below (Fig. 1):

The managers of hospitals that generated profit considered Mechanisms as having the highest impact on the financial result. Components were ranked the lowest by respondents. The managers of hospitals that suffered loss in the period concerned considered Components, Mechanisms and People as negatively affecting, i.e. decreasing, the financial result. The model for this group of respondents is presented below (Fig. 2):

The managers of hospitals that suffered loss in the year concerned considered organisational factors as negatively affecting, i.e. decreasing, the achieved financial result. When analysing the results between particular model variables: Components, Mechanisms and People, it should be noted that the lowest average value of responses was recorded for Mechanisms, meaning that they had the strongest negative impact on the financial result. They considered the People variable as having the lowest negative impact on the financial result.

Given the foregoing, it is assumed that the differences in hospital management models determined by the financial result differ significantly from one another as demonstrated below.

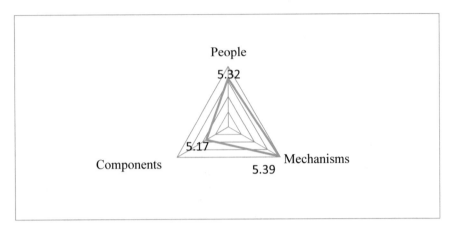

**Fig. 1.** The impact of organisational factors on the financial result – the model for hospitals that generated profit in 2014

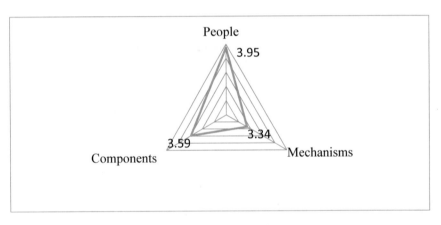

**Fig. 2.** The impact of organisational factors on the financial result – the model for hospitals that suffered loss in 2014

### 2.3   Comparative Model Analysis

To correctly interpret statistical data in order to find out the differences between the models and analyse their internal structure, it was necessary to first determine the type of distribution of the variables at issue in new data sets obtained: in hospitals with a positive (n = 59) and negative financial result (n = 45). For this purpose, the Shapiro-Wilk normality test was applied. The test revealed that the distribution of the People variable in the group of hospitals with a negative financial result had the characteristics of a normal distribution (p = 0.206 > 0.05). It should be noted that the Components variable also had such characteristics (p = 0.082 > 0.05), unlike the Mechanisms variable in the surveyed population (p = 0.025 < 0.05).

When examining the group of model variables obtained from the sample for hospitals with a positive financial result, it was shown that the distribution of the People variable in this group did not have the characteristics of a normal distribution ($p = 0.002 < 0.05$). The test revealed that the Mechanisms variable also did not have such characteristics ($p = 0.045 < 0.05$), unlike the Components variable in the surveyed population ($p = 0.883 > 0.05$).

Prior to comparing the models of hospitals with a positive and negative financial result, it was necessary to verify the existence of statistically significant differences between their components: HP5, HP6, HP7

$H_{S5}$: *The People model element in hospitals with a positive financial result is statistically significantly different from the element in hospitals with a negative financial result*

$$H_0 : \mu_{Ldod} = \mu_{Luj}$$

$$H_1 : \mu_{Ldod} \neq \mu_{Luj}$$

In order to verify the hypothesis, the non-parametric Mann-Whitney U test was applied to two independent samples.

Since $U = 423.500$; $p = 0.000 < 0.05$, hypothesis $H_0$ should be rejected in favour of alternative hypothesis $H_1$. It should therefore be assumed that the People model element differs between hospitals with a positive and negative financial result.

$H_{S6}$: *The Mechanisms model element in hospitals with a positive financial result is statistically significantly different from the element in hospitals with a negative financial result*

$$H_0 : \mu_{Mdod} = \mu_{Muj}$$

$$H_1 : \mu_{Mdod} \neq \mu_{Muj}$$

In order to verify the hypothesis, the non-parametric Mann-Whitney U test was applied to two independent samples. Since $U = 149.000$; $p = 0.000 < 0.05$, hypothesis $H_0$ should be rejected in favour of alternative hypothesis $H_1$. It should therefore be assumed that the Mechanisms model element differs between hospitals with a positive and negative financial result.

$H_{S7}$: *The Components model element in hospitals with a positive financial result is statistically significantly different from the element in hospitals with a negative financial result*

$$H_0 : \mu_{Sdod} = \mu_{Suj}$$

$$H_1 : \mu_{Sdod} \neq \mu_{Suj}$$

In order to verify the hypothesis, the parametric T-Student test was applied to two independent samples, assuming an unknown variance, since the variables in both groups had the characteristics of a normal distribution. Since $p = 0.000 < 0.05$, hypothesis $H_0$ should be rejected in favour of alternative hypothesis $H_1$. It should therefore be assumed that the Components model element differs between hospitals with a positive and negative financial result.

As demonstrated above, there are statistically significant differences between the model elements in hospitals that generated profit and suffered loss.

### 2.4    Identification of Differences in Economic Efficiency Models

In order to simplify an in-depth analysis of differences in both models, it was decided to present the average values of variables, reducing them to the value of absolute impact on the financial result of the hospital. The geometric approach to this transformation involves shifting the obtained results on the numerical axis from the 1–7 scale used during the research to the −3 to 3 scale (Table 4).

**Table 4.** The direct impact of the examined variables on the financial result of the hospital

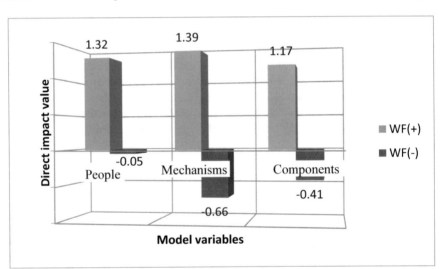

The People variable is at the top of the proposed organisational model. Having analysed the indications, presented in the surveys, of management staff in hospitals that generated profit in 2014, it can be concluded that they considered the People variable as having a small positive impact on the financial result. The managers of hospitals that demonstrated loss in the analysed period mostly considered the People variable, covering the Professional Qualifications, Skills, Knowledge, Experience and Motivation of the staff, as having no impact on the financial result.

This seems to be natural that, if managers consider all hospital staff groups as having a positive impact on, i.e. increasing, the financial result, the staff are perceived as the capital of the enterprise, thus confirming the adopted theoretical assumptions. The broadly understood management of human capital and investment in its development, as appropriate to tasks, functions and goals consistent with the goals of the organisation to be fulfilled by each hospital staff group, will contribute to the success of the entire organisation. The research results clearly showed that the managers of entities suffering loss did not see any potential in human capital to help achieve a positive financial result.

Mechanisms are the second variable of the proposed organisational model. The managers of hospitals that generated profit considered it as having the greatest impact from among all the three major model variables, i.e. as increasing that profit. The management staff of hospitals that suffered loss in the same period considered Mechanisms as negatively affecting the financial result. The analysis of the collected data is not enough to compare the internal structure of the Mechanisms variable for both surveyed groups. The hierarchical clustering algorithm used in the bottom-up approach revealed that the Mechanisms variable in the group of the positive financial result covered two elements, while in the group of the negative financial result – four elements.

Components are the third model variable. Just like with Mechanisms, it is not possible here to compare the internal structure because of a different number of factors in the group of a positive and negative financial result, being grouped in different indicator configurations.

## 3  Conclusions

The empirical studies on the model structure, as well as the analysis and assessment of the models determining the financial results of hospitals in the positive and negative group revealed significant differences. The presented model of impact of organisational factors on the financial result of the hospital, developed based on the results of surveys conducted among managers in hospitals that generated profit, differs from the model of impact of organisational factors on the financial result of the hospital developed based on the survey results collected from managers in hospitals that suffered loss at the same time.

The differences covered two aspects: firstly, the perception of the impact of organisational determinants on the financial result between the surveyed hospital groups; secondly, the internal structure of model variables in both groups. The comparative analysis of both models, preceded by a mathematical justification of its validity in the form of verification of research hypotheses, led to the conclusion that the financial result of the hospital, apart from tools resulting from the financial management policy adopted in the enterprise, was also influenced by organisational factors resulting from components, mechanisms and hospital staff attributes. It was also proved that managers in hospitals demonstrating loss in their balance sheet did not consider organisational factors as having any positive potential which, if managed well, could support profit-making, perceiving them instead as a cause of failure to achieve the set economic goal or as elements having no impact on the financial result.

# References

1. Dubas, K., Domagala, A.: The European hospitals' functioning determinants with special emphasis on the human resources issue. Sci. J. Health Care. Public Health Manag. **10**(3), 151 (2012)
2. Dubas-Jakóbczyk, K., Kamińska, K.: Private hospitals in Poland – importance to the health care system and development potential. Sci. J. Health Care. Public Health Manag. **15**(3), 236 (2017)
3. Golinowska, S., Kocot, E., Sowa, A.: Human resources in health care. Up-to-date trends and projections. Sci. J. Health Care. Public Health Manag. **11**(2), 126 (2013)
4. Gruca, T.S., Nath, D.: Health Care Manag. Sci. **4**, 91 (2001)
5. Hass-Symotiuk, M.: System for Measuring and Assessing Hospital Achievements, p. 70. Wolters Kluwer Polska Sp. z o.o., Warsaw (2011)
6. Kanownik, G.: Role of logistics in the rationalisation of hospital operating costs. Acta Universitatis Nicolai Copernici. Hum. Soc. Sci. Econ. **46**(2), 218 (2015)
7. Kautsch, M.: Can the county authorities effectively supervise their health care units? Publi Manag. **42**(4), p74 (2017)
8. Kautsch, M., Ster, M.: Operation and significance of boards of trustees in non-public local-government-owned hospitals. Sci. J. Health Care. Public Health Manag. **12**(2), 164 (2014)
9. Klich, J.: Transformation of independent health care units as a challenge for local governments. Sci. J. Health Care. Public Health Manag. **13**(1), 51 (2015)
10. Król, H., Ludwiczyński, A.: HR Management, p. 117. Wydawnictwo Naukowe PWN, Warsaw (2006)
11. Ludwig, M., Van Merode, F., Groot, W.: Eur. J. Health Econ. **11**, 302 (2010)
12. Mastromarco, C., Stastna, L., Votapkova, J.J.: Efficiency of hospitals in the Czech Republic: conditional efficiency approach. Prod. Anal. **51**, 74 (2019)
13. Mokrzycka, A.: Health care enterprise. Innovative legislative and systemic solution or reform astray? Sci. J. Health Care. Public Health Manag. **8**(2), 70–82 (2010)
14. Polish Language Dictionary, PWN, Warsaw (1996)
15. Rogala, M., Badora-Musial, K., Kowalska, I., Mokrzycka, A.: The role of marketing in the management of the children's university hospital of Kraków – case study. Sci. J. Health Care. Public Health Manag. **12**(2), 163–174 (2014)
16. Smigielska, M.: The influence of modernisation of the radiotherapy base on the financial condition of a health care unit based on the example of great Poland cancer centre. Contemp. Oncol. **12**(5), 244 (2008)
17. Sowada, C.: Debts of polish public hospitals in 2005–2014. The unsolved problem of overdue liabilities. Sci. J. Health Care. Public Health Manag. **12**(3), 267 (2014)
18. Szetela, A., Lichwa, K., Korniejenko, K.: Critical analysis of health policies concerning transformation of public hospitals into commercial law companies. Public Manag. **13**, 50, 62 (2011)

# The Need for Changes in the Management Processes of a Young Researcher's Career – Case Study

Radosław Ryńca$^{(\boxtimes)}$ and D. Robert Iskander

Wroclaw University of Science and Technology, Wroclaw, Poland
radoslaw.rynca@pwr.edu.pl

**Abstract.** The professional career of a young researcher associated with a higher education institution can be increasingly viewed as of system character, in which the individual is fully aware of the different processes occurring at each stage of their career. In the study, We defined the young researcher as a person that successfully passes through a series of stages from the beginning of their PhD programme to the time of obtaining the habilitation degree. Further, we analyzed the different stages of career development of the individual. We described its micro- and macro-environment in each of the considered stages of career development, and analyzed the level of competition within the industry using Porter's five forces. The main objective of the study that We set was to ascertain whether the rejection of the hitherto prevailing assumptions and structures of career management of a young scientist, is confirmed by the expectations of young research employees of Wroclaw University of Science and Technology. For this purpose, We conducted an anonymous survey in the form of a questionnaire on a group of 45 eligible individuals. The results systematise the expectations of young researchers and indicate the need for changes in the process of managing their careers.

**Keywords:** Processes management · University · Young researchers · Environment of young researchers

## 1 Introduction

From the beginning of the 1990s, the Polish higher education has been in a stage of transition resulting from economic, political, sociological, demographic and legal changes. Massification of education requires universities to diversify their educational offer and adapt it to new market expectations. In the subject literature attention is paid to the necessity of environmental debate about the directions of changes in the Polish higher education [10]. In the report: "*Strategia rozwoju szkolnictwa wyższego 2010–2020*" ("Higher Education Development Strategy 2010–2020") [9] prepared by a team appointed by the consortium involving the Conference of Rectors of Academic Schools in Poland, the Conference of Public Vocational Schools' Rectors and the Polish Rectors Foundation, attention was paid, for instance, to a weak position of Polish universities on the international stage as well as missing promotion of active attitudes towards self-development and continuous improvement, inadequate dynamics of

© Springer Nature Switzerland AG 2020
Z. Wilimowska et al. (Eds.): ISAT 2019, AISC 1052, pp. 97–110, 2020.
https://doi.org/10.1007/978-3-030-30443-0_9

academic career development, including small mobility of research employees. Thus the need for undertaking actions aimed to ensure professional development of young research employees is important. In the opinion of the authors of the article, these actions should be systemic. Career of a young research employee working at the university has more and more often a systemic form, in which an individual is fully aware of particular processes taking place at any stage of his or her professional career. The system for managing a young researcher's career and related processes is not universal and does not have uniform form in all countries. In some countries, alike in Poland, it is a result of a historical chain of events, traditions and beliefs not always related to the optimised effectiveness of career management processes.

For the purposes of this study, taking account of the Polish environment, a young research employee is defined as an individual who effectively goes through several paths of professional career development, from the beginning of the PhD programme to the time of obtaining the habilitation degree.[1] In line with the present assumptions, this process should last approximately 12 years (four years of the PhD programme and eight years until obtaining the habilitation degree), which, in the case of completing the master's degree studies at the age of approximately 24, gives a 36-year-old individual considered, in the academic community of Polish universities, as the so-called self-reliant individual.

When comparing this result with, for example, the system binding in Australia,[2] where individuals who begin PhD studies are most frequently graduates with Bachelor Degree with Honours aged 22,[3] and the PhD studies alone last three years, we obtain an individual with more than 10 years of professional experience [5]. During that time, in the Australian conditions as well as in other countries with the Anglo-Saxon culture (e.g. the UK, the United States of America or New Zealand), a given individual has real potential of a triple (namely every three years) shift at the levels of academic career, from the degree of *Lecturer/Assistant Professor*, through *Senior Lecturer*, to the degree of *Associate Professor/Professor*. Passages between particular degrees depend on research and teaching results of the individual and his or her involvement in the professional and social life at the university [4, 6, 7, 11]. Comparing the systems binding in Poland with the system implemented in Australia, it is worth deliberating on differing elements. In the opinion of the authors of the article, the initial stages of managing a young researcher's career (namely at the level of PhD studies and following directly the transitional stage before obtaining the diploma) are similar in both countries. Drastic differences between the systems become visible at the successive stages of professional career development [2]. In Australia, an individual after the PhD studies is treated as a self-reliant and responsible individual, including that he or she can supervise PhD dissertations. Additionally, he or she has clearly determined indicators of assessment and elements required for the development of his or her career at

---

[1] The habilitation degree, although sometimes translated as Doctor of Sciences, has no direct equivalent in the Anglo-Saxon system of academic career development.

[2] I have almost 20-year experience of working at universities in Australia.

[3] In Australia, children go to primary school at the age of five and complete high school at the age of seventeen.

each of the agreed three-year stages. On the contrary, in Poland, an individual after the PhD studies is not treated as a self-reliant and responsible individual. These are discouraging factors.

## 2 Professional Career of a Young Researcher

A typically understood path of the individual's professional career development includes a number of stages, starting from the period of exploration of personal potentials, the period of consolidation of professional career often with strong trend of development of vertical career, the period of potential dynamic changes, the period of stabilisation with strong trend of development of horizontal career, and, finally, the period of the individual's preparation for retirement. Figure 1 presents graphically these stages in the form of a chart of the level of the individual's professional involvement in the function of time. The chart is similar to charts known from literature (see, for example, [12] but it additionally considers the stage of dynamic changes taking place in the period of professional maturity. *The level of professional involvement* of the individual is defined in this study as a cumulative effect of professional satisfaction reflected in social and material status, and the possibility of unhindered fulfilment of one's personal needs and goals [1, 3]. The level of professional involvement is inseparably connected with dynamics of the individual's professional career.

In our opinion, this model, though generally adopted as correct in all economic spheres of the countries governed by market laws, does not fully reflect the course of typical career of a research and didactic employee connected with a university. To characterise this phenomenon closer, it is required to look previously at individual stages of the researcher's professional career and define possible options and conditions. And so, in the chronological order, the following was separated:

1. *The stage of exploration of personal potentials.* Apart from universities, this stage is completed, in principle, when the individual starts permanent professional work. In the case of academic career, this stage should cover the last year of the 2nd degree studies, when the individual decides to continue the studies on the 3rd degree of education (often in the form of PhD studies) or tries to combine responsibilities related to professional work with running more or less strictly related academic works.

2. *The stage of consolidation of professional career.* This stage has strong cultural, geographic and socio-economic conditions. Unfortunately, in the conditions of Polish universities, the majority of individuals who complete the 3rd degree studies do not have, in the micro- and macro-environment, systematic mechanisms that could lead them to the path of vertical career – which is so necessary in the currently existing conditions of the competitive international research industry. Additional factors that inhibit development of a young researcher's career in Poland is his or her poor mobility, and hierarchical and, in my opinion, anachronistic structures still prevailing at some universities.

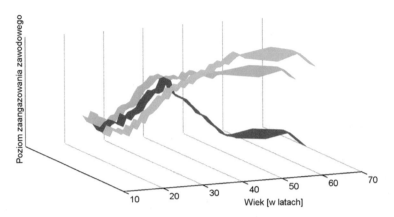

**Fig. 1.** Typical course of professional career including the stage of dynamic changes taking place in the period of professional maturity and the stages of vertical and horizontal career. Particular colours relate to particular individuals Source: prepared by the author

3. *The stage of potential dynamic changes in academic career.* Three main options of this stage can be separated. Namely, continuation of vertical professional career (green sash on Fig. 1), stabilisation of professional involvement (blue sash on Fig. 1) and decrease in professional involvement (red sash on Fig. 1), caused mainly by discouragement and unfulfilled expectations related to early academic career or changing dynamics of the individual's micro-environment.

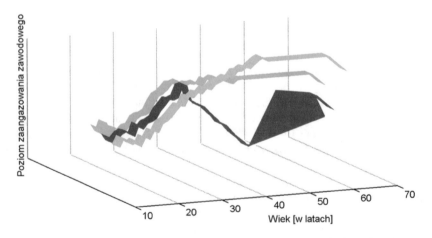

**Fig. 2.** Typical course of professional career including increase in professional involvement in the period of professional maturity caused by extended horizontal career (red sash) Source: prepared by the author

4. *The stage of career stabilisation.* This stage proceeds similarly for all options of the previous stage of potential dynamic changes, mainly due to a given individual's

maturity. It is characterised by a strong trend of development of horizontal career that, as a result, may contribute to significant increase in the level of professional involvement (red sash on Fig. 2).

5. *Final stage.* The last stage of each professional career, non-interrupted by acts of God, is the individual's preparation for retirement. Focus at the last stage would seem hardly justified in the context of deliberations over a young researcher's professional career. However, making the individual who starts research and didactic work aware of all elements of academic career is vital for optimising general level of professional involvement.

For the purpose of this study, *a young research employee* is defined as an individual who effectively passes through the first three stages of the path of professional career development, from the beginning of the PhD programme to the time of obtaining the habilitation degree.

## 3   Micro-environment and Macro-environment of a Young Researcher

In all of the above stages of academic career, the individual is not isolated, but is in the specific micro-environment (OB) and macro-environment (OD) which will be specified for each of the first three considered the stages (related directly to career of a young research employee). Regardless of the stage, OB includes always the closest relatives of the individual, such as parents, husband, wife, children or partner.

1. *The stage of exploration of personal potentials.* The early stage covering the choice of academic discipline, in which an individual undertakes to complete the PhD dissertation.
   a. OB: a group of potential tutors, PhD candidates and a team of direct employees related to a potential tutor, research infrastructure available directly to a given team, practical team mechanisms of controlling academic promotion. Environment of individuals directly connected with the entity during the studies who selected non-academic career.
   b. OD: formal university mechanisms of controlling academic promotion, academic community, as well as department and university infrastructure, university financing mechanisms for PhD studies.
2. *The stage of consolidation of professional career.* The stage in which the individual completes the PhD dissertation.
   a. OB: a tutor or a supervisor, PhD candidates and a team directly connected with a tutor/supervisor, research infrastructure directly available to a given team, practical mechanisms of controlling academic promotion.
   b. OD: international partners cooperating with an academic team, *formal* university mechanisms of controlling academic promotion, academic community as well as department and university infrastructure, national and international financing mechanisms for academic and research works.

3. *The stage of potential dynamic changes in academic career.* The stage, directly after obtaining the PhD degree, was often related to changing direction of research works, requiring the individual to show large self-involvement and expected certain degree self-reliance in further development of professional career.

   a. OB: a mentor or a group of mentors, a team directly connected with a mentor, research infrastructure directly available to a given team, *practical* team mechanisms of controlling academic promotion.

   b. OD: national and international financing mechanisms for academic and research works, international partners cooperating with an academic team, formal university mechanisms of controlling academic promotion, academic community as well as department and university infrastructure.

As it can be concluded from the above discussion, the macro-environment and the micro-environment of a young researcher are not always strictly subordinated to a given stage of professional career, but are characterised by dynamic changes in time.

## 4  The Impact of Porter's Five Forces on the Development of a Young Researcher's Career

At each of the analysed stages SWOT analysis can be conducted (SWOT: *Strengths, Weaknesses, Opportunities and Threats*). However, for the purpose of a more accurate description of the mechanisms optimising the level of professional involvement of a young research employee, the article makes use of the impact of Porter's five forces on the development of a young researcher's career.

Porter's competitive forces ([SD] - bargaining power of suppliers [MV] - bargaining power of buyers [NK] - threat of new entrants [ZS]- threat of substitutes [RO] - industry rivalry) can be referred to particular stages of development of a young researcher's career, but they are applicable mainly at its first two stages, namely at the stage of exploration of personal potentials and at the stage of consolidation of a given professional career.

The analysis of the impact of Porter's forces [8] **on the development of a young researcher's career** should enable better understanding of the individual's position in the micro- and macro-environment at different stages of career development which, in the future, may be used in creating new procedures for supporting and streamlining activities, in the form of recommendations that would apply to the closest and macro-environment of a young researcher.

1. *The stage of exploration of personal potentials.* For a future PhD candidate, as a future buyer of formal professional qualifications (the granted PhD degree), the following forces can be examined:

   a. SD: potential academic, professional, organisational as well as interpersonal competences of the future tutor (tutors, in the case of interdisciplinary dissertations) or the supervisor/supervisors of the dissertation, potential academic competences of the faculty members, international status of university, academic and research infrastructure, links between the university and industry, university financing mechanisms for PhD studies, degree of subsidy for PhD studies.

b. SN: relatively low number of very well educated candidates for PhD studies, relatively high number of offered places, often low degree of subsidies for PhD studies.

c. NK: number of high PhD scholarships offered under research grants, increasing number of job offers for skilful university graduates without PhD.

d. ZS: increasing number of academic projects employing assistants without the requirement of supervising PhD dissertations, increasing number of job offers for skilful university graduates without PhD.

e. RS: number of willing candidates who consider the beginning of the PhD programme as a prolongation of the period of studies or who are not at this initial stage sure about choosing academic career, rivalry of universities.

2. *Stage of consolidation of a given professional career.* The individual who started the PhD programme may be considered as the buyer of specific real skills (as opposed to formal qualifications) distinguishing the individual from his or her lateral professional environment or as the holder of membership in specialised national and international academic networks. Real skills include proficiency in using advanced mathematic apparatus or possession of advanced technical and academic knowledge in a given domain.

a. SD: actual academic, professional, organisational as well as interpersonal competences of the tutor or the supervisor of the dissertation, linkages between the tutor and the industry and top international academic centres, effectiveness of the tutor/supervisor in obtaining external funds for financing academic research in the team.

b. SN: individual's motivation and involvement in gaining knowledge.

c. NK: new PhD candidates and young research employees from beyond the university who want to work with a given tutor/supervisor or academic team.

d. ZS: impact of the individual's micro-environment. The second stage of career development is very dependent on the individual's micro-environment. Pressure of immediate family or close colleagues/friends of the individual may be motivating or discouraging, which may result in searching for other opportunities for career development beyond research work.

e. RS: a great number of PhD candidates per one supervisor. Like for NK, examples may concern new, but also existing relations between the supervisor and the PhD candidates in the team.

3. *Stage of potential dynamic changes in academic career.* At this stage, a young research employee may be treated to a decreasing degree as the buyer of skills, but instead he or she should be treated as the buyer of a good academic position.

a. SD: actual academic, organisational as well as interpersonal competences of *the mentor*, international status of the university, academic and research infrastructure.

b. SN: mobility of the individual.

   c. NK: dynamic free labour market for individuals with PhD dissertation.

   d. ZS: impact of the individual's micro-environment.

   e. RO: efforts of both parties to employ the best individual and in the best academic team.

To sum up the Porter's five competitive forces discussed above, it can be concluded that at the stage of exploration of personal potentials, the individual who starts academic work is strongly motivated, but he or she remains mostly in the phase of unconfirmed expectations termed in the paper as the phase of *dreams*. This phase is connected mainly with ignoring relations between the tutor or the supervisor of the PhD dissertation and the PhD candidate, which is most often individual for each pair of such relation. For this reason, bargaining power of suppliers at this stage is strictly correlated with final results related to the completion of the PhD dissertation.

The stage of consolidation of a given professional career, namely the time of PhD studies is a period when the individual acquires real skills. This phase is termed as the phase of *reality*, in which early dreams of the individual can either come true or, most often, be verified.

In the third stage of potential dynamic changes in academic career, namely after obtaining the PhD degree, a young research employee is to a decreasing degree perceived as the buyer. As a result, at this stage, Porter's forces are applied to a decreasing degree. However, it is worth emphasising that they will apply in the case of each change in the individual's workplace.

## 5  Case Study

### 5.1  Research Arrangement

The purpose of the carried out research is to check whether the below defined assumptions and the proposed result around which carried management processes should be arranged are confirmed by expectations of young researchers working at the Wrocław University of Technology.

In the paper the following assumptions and result were assumed:

**Assumption 1.** A young research employee before obtaining the habilitation degree is not a self-reliant individual.

**Assumption 2.** A young research employee obtains self-reliance upon receipt the habilitation degree.

**Result:** A young research employee employed at the university in Poland is a motivated individual who is successful on the international stage.

To fulfil the above objectives, an interview in the form of questionnaire was conducted. The questionnaire consisted of closed questions grouped into thematic fields, such as: boundary conditions, competences of the tutor (mentor), academic and teaching infrastructure, motivational factors, communication with the tutor (mentor), organisational competences of the university. Data received from the research was subjected to a statistical analysis, which may be used to create a procedure for supporting and streamlining activities that would apply to the closest and macro-environment of a young researcher in the form of recommendations.

## 5.2  Description of Results

The research covered 45 young students and researchers of the Wrocław University of Technology, including 19 women and 26 men. Nine of the surveyed people were at the first stage of career development, before starting the PhD programme. 22 and 14 respondents were at the second and third career stage, respectively. This latter group covered both individuals directly after the PhD programme (up to one year) and one that completed the PhD programme 10 years ago. Average period after the PhD programme in this group (one standard deviation) was $3.5 \pm 2.9$ year.

The respondents' answers concerning boundary conditions determining the potential application of reengineering elements aroused large interest of the authors of the article. The results are presented in Table 1. For the whole examined group it was also checked whether there was some correlation between the respondents' answers and their age or present career stage. The results are presented in Table 2. Statistically significant correlation at the level of confidence $\alpha \leq 0.05$ was recorded for question A in combination with age of the respondent and for question B in combination with career stage.

| Competences of the tutor/supervisor/mentor | Definitely agree | Rather agree | Rather disagree | Definitely disagree |
|---|---|---|---|---|

**Table 1.** Group results of the respondents in answers to questions related to boundary conditions of the research. Specification.

| Boundary conditions | Definitely agree | Rather agree | Rather disagree | Definitely disagree |
|---|---|---|---|---|
| **A:** A young research employee before obtaining the habilitation degree is not a self-reliant individual. | 8 | 37 | 0 | 0 |
| **B:** A young research employee *obtains* self-reliance upon receipt of the habilitation degree | 0 | 17 | 27 | 1 |
| **C:** A young research employee employed at the university in Poland should be *a motivated* individual *who is successful* on the international stage | 27 | 18 | 0 | 0 |

Source: prepared by the author on the basis of the conducted research

**Table 2.** Group results of the respondents in answers to questions related to boundary conditions of the research. Correlation with age or career stage. Bold means statistically significant correlation.

| Boundary conditions | Age | Career stage |
|---|---|---|
| **A:** A young research employee before obtaining the habilitation degree is not a self-reliant individual. | **r = 0.412, p = 0.005** | r = 0.174, p = 0.253 |
| **B:** A young research employee *obtains* self-reliance upon receipt of the habilitation degree | r = 0.275, p = 0.067 | **r = 0.528, p = 0.0002** |
| **C:** A young research employee employed at the university in Poland should be *a motivated* individual *who is successful* on the international stage | r = 0.013, p = 0.931 | r = 0.250, p = 0.088 |

Source: prepared by the author on the basis of the conducted research

The respondents' answers to questions concerning competences of the tutor/supervisor/mentor are grouped in Table 3. A number of questions related to specific qualifications obtained a very large (more than 80%, and in some cases 100%) number of the answers "definitely agree". Interestingly, in the opinion of the respondents, the title or public image of the tutor/supervisor/mentor were not important. The correlation analysis of this question with age and career stage is presented in Table 4. Statistically significant correlations were obtained for a number of questions.

**Table 3.** Group results of the respondents in answers to questions related to competences of the tutor/supervisor/mentor. Specification.

| | | | | |
|---|---|---|---|---|
| A: Title/degree of the tutor/supervisor/mentor is important for me | 0 | 8 | 9 | 28 |
| B: Public image (perceived beyond the university) of the tutor/supervisor/mentor is important for me. | 11 | 12 | 13 | 9 |
| C: Experience of the tutor/supervisor/mentor in conducting academic research is important for me | 39 | 6 | 0 | 0 |
| D: The results of academic research of the tutor/supervisor/mentor are important for me | 27 | 17 | 1 | 0 |
| E: International status of the tutor/supervisor/mentor is important for me | 12 | 32 | 1 | 0 |
| F: Experience of the tutor/supervisor/mentor in supervising PhD dissertations | 45 | 0 | 0 | 0 |

Source: prepared by the author on the basis of the conducted research

The respondents' answers to questions concerning academic and research infrastructure are grouped in Table 5 and correlations with age or career stage are grouped in Table 6. Similarly, Tables 7 and 8 present the respondents' answers to questions concerning motivational factors. Definitely positive answers to questions A, B, E and F give a picture of a very determined and motivated young group of research employees. Answers concerning friendly atmosphere at work and employment stability had statistically significant correlations with age and career stage.

**Table 4.** Group results of the respondents in answers to questions related to competences of the tutor/supervisor/mentor. Correlation with age or career stage. Bold means statistically significant correlation.

| Competences of the tutor/supervisor/mentor | Age | Career stage |
|---|---|---|
| A: Title/degree of the tutor/supervisor/mentor is important for me | **r = 0.379, p = 0.010** | **r = 0.802, p = 0.000** |
| B: Public image (perceived beyond the university) of the tutor/supervisor/mentor is important for me | r = 0.153, p = 0.315 | r = 0.125, p = 0.415 |
| C: Experience of the tutor/supervisor/mentor in conducting academic research is important for me | r = 0.231, p = 0.127 | **r = 0.432, p = 0.003** |
| D: The results of academic research of the tutor/supervisor/mentor are important for me | **r = 0.450, p = 0.002** | r = 0.169, p = 0.266 |
| E: International status of the tutor/supervisor/mentor is important for me | r = 0.109, p = 0.475 | **r = 0.380, p = 0.010** |
| F: Experience of the tutor/supervisor/mentor in supervising PhD dissertations | – | – |

Source: prepared by the author on the basis of the conducted research

**Table 5.** Group results of respondents in answers to questions concerning academic and research infrastructure. Specification

| Academic and research infrastructure | Definitely agree | Rather agree | Rather disagree | Definitely disagree |
|---|---|---|---|---|
| A: Uniqueness of research apparatus is necessary for effective research work | 8 | 37 | 0 | 0 |
| B: Access to the most recent technologies is critical for effective research work. | 0 | 17 | 27 | 1 |
| C: Access to specialised software is necessary for effective research work | 27 | 18 | 0 | 0 |

Source: prepared by the author on the basis of the conducted research

**Table 6.** Group results of respondents in answers to questions concerning academic and research infrastructure. Correlation with age or career stage. Bold means statistically significant correlation.

| Academic and research infrastructure | Age | Career stage |
|---|---|---|
| A: Uniqueness of research apparatus is necessary for effective research work | **r = 0.410, p = 0.005** | r = 0.174, p = 0.254 |
| B: Access to the most recent technologies is critical for effective research work. | r = 0.275, p = 0.067 | **r = 0.528, p = 0.002** |
| C: Access to specialised software is necessary for effective research work | r = 0.013, p = 0.931 | **r = 0.257, p = 0.008** |

Source: prepared by the author on the basis of the conducted research

**Table 7.** Group results of respondents in answers to questions related to motivational factors. Specification.

| Motivational factors | Definitely agree | Rather agree | Rather disagree | Definitely disagree |
|---|---|---|---|---|
| A: I am willing to gain knowledge and develop continuously | 45 | 0 | 0 | 0 |
| B: I am aware of selecting professional career. | 45 | 0 | 0 | 0 |
| C: I am mobile | 17 | 13 | 11 | 4 |
| D: Friendly atmosphere at work is important for me | 5 | 37 | 3 | 0 |
| E: Possibility of professional development (promotions) is important for me. | 45 | 0 | 0 | 0 |
| F: Possibility of cooperation with other research centres in the country/abroad is important for me. | 45 | 0 | 0 | 0 |
| G: Employment stability is important for me. | 5 | 25 | 15 | 0 |

Source: prepared by the author on the basis of the conducted research

**Table 8.** Group results of respondents in answers to questions related to motivational factors. Correlation with age or career stage. Bold means statistically significant correlation.

| Motivational factors | Age | Career stage |
|---|---|---|
| A: I am willing to gain knowledge and develop continuously | - | - |
| B: I am aware of selecting professional career | - | - |
| C: I am mobile | $r = 0.238$, $p = 0.115$ | $r = 0.216$, $p = 0.154$ |
| D: Friendly atmosphere at work is important for me | **$r = 0.296$, $p = 0.048$** | **$r = 0.392$, $p = 0.008$** |
| E: Possibility of professional development (promotions) is important for me | - | - |
| F: Possibility of cooperation with other research centres in the country/abroad is important for me | - | - |
| G: Employment stability is important for me | **$r = 0.312$, $p = 0.037$** | **$r = 0.356$, $p = 0.016$** |

Source: prepared by the author on the basis of the conducted research

The survey covered also questions about communication with the tutor, supervisor or mentor. As expected, the frequency of the expected meetings with the tutor/supervisor/mentor was statistically firmly correlated with age ($r = 0.652$, $p < 0.001$) and the career stage ($r = 0.702$, $p < 0.001$).

In the case of the relation between the tutor/supervisor/mentor and the tutee, the answers were divided into two groups: 15 people preferred partner model and 30 persons preferred tuition model. Nobody preferred supervisor/subordinate model. This result was statistically firmly correlated with the career stage ($r = 0.556$, $p < 0.001$).

## 5.3   Discussion of Results

The research on the group of 45 young students and researchers of the Wrocław University of Technology is treated by the authors of the article as a pilot research. Despite relatively small size of the sample, a relatively large number of statistically important correlations was obtained. It is worth noting that this result was obtained despite uneven proportion of groups related to age, sex and young researcher's career development stage.

Questions related to competences of the tutor/supervisor/mentor obtained the greatest number of answers "definitely agree". Similar results were obtained for questions related to communication with the tutor/supervisor/mentor and questions concerning organisational competences of universities. These results clearly indicate high maturity of respondents and their high expectations, both from tutors/supervisors/ mentors and the university environment. In many cases age, classified to 4 groups for the purposes of the paper, and stage of respondent's professional career had statistically significant correlations with answers. It is justified to assume the causality of these results.

A number of questions in the survey received unambiguous result (regardless of age and career stage or sex), especially those concerning university competences. This result could suggest improper selection of questions in terms of the examined population. On the other hand, it may mean determination of views respondents who are well acquainted in the administrative environment of the university.

Finally, it is worth emphasising the local character of the examined group that is not representative from the point of view of the whole university.

## 6   Summary

The study defines a young research employee as a person that successfully goes through a number of professional stages from the beginning of the PhD programme to the time of obtaining the habilitation degree. The article presents the environment of a young researcher and the impact of Porter's five forces on the development of his or her professional career. In the paper attention was paid to the need to verify so far prevailing assumptions and structures of managing professional career of a young researcher, which is confirmed in expectations of young employees of the Wrocław University of Science and Technology. For this purpose, an anonymous pilot survey was conducted in the form of a questionnaire on the group of 45 people. The results systematise the expectations of young researchers and indicate the need for changes in the process of managing their careers.

The article presents deliberations concerning professional career of a young research employee, working at the university in Poland. After defining a young research employee, the authors analysed particular stages of his or her career development, beginning from the stage of exploration of personal potentials, through the stage of consolidation of a given professional career, to the stage of consolidation of a given professional career, completed with obtaining the habilitation degree.

In order to separate factors affecting development of a young researcher's career, factors affecting development of a young employee's career on the basis of Porter's five forces were identified. The authors of the article do not know about any examples of applying this analysis with regard to relations between the tutor/supervisor/mentor and the tutee. The identification of factors under Porter's five forces made it possible to construct a questionnaire whose main objective was to check "the hypothesis" about the need for changes in the management processes of young researcher's career.

## References

1. Bartkowiak, G.: Człowiek w pracy. Od stresu do sukcesu w organizacji, Polskie Wydawnictwo Ekonomiczne Warszawa (2009)
2. Bazley, P.: Defining 'early career' in research. High. Educ. **45**, 257–279 (2003)
3. Bochyńska-Śmigielska, E.: Determinanty zaangażowania i efektywności zawodowej. Zeszyty Naukowe Towarzystwa Doktorantów UJ. Nauki Społeczne **4**(1/2012), 97–113 (2012)
4. Dziedziczak-Foltyn, A.: Strategie uczelniane - antycypacja, translacja czy pseudokonformizacja celów polityki rozwoju szkolnictwa wyższego? Prace Naukowe Wyższej Szkoły Bankowej w Gdańsku **14**, 183–202 (2011)
5. Kemp, L., Stevens, K., Asmar, C., Grbich, C., Marsh, H., Bhathal, R.: Waiting in the Wings: A Study of Early Career Academic Researchers in Australia. Australian Government Publishing Service, Canberra (1996)
6. Matthews, K.E., Jason, M., Lodge, J.M., Bosanquet, A.: Early career academic perceptions, attitudes and professional development activities: questioning the teaching and research gap to further academic development. Int. J. Acad. Dev. **19**(2), 112–124 (2014)
7. McAlpine, L., Amundsen, C., Turner, G.: Identity-trajectory: Reframing early career academic experience. Br. Educ. Res. J. **40**(6), 952–969 (2014)
8. Porter, M.E.: Strategia konkurencji. Metody analizy sektorów i konkurentów, PWE, Warszawa (1992)
9. Raport Strategia rozwoju szkolnictwa wyższego 2010–2020 - Projekt środowiskowy. http://www.krasp.org.pl/pl/strategia/strategia. Accessed 1 May 2016
10. Raport, Bank Światowy, Europejski Bank Inwestycyjny, Szkolnictwo wyższe w Polsce, Raport nr 29718, 13 lipca 2004
11. Thomas, J.D., Lunsford, L.G., Rodrigues, H.A.: Early career academic staff support: evaluating mentoring networks. J. High. Educ. Policy Manag. **37**(3), 320–329 (2015)
12. Zając, C.: Zarządzanie zasobami ludzkimi, Wydawnictwo Wyższej Szkoły Bankowej, Poznań (2007)

# Artificial Neural Networks in Forecasting Cancer Therapy Methods and Costs of Cancer Patient Treatment. Case Study for Breast Cancer

Zbigniew Leszczyński⬤ and Tomasz Jasiński$^{(\boxtimes)}$⬤

Lodz University of Technology, Piotrkowska 266, 90924 Lodz, Poland
tomasz.jasinski@p.lodz.pl

**Abstract.** Forecasting of cancer therapy and its cost is a key factor in the process of managing the oncological patient's treatment. This research is based on two predictive models of artificial neural networks (ANNs). The paper scope also includes a statistical analysis of correlation of the ANN variables. Research shows that ANNs are a viable alternative to traditional method of estimating the cost of a cancer therapy, especially in situations of poor recognition of the nature of medical relations between costs and their cost drivers, or in situations of non-linear, multidimensional relations between variables. Studies have also shown that ANN is an adequate method for predicting cancer therapy. The first contribution of this study is application of an innovative, complex approach based on two ANN models in the management of oncological patient treatment. The second contribution is the use of medical specifications required by the oncological patient directly to estimate the costs of cancer therapy.

**Keywords:** Forecasting · Cost management · Cancer therapy · Neural network

## 1 Introduction

The use of artificial neural networks (ANNs) in biological and medical research already at the end of the last century has become extremely important. An increase in medical use of ANN has been observed in many fields such as anaesthesiology [1], radiology [2], cardiology [3, 4], psychiatry [5] and neurology [6]. Although the cancer treatment process may be predictable, in the sense that oncologists and pathologists are highly likely to identify the existence and spread of cancer. Reliable evaluation of cancer development is necessary to determine in a short period of time adequate methods of cancer therapy. Factors related to prediction of cancer, which indicate the potential response of an oncological patient to the planned treatment, are important for the choice of cancer therapy method [7]. Research in the area of molecular prognostic factors in breast cancer was carried out, inter alia, in [8]. Identification and examination of many potential predictive factors in a specific oncological patient is an extensive information material for the medical council in selection of adequate cancer therapy methods. As a result, doctors have to deal with more information on the disease and its advancement for specific oncological patients. A large amount and variety of information makes it

© Springer Nature Switzerland AG 2020
Z. Wilimowska et al. (Eds.): ISAT 2019, AISC 1052, pp. 111–120, 2020.
https://doi.org/10.1007/978-3-030-30443-0_10

difficult for oncologists to interpret it and, consequently, to use it in the process of selecting cancer therapy methods. However, in the decision-making process, the model can play a major role for the payer of medical services (National Health Fund – NHF). It standardizes the selection of therapy (and its costs), which facilitates and accelerates the NHF decision-making process in the scope of payments for medical services (in the area of oncological therapy). The above predisposes the presented model for its use by the NHF, much more than by individual oncologists.

The way to increase the effectiveness and shorten the time for selection of adequate cancer therapy methods is the use of ANNs. These models are particularly useful in solving complex problems where algorithmic solutions are time-consuming and difficult to define [9]. ANNs use predictive factors as input neurons and are the perfect way to combine input signals, thus the output neurons can form the predictive basis for determining oncological patient therapy. Basically, the two features that determine the reliability of ANNs are selection and reliability of input variables, as well as the correct classification of cancer therapy methods from the point of view of a combination of predictive factors [10].

The management of health care facilities under different operating conditions has fundamentally changed the need for information on the entity activities, both by external users and management staff [11]. Procedures of oncological patient treatment are among the most cost-intensive and complex medical procedures.

The improvement level of traditional forecasting methods regarding the costs of oncological therapy is not commensurate with changes in medical technologies, and their accuracy and adequacy is insufficient (traditional methods of forecasting the costs of oncological therapy are often called cost calculation of cancer therapy in the literature and practice). Current forecasting methods of the costs of oncological therapies remained unchanged in relation to the changing medical environment (advanced medical technologies, complex medical and pharmacological procedures, digitalization of procedures, complex biomedical technologies). Complex oncological procedures determine the use of a large number of cost drivers [12, 13].

ANNs have the potential to overcome disadvantages and limitations of traditional forecasting methods of oncology therapy costs [14]. ANNs can be complex models, non-parametric forecasting of oncology therapy costs with unlimited number of cost drivers and real time of obtaining information on the therapy costs of an oncological patient [15]. Forecasting treatment costs using ANN is one of the first such studies in Poland.

The aim of this paper is to present, in application terms, ANNs in forecasting of cancer therapy and the cost of cancer therapy. Both elements are a prerequisite for an optimal oncology management system.

## 2   Materials and Methods

ANNs are increasingly used to identify and classify breast cancer (see [16] and [17]). Research usually uses medical images and can only indicate the identification of cancer or its lack [18], as well as recognition of mild lesions [19] and in situ [20]. In the subject literature, there is a research gap in the area of ANN's indication of appropriate cancer therapy and its costs.

Two ANN models were used in the empirical study. The first model was used to predict the type of breast cancer therapy, and the second to estimate the treatment costs of an oncological patient. Forecasting the costs of oncological treatment is based on the results obtained from the first ANN model, thus the whole model is cascading (Fig. 1).

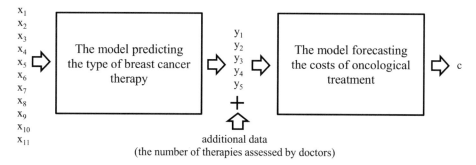

**Fig. 1.** Diagram of the cascade cost model of oncological treatment.

## 2.1   Artificial Neural Networks in Forecasting Cancer Therapy

The empirical study covered the forecast of breast cancer therapy methods on the basis of selected predictive factors, which are important for selection of the path of onco-logical treatment of a specific patient. The most important predictive factor is the presence of estrogen (ER), progesterone (PgR) and HER2 receptors in breast cancer cells [21]. Predictive factors were selected based on the analysis of recommendations of the Polish Clinical Oncology Association concerning the methods of breast cancer diagnosis and treatment.

In order to develop a set of training data, the values of predictive parameters and data on applied therapies were obtained for 160 patients, who underwent oncological treatment.

The set of independent variables – predictive factors included eleven elements. Applied cancer therapies represent a dependent variables expressed as a logical values (see Table 1).

The data were randomly grouped into three sets: training (100 samples), validation and test (30 elements each). Such a division was related to the number of available data and constituted a compromise between the desired high number of training set (pro-viding the possibility of more effective learning) and the required number of cases in the test set (enabling reliable evaluation of the model operation). Various models of ANNs were tested. Figure 2 shows the architecture (11-9-7-5, logistic activation functions in hidden layers) of the best ANN model.

In case of ANN, both the increase in the number of hidden layers and the number of nodes in them did not improve the accuracy of obtained results, but only significantly prolonged the waiting time for results. In order to improve the quality of ANN, it is necessary to collect data prepared for training ANN. As a result of limited amount of test data, the model presented may be affected by a large error due to data distortion. Different treatment centres in Poland apply slightly modified medical procedures. Also, the choice of prognostic factors chosen by researchers may omit some of the important data that are taken into account when determining the therapeutic methods for individual patients.

**Table 1.** Independent and dependent variables.

| Name | Description | Type* | Details |
|------|-------------|-------|---------|
| Independent variables | | | |
| $x_1$ | Patient's age | N | Expressed in years |
| $x_2$ | Burdened family interview (someone in the family had cancer) | L | Yes – 1, no – 0 |
| $x_3$ | T classification (describing the severity and malignancy of cancer) | N | TX – 0, T0 – 1, Tis (DCIS) – 2, Tis (Paget) – 3, T1mi – 4, T1a – 5, T1b – 6, T1c – 7, T2 – 8, T3 – 9, T4a – 10, T4b – 11, T4c – 12, T4d – 13 |
| $x_4$ | N classification (describing the severity and malignancy of cancer) | N | NX – 0, N0 – 1, N1 – 2, N2a – 3, N2b – 4, N3a – 5, N3b – 6, N3c – 7 |
| $x_5$ | M classification (describing the severity and malignancy of cancer) | L | MX – 0 (no distant metastasis, cancer has not spread to other parts of the body), M0 – 1 (distant metastasis) |
| $x_6$ | Breast cancer subtype (descriptive value) | N | luminal A – 1 luminal B – 2, HER2 – 3, positive special types – 4 |
| $x_7$ | Expression rate of estrogen receptors – ER | P | |
| $x_8$ | Expression rate of progesterone receptors – PgR | P | |
| $x_9$ | Overexpression of receptors and HER2 amplification | N | |
| $x_{10}$ | Cellular marker of Ki-67 proliferation | P | |
| $x_{11}$ | VNPI index | N | |
| Dependent variables | | | |
| $y_1$ | Radical surgical treatment | L | Radical – 1, saving – 0 |
| $y_2$ | Chemotherapy | L | Yes – 1, no – 0 |
| $y_3$ | Hormonotherapy | L | Yes – 1, no – 0 |
| $y_4$ | Radiotherapy | L | Yes – 1, no – 0 |
| $y_5$ | Supplementary treatment with trastuzumab | L | Yes – 1, no – 0 |

* N – numerical value (integer), P – percentage value, L – logical value (binary)

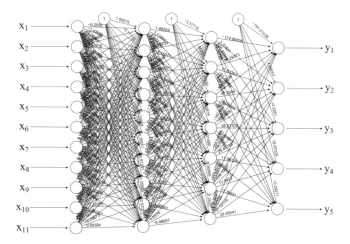

**Fig. 2.** Structure of the best MLP forecasting breast cancer therapy methods.

## 2.2 Artificial Neural Networks in Forecasting Costs of Oncological Treatment

Empirical research included the estimated cost of oncological treatment of a patient with breast cancer on the basis of selected medical data. The authors of the study obtained data for 100 oncological patients, including the actual costs of oncological treatment of a patient, which formed the explanatory variable of the model. The set of data included five explanatory variables ($y_1$, $y_2$, $y_3$, $y_4$, $y_5$) and one response variable (c). The data were randomly grouped into three sets: training (50 samples), validation (10 elements) and test (40 cases). Two groups of ANNs were tested: MLP and RBF (see [22] and [23]). More than 20,000 different models were tested in each case. Only the best ANN is presented in the study.

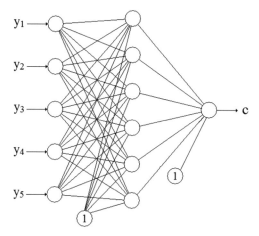

**Fig. 3.** Structure of the best MLP forecasting the cost of oncological treatment.

Studies have shown that MLP-type networks generate a significantly better cost estimation for cost of oncological treatment than RBF networks. MAPE for the best MLP is 16.59%. The determination coefficient $R^2$ is 0.96.

Figure 3 presents a construction scheme of the successful MLP model (architecture: 5-6-1, activation functions: hidden and output layers – logistic). Figure 4 shows the histogram of prediction errors for data from the test set. Table 2 presents examples of the test set data with values of the estimated cost.

Extending MLP by additional hidden layers did not translate into an increase in the quality of cost estimation. This is confirmed by the necessity of keeping the model possibly uncomplicated, a need often postulated in research. Empirical studies have confirmed that the structure of the model should be selected according to the analysed issue without any tendency for its excessive complexity.

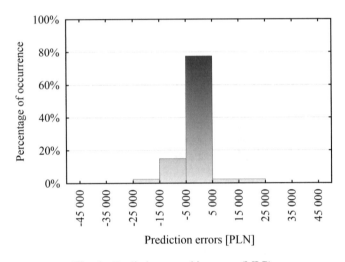

**Fig. 4.** Prediction error histogram (MLP).

**Table 2.** Examples of the test set data with values of the estimated cost.

| Patient no. | $y_1$ | $y_2$ | $y_3$ | $y_4$ | $y_5$ | Expected cost [PLN] | Estimated cost [PLN] |
|---|---|---|---|---|---|---|---|
| 9 | 1 | 12 | 12 | 1 | 12 | 88813.56 | 83836.01 |
| 17 | 0 | 4 | 4 | 1 | 4 | 17355.29 | 15604.77 |
| 18 | 1 | 4 | 4 | 1 | 4 | 19728.24 | 20128.88 |
| 23 | 1 | 6 | 4 | 1 | 6 | 19501.68 | 24693.49 |
| 29 | 1 | 12 | 12 | 0 | 12 | 72560.96 | 74284.28 |

## 3   Regression Analysis

A regression model with a single explanatory variable shows that independent variables $y_2$, $y_3$, $y_5$ have (to some extent) linear impact on the level of the expected cost of oncological treatment (c). Variables $y_1$, $y_4$ do not show a linear relationship with the variable c. Independent variables $y_2$, $y_3$ and $y_5$ are positively correlated with the expected cost of oncological treatment – c (see Fig. 5). Determination coefficients $R^2$ for variables $y_2$, $y_3$ and $y_5$ are within the range of 0.65–0.79.

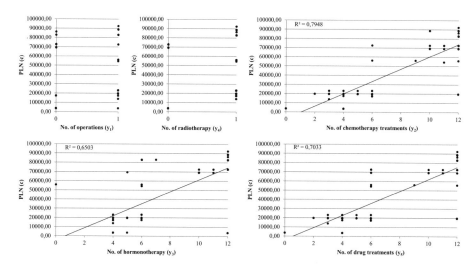

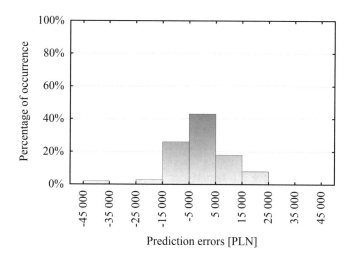

**Fig. 5.** Dependencies between explanatory variables and response variable.

**Fig. 6.** Prediction error histogram (multiple regression).

Multiple regression was used as a comparative model for ANNs. In order to confirm the lack of correlation between the explanatory variables, the variance inflation factor (VIF) was calculated for each of them. In all cases, their values were less than 10 ($VIF_{Y2} = 7.23$, $VIF_{Y3} = 1.93$, $VIF_{Y5} = 6.16$), which indicates an acceptable level of autocorrelation and justifies the use of the three variables in the multiple regression model. The determination coefficient $R^2$ is 0.87. Figure 6 shows the histogram of prediction errors.

## 4   Discussion and Conclusions

The use of ANNs in the forecasting of cancer therapy methods improves the work of medical council, which is an element of the oncological package (the medical council as an element of the oncological package has been implemented by the legislator). The medical council should meet within a maximum of two weeks after the oncological patient's admission to hospital. This element of the oncological package involves doctors of many specialties, which generates high costs and delays the implementation of adequate treatment. Procedures for selection of cancer therapy methods by the medical council are characterized by a high degree of repeatability due to the limited number of effective and possible to carry out cancer therapy. The stage of oncological disease progression is quantified and relatively easy to determine on the basis of cancer description. All these factors predestine the procedures for selection of cancer therapies to be subjected to the automation process.

ANNs can be practically applied in the optimal and fast selection of the most effective method of cancer therapy. Thanks to this, the same treatment can be implemented much faster, and the elimination of procedure for selection of cancer therapies by the medical council will reduce the costs of oncological package. The automation effect of this procedure will also reduce the workload of oncologists, which will increase the availability of time for the patient.

The oncological patient will be able to obtain information on cancer therapy methods as soon as all examination results are provided to the oncologist and data are entered into the program using the ANN. During the same visit, when the patient provides a complete set of tests, they will receive information on the development stage of disease and adequate therapy methods.

Currently, the severity of disease and the choice of cancer therapy methods are determined only once, during the first diagnosis. Even if the cancer increases its range or spreads, the classification of cancer severity and selected therapies does not change. Thanks to the use of ANNs, the oncologist will be able to continuously evaluate the progress of cancer and adapt therapy methods appropriate to the progress of cancer without convening a medical consultation. However, it would be reasonable in the future to extend the presented models with gene analysis.

ANNs do not eliminate all the difficulties associated with traditional methods of forecasting the cost of oncological therapy. They are a viable alternative to the traditional method of forecasting the cost of oncological therapy. Oncological therapies are very complex in terms of medical procedures and advanced technologies, making ANNs one of the most optimal tool for forecasting the costs of treatment of an oncological patient. The use of ANNs in forecasting the costs of oncological therapy reduces the need for in-depth medical knowledge among cost planners. From a practical point of view, the ANNs are capable of analyzing a much larger number of potential cost carriers (even several dozen) than the traditional calculation of the treatment cost of oncological patient. The ANN model can forecast the treatment cost of an oncological patient without the need for a mathematical description of the functional dependence between independent variables (cost carriers) and dependent variables (cost of treatment of an oncological patient) [24]. In case of ANNs, the time of access to cost information, as well as the information structure are of critical importance, namely: using medical specifications required by the oncological patient directly to estimate the cost of cancer therapy.

In contrast, computer-aided ANN is not simple to use and requires specific IT knowledge. The disadvantage of ANNs is a significant risk in the process of building the ANN model with a small number of data sets. As indicated by the results of both literature [25] and empirical studies conducted by the authors of the paper, the number of data samples is one of the basic elements influencing the quality of forecast.

Theoretical and empirical analysis carried out in the research fully confirm the statement that ANN models are one of the most innovative tool for forecasting cancer therapy methods and the costs of implementing these methods for a specific oncological patient. Implementation of these models into oncological practice will ensure optimization of the treatment system for oncological patients. The concept and procedures of building ANN models presented in this paper constitute a reference point for further theoretical and practical research in this field.

# References

1. Narus, S.P., Kuck, K., Westenskow, D.R.: Intelligent monitor for an anesthesia breathing circuit. In: Gardner, R.M. (ed.) Proceedings of the Annual Symposium on Computer Application in Medical Care, pp. 96–100. Hanley & Belfus, Philadelphia (1995)
2. Wn, Y.C., Doi, K., Giger, M.L.: Detection of lung nodules in digital chest radiographs using artificial neural networks: a pilot study. J. Digit. Imaging 8(2), 88–94 (1995)
3. Keem, S., Meadows, H., Kemp, H.: Hierarchical neural networks in quantitative coronary arteriography. In: Proceedings of the 4th International Conference on Artificial Neural Networks, pp. 459–464. IEEE, London (1995)
4. Andreae, M.H.: Neural networks and early diagnosis of myocardial infarction. Lancet 347(8998), 407–408 (1996)
5. Dumitra, A., Radulescu, E., Lazarescu, V.: Improved classification of psychiatric mood disorders using a feedforward neural network. Medinfo 8(1), 818–822 (1995)
6. Moreno, L., Pifiero, J.D., Sanchez, J.L., Mafias, J., Acosta, L., Hamilton, A.: Brain maturation using neural classifier. IEEE Trans. Biomed. Eng. 42(4), 428–432 (1995)
7. Webber-Foster, R., Kvizhinadze, G., Rivalland, G., Blakely, T.: Cost-effectiveness analysis of docetaxel versus weekly paclitaxel in adjuvant treatment of regional breast cancer in New Zealand. PharmacoEconomics 32(7), 707–724 (2014)
8. Angus, B., Lennard, T.W.J., Naguib, R.N.G., Sherbet, G.V.: Analysis of molecular prognostic factors in breast cancer by artificial neural networks. In: Naguib, R.N.G., Sherbet, G.V. (eds.) Artificial Neural Network in Cancer Diagnosis, Prognosis, and Patient Management, pp. 9–132. CRC Press LLC, New York (2001)
9. Karakış, R., Tez, M., Kılıç, Y.A., Kuru, Y., Güler, I.: A genetic algorithm model based on artificial neural network for prediction of the axillary lymph node status in breast cancer. Eng. Appl. Artif. Intell. 26(3), 945–950 (2013)

10. Karczmarek-Borowska, B.: Czynniki prognostyczne i predykcyjne dla raka piersi. Przegląd Medyczny Uniwersytetu Rzeszowskiego **4**, 350–355 (2009)
11. Young, D.W.: Management Accounting in Health Care Organizations. Jossey-Bass, San Francisco (2008)
12. Lawson, R.: Costing practices in healthcare organizations: a look at adoption of ABC. Healthcare Financ. Manag. 12 (2017)
13. Wahab, A., Mohamad, M., Said, J.: The implementation of activity-based costing in the Accountant General's Department of Malaysia. Asian J. Account. Gov. **9**, 63–76 (2018)
14. Setyawati, B.R., Sahirman, S., Creese, R.C.: Neural networks for cost estimation. Association for the Advancement of Cost Engineering International Transactions. AACE International, Morgantown (2002)
15. Cokins, G., Lawson, R., Tholemeier, R.: Patient-level costing and profitability: making it work. Healthcare Financ. Manag. 1 (2019)
16. Lafta, H.A., Hasan, Z.F., Ayoob, N.K.: Classification of medical datasets using back propagation neural network powered by genetic-based features elector. Int. J. Electr. Comput. Eng. **9**(2), 1379–1384 (2019)
17. Basha, S.M., Rajput, D.S., Iyengar, N.C.S.N., Caytiles, R.D.: A novel approach to perform analysis and prediction on breast cancer dataset using R. Int. J. Grid Distrib. Comput. **11**(2), 41–54 (2018)
18. Kaymak, S., Helwan, A., Uzun, D.: Breast cancer image classification using artificial neural networks. Procedia Comput. Sci. **120**, 126–131 (2017)
19. Ting, F.F., Tan, Y.J., Sim, K.S.: Convolutional neural network improvement for breast cancer classification. Expert Syst. Appl. **120**, 103–115 (2019)
20. Vo, D.M., Nguyen, N.-Q., Lee, S.-W.: Classification of breast cancer histology images using incremental boosting convolution networks. Inf. Sci. **482**, 123–138 (2019)
21. Baselga, J.: Herceptin alone or in combination with chemotherapy in the treatment of HER2-positive metastatic breast cancer: pivotal trials. Oncology **61**(Suppl. 2), 14–21 (2001)
22. Rosenblatt, F.: The perceptron: a probabilistic model for information storage and organization in the brain. Psychol. Rev. **65**(6), 386–408 (1958)
23. Broomhead, D.S., Lowe, D.: Multivariable functional interpolation and adaptive networks. Complex Syst. **2**, 321–355 (1988)
24. Liu, H.: Cost estimation and sensitivity analysis on cost factors: a case study on Taylor Kriging, regression and artificial neural networks. Eng. Econ. **55**(3), 201–224 (2010)
25. Bode, J.: Neural networks for cost estimation: simulations and pilot application. Int. J. Prod. Res. **38**(6), 1231–1254 (2000)

# Modeling and Stability Investigation of Investment of Health Sector on Regional Level

V. P. Martsenyuk[1] , D. V. Vakulenko[2(✉)] , S. M. Skochylyas[3] ,
and L. O. Vakulenko[4]

[1] University of Bielsko-Biala, 2, Willowa str., Bielsko-Biala, Poland
vmartsenyuk@ath.bielsko.pl
[2] I. Horbachevsky Ternopil State Medical University, 1, m.Voli,
Ternopil 46001, Ukraine
dmitro_v@ukr.net
[3] Ternopil National Economic University, 11, Lvivska str.,
Ternopil 46001, Ukraine
skochulyas@gmail.com
[4] Ternopil Volodymyr Hnatiuk National Pedagogical University,
2 Maxyma Kryvonosa str., Ternopil 46027, Ukraine

**Abstract.** The current state of investment activity is characterized by low innovation, institutional incompleteness, inconsistency and imbalance of technological, economic and social aspects, thus causing us, on the basis of scientific and practical works, to update the importance of systematization of knowledge in the investment sphere of the healthcare sector in Ukraine with the identification of the main aspects.

We have also considered the conceptual foundations of the health sector investment capacity, generalization of which will make it possible to form the proper theoretical and methodological basis of investment development technology based on the research of the medical industry business entities' investment activity tendencies. It has been found that investments in health care should correlate with the scientific substantiation and practical implementation of new management methods that would improve the infrastructure capacity of the healthcare sector, modernize the medical facilities, create new means of production based on modern technology and advanced technologies, and intensify socio-economic investment strategy development.

The research proposes an investment model, aimed at maintaining high investment activity and relative improvement of the region's competitiveness.

Based on the devised management model of the regional medical industry development using a system of nonlinear ordinary differential equations with two distributed delays, the authors conducted a qualitative analysis, found three stationary states and presented their economic interpretation. The presented model is investigated for the stability of its solutions and the quantitative research on Ternopil region medical industry management model has been carried out.

© Springer Nature Switzerland AG 2020
Z. Wilimowska et al. (Eds.): ISAT 2019, AISC 1052, pp. 121–131, 2020.
https://doi.org/10.1007/978-3-030-30443-0_11

**Keywords:** Investment · Investment potential of the region ·
Investment process · Investment attractiveness · Investment infrastructure ·
Mathematical model

# 1 Introduction

In the context of reforming Ukraine's economy and European integration, investments are becoming the most important factor of economic growth, which determines the level of competitiveness of the national economy and ensures sustainable development of the country. Investments are the key characteristic of modern scientific, technical, industrial, socio-economic and all social processes.

Today in Ukraine, the level of investment activity and the reproduction of innovative products market segment remains insufficient to provide an innovative breakthrough in the national economy. Under such conditions the implementation of an innovative model of structural adjustment should become the core of modern Ukrainian economic competitiveness model, whereas it is the knowledge economy strategy that needs to be implemented in the future. Investment activity is a factor ensuring the competitiveness of products and the efficiency of the use of resources, increases the degree of adaptability of the enterprise to the environment, and creates the prerequisites for a stable future prospect. The attraction of investments into economic circulation is due to the system of qualitative and positive effects that they create for both separate branches of the national economy and for the economy as a whole. At the present stage of the reform of the health care sector of Ukraine, the problem of regulating the investment provision of the medical sector development is of great importance.

The question of meaningful content of the concept of "investments", "investment attractiveness", the ambiguity of the essence and features that determine the properties, were considered in the works of a significant number of domestic and foreign scientists. A great contribution to the study of these problems was made by I. Blank, A. Pavlyuk, O. Savluk, O. Skibitsky, V. Fedorenko, F. Kotler, L. Gitman W. Sharpe, J. Gordon and others.

There are many modifications to the notion of "investment", the emergence of which is caused by the specifics and traditions of various economic schools and trends. The Nobel Prize winner in economics W. Sharpe in his monograph "Investments", written in cooperation with other American scholars, states: "In the broadest sense, the term "invest" means to part with the money today in order to get them in a larger sum in the future" [22]. In their monograph "Basics of Investing" American economists L. Guitman and M. Johck state that "Investing is a way of placing capital that should ensure the preservation or growth of capital." [7]. In authors' research the notion of "investment" is considered together with "investment attractiveness" of the region or administrative unit.

Investment attractiveness of the region, according to the method of I. Blank is estimated on the basis of ranking according to the following indicators: level of general economic development of the region; the level of development of the investment infrastructure of the region; demographic characteristics of the region; level of development of market relations and commercial infrastructure of the region; level of criminogenic, ecological and other risks [3]. We agree with the opinion [17, pp. 386–387] that it is expedient to estimate the investment attractiveness of the region on the basis of analysis of 5 generalized indicators: the level of eneral economic development of the region; the region infrastructure development investment level; demographic factor; level of development of market relations and commercial infrastructure of the region; level of criminogenic, ecological and other types of risks. O. Mikhailovska in her writings defines the investment attractiveness of industries as an integral indicator, combining the characteristics of each type of industry from the point of the operational efficiency and profitability. To assess the current state of investment attractiveness and its forecasting, the following principles are used:

– level of perspective development, profitability of enterprise activities of certain branches and their dynamics;
– the investment climate in the industry and the level of investment risk [5].

According to O. Rivak, investment attractiveness is reflected by the capital market, whose economic leverage can lead to revitalization of investment processes in industries [5]. Despite the large range of research carried out and significant scientific and practical developments, a number of issues remain in the area of increasing the investment attractiveness of the region, the areas in need of further research. The assessment of investment attractiveness increases the reasonableness of management decisions, creates conditions for efficient investment activity and requires modernization of the state management system of investment processes.

**Article Objective Statement.** Objectives of the article include the following: systematization of conceptual approaches to the assessment and modeling of regional investment activities in the field of health care and the identification of the most valuable elements, the synthesis of which will make it possible to form the appropriate theoretical and methodological basis; construction of a mathematical model for the reorganization of the investment potential of the healthcare sector of the region, conducting its qualitative analysis, studying stationary solutions of its stability and conducting a numerical experiment.

**Literature Review.** It is advisable for the national state regulation system to combine the interests of the state and private business in medicine, to ensure the diffusion of Ukrainian goods and services to world markets, and support the priority directions of the medical industry development. Of crucial importance is the problem of regulation of development-aimed medical sector investment provision at the level of the legislative framework, in particular: concept of the national program "Health 2020: Ukrainian dimension", which emphasizes the need for interaction between the public and private sectors, the law of Ukraine "On amendments to certain legislative acts of Ukraine on the improvement of health care legislation" [19], which provides for the autonomy of medical institutions and their transformation into a kind of non-profit enterprises, and the law "On state financial guarantees of medical care of the population " [20]. In the law of Ukraine "On the State Budget of Ukraine for 2018" the total health care expenditures will amount to UAH 86 billion, or 9.1% of the state budget or 2.6% of GDP [19.21].

Let us consider the investment attractiveness of Ternopil region medical industry infrastructure as an example. According to the results of the analysis of statistical data for 2016–2017, Ternopil region ranked 2nd in terms of infrastructure development among all regions of Ukraine. The status analysis of certain infrastructural components of Ternopil region revealed the presence of certain obstacles, which upon overcoming will contribute to the investment attractiveness of the region: the development of special programs and investment attraction to implement measures to replenish regional medical institutions with modern diagnostic equipment; the formation of a single electronic medical space that needs to be solved through the intensive use of information technology. In addition, there is a problem of the growth of information flow volumes, therefore, it is necessary to improve state mechanisms for the processing and use of medical information, its rational collection and analysis. In order to solve the problems of creating a single medical information space, a sectoral system of scientific medical service and medical statistical information with the appropriate infrastructure, personnel potential, scientific and methodological support, financial resources, etc., is needed for the development of new methods; continued creation of conditions for the work of a family doctor; provision of family medicine facilities with modern diagnostic equipment, transport and housing; optimization of the network of medical establishments within the region; completion of the "medical reform" within the state. An assessment of the investment attractiveness of Ternopil region medical industry infrastructure is proposed to be carried out in the following areas:

– Assessment of general economic development. The higher the indicators of economic development, the more opportunities the medical institution has to attract health care users.
– Demographic characteristics of the region. The number and composition of the population is instrumental in shaping the demand for health services.
– Level of development of market relations and market infrastructure of the region. The well-developed market infrastructure of the region with the presence of construction organizations, enterprises for the maintenance of medical equipment, suppliers of medicines and dressings materials, creates favorable conditions for the medical institution development and operation.
– The level of investment activity security is estimated by the indicator of economic and criminal delinquency, and the number of bankruptcies in the region among the enterprises providing medical services.
– The level of medium-sized investment risks.
– Indicators characterizing: specific capital intensity of services or products, normative profitability of products (services), etc., fixed assets through which services are provided or products are produced, in particular current book value, depreciation and obsolescence, anticipated decommissioning, standard service life, duration of the investment lag, etc., demand or products and services of the industry; state support for the development of the medical sector.
– Improving the quality and efficiency of providing medical services and the degree of satisfaction of the population's needs in these services will ensure the preservation and strengthening of public health, increase in life expectancy and decrease in the level of morbidity, disability and mortality. Improvement of these regulatory instruments and methods, with their further introduction into practical activities, will contribute to ensuring the effective functioning of the healthcare sector.

These areas are the basis for developing recommendations for improving the organization of planning and analytical work of the relevant state authorities and local self-government, which promotes an increase in the degree of efficiency of providing potential investors with information about the investment attractiveness of the structural components of the sectoral health care system and increasing the completeness and reliability of information through the improvement of statistical reporting and statistical processing methods.

## 2  Materials and Methods

Literature revue of complex compartments of regional model health care sector on regional level are described. For describing investigation of investment of health sector on regional level are used determinal mathematical model. Using the system of nonlinear ordinary differential equations with two delayed distributions, we construct a model of investment of the health sector on regional level. In this paper, an approach to the construction and analysis of models of investment of health sector on regional level in the classes of balance models, functional-differential equations, called population dynamics equations, is proposed. In this case, the phase coordinates, which are the numbers of the compartments model are described by the average value.

The presented model is investigated on the stability of solutions at the first approximation and a quantitative study of the model of investment of the health sector on regional level has been carried out. Investigation of nonlinear dynamics of investment of health sector on regional level has been carried out by analysis of phase portraits [11, 12, 16].

## 3  Experimental

Construction of a mathematical model for reorganization of the investment potential region healthcare sector. Let us assume that the main factors influencing the investment attractiveness of the health sector are the following [11, 12]: $C(t)$ - the consumer potential of investment attractiveness of the investment object (healthcare sector), which depends on many factors, in particular: the prospects for the development of the medical services sector, the availability of modern medical technologies, scientific advances in the industry and the possibility of their application in the work of a medical institution, the need in this type of medical care and the level of the satisfaction of this need, the share of paid medical services in the structure of all services provided by public and private medical institutions; improving the quality and efficiency of providing medical services and the degree of satisfaction of the population's needs in these services. $B(t)$ - production-resource potential (PRP) of investment attractiveness of investment objects (health care), in particular such crucial components as organizational and managerial, logistical, personnel, financial potential and potential of territorial development; $K(t)$ - investment risks of investment attractiveness of investment objects healthcare) to which we refer: the level of competition in the industry; the average level of profitability of medical institutions; an increase in prices for medical services and an inflation index;

$M(t)$ - investment potential of attractiveness of investment objects (healthcare). Indication of the coefficients used in the model: the increase in the level of attractiveness of consumer services, $\beta_{CB}$, $d_C$ – the coefficient of decline in the competitiveness of consumer services, $\beta_B$ – the coefficient of PRP growth. At the same time, the PRP involved in the formation of investment potential requires time $\tau_B$, in particular for the formation of personnel, reorganization of the infrastructure of the region, etc. $\gamma_{BM}$ - coefficient indicating the PRP share in the formation of the investment potential of the region. $\beta_K$ we will call the coefficient of investment risks, $\alpha_{KM}$ – a coefficient indicating the share of investment risks in the investment potential. Time required for investment risks. $\tau_K$. $k_C$ - coefficient of formation of investment potential at the expense of consumer potential. C. $k_B$ - coefficient indicating the probability of forming the investment potential of the PRP. $k_M$ - coefficient of proportionality, which indicates the probability of lowering the investment potential. $k_{KM}$ - coefficient indicating the probability of reducing the investment potential by investment risks.

Thus, we come to the following system of nonlinear ordinary differential equations with distributed delay:

$$
\begin{aligned}
\frac{dC(t)}{dt} &= \beta_{BC}B(t) - d_C C(t), \\
\frac{dB(t)}{dt} &= \beta_B B(t) - d_B B(t) + \gamma_{BM} M(t - \tau_B)B(t - \tau_B) \\
\frac{dK(t)}{dt} &= \beta_K K(t) - d_K K(t) + \alpha_{KM} K(t - \tau_K)B(t - \tau_K) \\
\frac{dM(t)}{dt} &= k_C C(t) - k_B B(t)M(t) + k_M M(t) - k_{KM} M(t)K(t)
\end{aligned}
\tag{1}
$$

For the system of Eqs. (1) we specify the initial conditions:

$$
\begin{aligned}
C(t) &= C^0, K(t) = K^0 \\
B(t) &= B^0, M(t) = M^0, \ t \in [t^0 - \max\{\tau_B, \tau_K\}, t_0].
\end{aligned}
\tag{2}
$$

The system of Eqs. (1) with the initial conditions (2) will be referred to as the management mathematical model of the regional medical industry development [11].

**Remark.** In (1), (2) we take $t^0 = 0$ as the initial time instant; in what follows we assume that the initial conditions are positive and all parameters of the model are constant and positive quantities.

**Calculation.** By the developed computer program one has carried out the quantitative research of management model of the regional medical industry development in case when $\beta_{CB} = 0,78$, $d_C = 0,8$ $\beta_B = 21$, $d_K = 1$, $\gamma_{BM} = 0,47$, $d_B = 0,8$, $\beta_K = 10000$, $\alpha_{KM} = 10$, $k_B = 0,12$ $k_M = 0,12$ $k_{KM} = 0,12$. At $\tau \in [-\tau, 0]$ for the initial set the following initial conditions $C(t) = 1, B(t) = 1$, $K(t) = 0$, $M(t) = 1$ are valid. The modeling demonstrates the qualitative dependence of a regional medical industry development on $k_C$, - the coefficient of formation of investment potential at the expense of consumer potential and required time $\tau_K$ - time required for investment risks (in days).

We now consider the system behavior when the delay $\tau_B = 25$, as the bifurcating parameters we consider $k_C \geq 0$ and their delay $\tau_K \geq 0$. At values $k_C = 0, 31$, $\tau_K = 0$, (Fig. 1a) the attractor is a stable node. At $k_C = 0, 31$, $\tau_B = 25, 3454212$ there appears a pair of complex conjugate roots with positive real parts. The attractor in the system is the unstable limit cycle (Fig. 1b). At $k_C = 0, 32$ $\tau_B = 24, 023215306$, the characteristic quasipolinomial gets a pair of purely imaginary roots that correspond to Hopf bifurcation (Fig. 1c). At $k_C = 0, 35$, $\tau_B = 20, 390995$, the attractor passes into the stable focus that associates with the presence of roots of characteristic quasipolinomial with negative real parts (Fig. 1d).

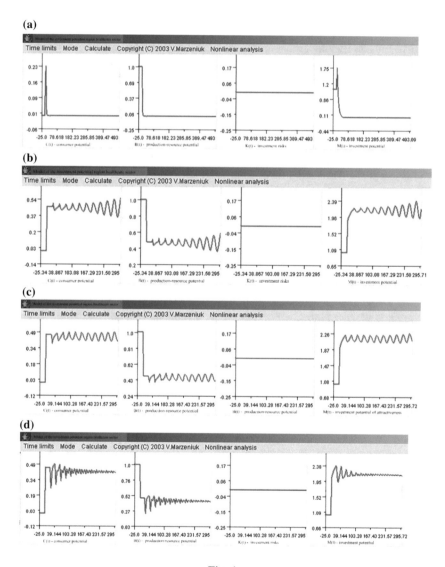

**Fig. 1.**

## 4 Results

The conceptual foundations of the investment capability of the health care sector, the generalization of which will allow the forming of a proper theoretical and methodological basis of investment development technology based on the research of tendencies of investment activity of the business entities of the medical industry were considered.

The main factors influencing the investment attractiveness of the health sector are the following: the consumer potential of investment attractiveness of the investment object (healthcare sector); production-resource potential (PRP) of investment attractiveness of investment objects (health care); investment risks of investment attractiveness of investment objects healthcare); investment potential of attractiveness of investment objects (healthcare).

An investment model is proposed, which is aimed at maintaining high investment activity and the relative improvement of the region's competitiveness.

Based on the developed model of management of the development of the medical industry with the help of a system of nonlinear ordinary differential equations with two distributed delay, a qualitative analysis was carried out, three stationary states were found, economic interpretation of each was presented. This condition can be interpreted as one of the conditions of investment attractiveness of the health sector with a sufficient GRP, in particular, such critical components: organizational, managerial, material and technical, personnel, financial potential and potential of territorial development. With the help of the developed computer program quantitative research of management model of the medical field development of the region is carried out, the behavior of the system under various bifurcating parameters is considered and the stationary states are shown – a stable node, an unstable boundary cycle and a steady focus. The presented model is investigated on the stability of solutions and its quantitative research on the model of management of the development of the medical sector has been carried out.

It was found that investments in health care should relate to the scientific substantiation and practical introduction of new management methods that would improve the infrastructure capacity of the healthcare sector, modernize the medical facilities, create new means of production based on modern technology and advanced technologies, intensify development investment strategy for socio-economic development.

Discussion. The main factors influencing the investment attractiveness of the health sector are the following [11, 12]:

- the consumer potential of investment attractiveness of the investment object (healthcare sector), which depends on many factors, in particular: the prospects for the development of the medical services sector, the availability of modern medical technologies, scientific advances in the industry and the possibility of their application in the work of a medical institution, improving the quality and efficiency of providing medical services and the degree of satisfaction of the population's needs in these services, the need in this type of medical care and the level of the satisfaction of this need, the share of paid medical services in the structure of all services provided by public and private medical institutions;

- production-resource potential (PRP) of investment attractiveness of investment objects (health care), in particular such crucial components as organizational and managerial, logistical, personnel, financial potential and potential of territorial development;
- investment risks of investment attractiveness of investment objects healthcare) to which we refer: the level of competition in the industry; the average level of profitability of medical institutions; an increase in prices for medical services and an inflation index;
- investment potential of attractiveness of investment objects (healthcare). Indication of the coefficients used in the model: the increase in the level of attractiveness of consumer services.

An investment model is proposed, which is aimed at maintaining high investment activity and the relative improvement of the region's competitiveness.

In the future, for support optimal design making by managers' healthcare sectors of region are recommended solve optimal control task with phase constraints.

The proposed method can be used both to check the optimally of existing solutions (projects), and to develop new ones. We note that the system under consideration can also be used to model high investment activity and relative improvement of the region's competitiveness through n solutions (projects) by introducing additional variables.

## 5   Conclusion

The current state of investment activity is characterized by low innovation, institutional incompleteness, inconsistency and imbalance of technological, economic and social aspects, thus causing us, on the basis of scientific and practical works, to update the importance of systematization of knowledge in the investment sphere of the healthcare sector in Ukraine with the identification of the main aspects.

The research proposes an investment model, aimed at maintaining high investment activity and relative improvement of the region's competitiveness.

Based on the devised management model of the regional medical industry development using a system of nonlinear ordinary differential equations with two distributed delays, the authors conducted a qualitative analysis, found three stationary states and presented their economic interpretation. This condition can be interpreted as one of the conditions of investment attractiveness of the health sector with a sufficient GRP, in particular, with such crucial components: organizational, managerial, material and technical, personnel, financial and territorial development potential. Using the designed computer program, we carried out quantitative research of the regional medical industry management model, considered the behavior of the system under different bifurcating parameters and presented the following stationary states: a stable node, an unstable boundary cycle and a steady focus. The presented model is investigated for the stability of its solutions and the quantitative research on Ternopil region medical industry management model has been carried out.

The prospect of this study is to reseach the management model of the regional medical industry development for optimal control management, investment and legislative factors.

# References

1. Akhlamov, A.H., Kusyk, N.L.: Economics and health financing: teaching method. Manual, ORIDU NADU, Odesa (2011). http://academy.gov.ua/NMKD/library_nadu/Biblioteka_Magistra/3b63646e-4026-4df2-826d-c561b5313180.pdf
2. Baieva, O.V.: Healthcare Management: Teach. Manual, K.: Center for Educational Literature (2008). https://medic.studio/zdravoohranenii-menedjment/menedjment-galuzi-ohoroni-zdorovya-navch.html
3. Blank, Y.A.: Investment management: study. course. K.: Элha–N, Nyka–Tsentr (2001). http://www.alleng.ru/d/econ-fin/econ-fin173.htm
4. Dieudonne, J.: Foundation of Modern Analysis. Academic Press, New York (1960)
5. Farat, O.V.: Investment attractiveness of the sectors of the national economy of Ukraine: the current state and prospects of development. Innovative economy: allukr. sciences - production journ – Ternopil **43**(5), 34–40 (2013). http://ie.at.ua/IE_2013/InnEco_5-43-2013.pdf
6. Fedorenko, V.H.: Investing in the national economy: monogr. K.: IPK DSZU (2011). http://ipk.edu.ua/journal/El-librari/Fedorenko%202011%20Invest.pdf
7. Gitman, L.J., Joehnk, M.D., Smart, S.B.: Fundamentals of Investing. Prentice Hall, Upper Saddle River (2011). 013611704X, 9780136117049
8. Alexander, G.J., Sharpe, W.F., Bailey, J.V.: Fundamentals of Investments. Prentice Hall, Upper Saddle River (2001)
9. Hale, J.K.: Theory of Functional Differential Equations. Springer, Heidelberg (1977). https://doi.org/10.1007/978-1-4612-9892-2
10. Holubiatnikov, V.T.: Investments and regional project activity: training. Manual. LRIDU NADU, Lviv (2011)
11. Marchuk G.I.: Mathematical Models in Immunology. Nauka, Moscow (1980)
12. Martsenyuk, V.P., Vakulenko, D.V.: On model of interaction of cell elements at bone tissue. Remodeling J. Autom. Inf. Sci. Begellhouse **39**(3), 68–80 (2007). https://doi.org/10.1615/JAutomatInfScien.v39.i3.70
13. Martsenyuk, V.P., Vakulenko, D.V.: Optimal control of regimens of medical therapy and physiotherapy in the task of reconstruction of bone tissue. J. Syst. Res. Inf. Technol. **3**, 108–122 (2011)
14. Mykhalchuk, V.M.: Economic substantiation of introduction of new medical technologies in health protection. Bull. Soc. Hyg. Health Care Organ. Ukraine **67**(1), 45–49 (2016). https://ojs.tdmu.edu.ua/index.php/visnyk-gigieny/issue/view/428
15. Pavliuk, A.P.: Priorities of investment policy in the context of modernization of the Ukrainian economy: analytical report. K.: NISD (2013). http://www.niss.gov.ua/content/articles/files/Invest_polit.indd-c9ae5.pdf
16. Ruan, S.: On the zeros of transcendental function with applications to stability of delay differential equations. Preprint/Oxford University, Oxford (1999)
17. Savluk, O.: Estimation of investment attractiveness of regions of Ukraine. Bull. Kiev Natl. Trade Econ. Univ. **5**, 31 (2013). http://visnik.knteu.kiev.ua/index.php?option=com_content&view=article&id=123&Itemid=526&lang=uk
18. Skibitskyi, O.M.: Innovation and Investment Management: Teaching. manual. K.: Center for Educational Literature (2009). https://subject.com.ua/pdf/122.pdf

19. The Law of Ukraine "On amendments to certain legislative acts of Ukraine on the improvement of health care legislation" from 06.04.2017 №2002-VII (2017). http://oz.zt.gov.ua/index.php?option=com_content&view=article&id=903&Itemid=20
20. The Law of Ukraine "On state financial guarantees of medical care of the population" 19.10.2017 №2168-VIII (2017). http://oz.zt.gov.ua/index.php?option=com_content&view=article&id=903&Itemid=20
21. Tsarenko, M.O.: Regional investment programs as a tool for improving the efficiency of investment activity management in the regions. In: Proceedings. Sir: Economy, vol. 20, pp. 41–46 (2012). http://nbuv.gov.ua/UJRN/Nznuoa_2012_20_11
22. Sharpe, W., Alexander, G.J., Bailey, J.W.: Investments, 6th edn. Prentice-Hall (1999). ISBN 10:013011507X

# Bibliomining the Pareto Principle of Public Libraries

Yi-Ting Yang and Jiann-Cherng Shieh[✉]

Graduate Institute of Library and Information Studies,
National Taiwan Normal University, Taipei, Taiwan
jcshieh@ntnu.edu.tw

**Abstract.** The operation management or readers marketing of libraries cannot be only based on the experience and intuition of the library administrators. There must be some practical evidence to provide them as a reference for decision-making. Bibliomining is the use of statistics, bibliometrics, data mining, and other techniques to analyze the data generated by library operations, to explore the information available to library managers, and to assist their decision-making. From the application of the Pareto principle in business management, it is known that analyzing vital customers can help the growth of profit of product by marketing. Also, the main profit analysis helps in effective production management. This can, of course, be used analogously to the relationship of library readers and their circulations. Whenever using library readers borrowing data to discover the 80/20 rule (or Pareto Principle) existed in library circulation, we can further discuss the issues of operation management and reader marketing of the library from three perspectives: vital readers, major collection and both. In this study, we focus on the vital readers and the major collection as research topics. Under the situation of 80/20 in library circulation, we apply data mining techniques to further explore what characteristics of vital readers, useful collections and their impacts on the library.

**Keywords:** Data mining · Pareto principle · 80/20 rule · Public libraries · Bibliomining

## 1 Introduction

The Pareto Principle concept has been recognized as a popular guideline or tool in business and social context. The principle known as the 80/20 rule indicates that there is an unbalanced relationship between causes and results or between efforts and rewards: specifically, it maintains that 80% of rewards usually come from 20% of efforts, and the other 80% of efforts only produce 20% of the results. Therefore, if you can recognize, focus on, and control the vital 20% of efforts, you will obtain greater profits or efficiency.

The Pareto principle has been adapted to many applications such as economics, sports, computing, occupational health and safety, engineering control, quality control, health care, etc. Is there the Pareto Principle in library context? In recent years, a great deal of research has analyzed library data and uncovered a variety of trends, patterns,

© Springer Nature Switzerland AG 2020
Z. Wilimowska et al. (Eds.): ISAT 2019, AISC 1052, pp. 132–140, 2020.
https://doi.org/10.1007/978-3-030-30443-0_12

and relationships. Bibliomining, or data mining in libraries, is the application of data mining techniques to data produced from library services [1, 2]. By applying statistic, bibliometrics, or data mining tools, libraries can better understand usage patterns and rules, enabling library managers to make decisions to meet reader needs based on those mining results [3, 4]. However, quality decisions must be based on quality data. Data processing is an important step in the knowledge discovery process [5, 6]. Identifying vital data and reducing the data to be analyzed can lead to huge decision-making payoffs. Similarly, identifying vital readers and core collections would allow library managers to provide better services and popular materials to promote library utilization and reader satisfaction.

Investigating distributions of circulation data can help libraries understand reader behaviors. Previous discussions of data mining applications in academic libraries have emphasized usage analysis [7]. There have been a number of studies that have inspected usage analysis of academic libraries [8, 9]. Renaud, Britton, Wang, and Ogihara [10] analyzed data from a university library and revealed the distributions of check-out activities based on reader type, academic department, LC classification, material type, life span, and so on. They also correlated the findings with student grade point averages. Goodall and Pattern [11] analyzed the usage data of electronic resources, book loans, and visits in an academic library and correlated these data points with academic achievement.

However, in comparison with academic libraries, public libraries service all kinds of readers, which is much complex in real practice. Before data analyzing or data mining, segmenting these various readers and collections would conduct meaning results. Less research was to focus on this issue. Yang and Shieh [12] proposed their study that they analyzed the circulation data of one public library and uncovered the existence of Pareto principle. They took the 80/20 result to identify vital few readers and useful many circulation collections. The consequence is much helpful and effective in library target marketing its readers, and even in promoting their collection development.

Following the discovery of the public library 80/20 rule from its circulation data, through bibliomining technique, this study will further explore public libraries borrowing patterns or characteristics from different viewpoints of vital readers, major collections, or both, and their impacts on the public library.

## 2   Literature Review

### 2.1   The 80/20 Rule

The 80/20 rule originated from the Pareto Principle, named for the Italian economist Vilfredo Pareto who identified a general imbalance in property allocation: most (80%) wealth belongs to a few (20%) people. This model of imbalance has been observed repeatedly. In the late 1940 s, Juran named his notion of "the vital few readers and the trivial many" as Pareto Principle after the Italian economist. The 80/20 rule is an extension of the Pareto Principle developed by Richard Koch based on a theoretical view of Pareto and Juran. Koch [13] pointed out that the 80/20 rule applies in various fields. In business, 80% of a company's profits come from 20% of its customers, 80% of revenues come from 20% of the products, and 80% of sales come from 20% of the sellers. In quality management, 80% of the problems come from 20% of the faults. In computer science, most software takes 80% of the time to run 20% of the programs. The unbalanced relationship between efforts and rewards or causes and results makes delineating the vital few readers very important. Several studies also have suggested the benefit of applying the 80/20 rule [14, 15]. Concentrating on the groups of customers and the specific markets that are profitable can substantially improve a company's bottom line, having insight into the vital few readers is an important issue.

### 2.2    Applications of the 80/20 Rule in Libraries

The study of the Pareto Principle in libraries was initialized by Trueswell [15] who applied the 80/20 rule to address the relationship between collections and circulation numbers. Trueswell [15] noted that about 20% of collections bring 80% of circulation numbers [16]. Hardesty [17] traced the book acquisitions and circulations of a university for five years. He found that 30% of books accounted for 80% of circulation. In recent years, Singson and Hangsing [18] analyzed usage patterns of electronic journals academic consortia. They found the reader downloads for some publishers follow the 80/20 rule. The few, core journals were downloaded the most. Some research suggested the 80/20 rule could be used to identify the core collections within libraries. Burrell [19] investigated the circulation data of university libraries and public libraries and found between 43% and 58% of circulating collections are required to account for 80% of borrowings. He developed a theoretical model of library operations to help libraries identify their core collections. Nisonger [20] examined the 80/20 rule in relation to the use of print serials, downloads from electronic databases, and journal citations, concluding that the 80/20 rule is a valid method for determining core concepts in journal collection management.

However, sporadic studies have explored the distributions of circulation in public libraries. Recently, Yang and Shieh's study [12] examined a circulation dataset from a public library in Taiwan and analyzed usage patterns to understand the distributions of readers and circulations. Their research explored the existence of 80/20 rule on circulation data of the public library.

Considering the importance of the user-centered concept today, this study will apply bibliomining techniques to further analyze the characteristics of a few vital readers, the types of useful many library collections, and their associations. The results will provide important decision-making information about library marketing and collection development.

## 3    Research Methods

### 3.1    Data Description

The data collection for analyzing purpose includes circulation data, reader data, item data, and branch location data. Circulation dataset contains more than 18 million transactions conducted over two years, and reader dataset contains data about 460 thousand readers.

### 3.2   Preprocessing Data Privacy

To preserve privacy, the data had been adopted through preprocessing before we got and processed it. Columns that may have identified someone by providing information such as reader names, addresses, or phone numbers have been deleted. Furthermore, reader corresponding data and branch related data had been translated into substituted codes by some one-way hash functions respectively [21].

### 3.3   Data Processing Procedure

Following the 80/20 findings (75/25 for the case public library) of Yang and Shieh's, this paper will further focus on the vital few readers and useful many collections to apply data mining techniques to investigate their uncover information patterns. We will first extract the relevant data of about 120 thousand vital readers and the 13,657 thousand circulation books that they borrowed. Then we will further apply data mining techniques to analyze the vital readers and useful circulation collections to explore information that is implicit in it, that can be used as a reference for library marketing or management decisions.

### 3.4   Analysis Tools and System Environment

The database system used in this study is Microsoft SQL Server 2014. And we applied Microsoft SQL Server Data Tools (SSDT) and Microsoft Excel 2016 as a data mining tool to analyze data. We adopted a PC workstation with Intel Core i7-7700 CPU, 16G memory and 1T SSD to support database system operation, data processing, and analysis tasks.

## 4   Data Mining and Findings

Data mining results and findings of this research are presented in the following three sections.

### 4.1   The Vital Few Readers

According to the number of readers, the **top five** regions with the largest number of vital readers are Location 15, Location 3, Location 26, Location 25 and Location 9. Totally, there are 65966 vital readers (14.35% of all readers). The number of circulation items they borrowed is total7893286 (43.70% of all collections) and is depicted in the following Table 1.

**Table 1.** The numbers of vital readers of locations and their borrowed items

|  | Readers | | Borrowed item | |
|---|---|---|---|---|
|  | Number | % | Number | % |
| All | 459755 | 100.00 | 18062800 | 100.00 |
| Pareto Principle | 114727 | 24.95 | 13657491 | 75.61 |
| Vital Readers (Location 15) | 22370 | 4.87% | 2706225 | 14.98% |
| Vital Readers (Location 3) | 13329 | 2.90% | 1559025 | 8.63% |
| Vital Readers (Location 26) | 12549 | 2.73% | 1518296 | 8.41% |
| Vital Readers (Location 25) | 9779 | 2.13% | 1119487 | 6.20% |
| Vital Readers (Location 9) | 7939 | 1.73% | 990253 | 5.48% |

Since this public library service area is quite extensive, it is worthy of further analyzing whether the characteristics of readers in different regions are different. Thus we apply data mining of the clustering algorithms to vital readers on location and age. There are 6 clusters resulted as shown in Table 2.

**Table 2.** Clustering analysis of the vital readers

|  | Cluster 1 | Cluster 5 | Cluster 3 | Cluster 2 | Cluster 4 | Cluster 6 |
|---|---|---|---|---|---|---|
| Size | 16413 | 13537 | 12668 | 11008 | 6696 | 5644 |
| Age | Youth (59.2%) | Kid (65%) | Youth (62.4%) | Youth (100%) | Middle Age (92%) | Kid (58.8%) |
|  | Middle Age (40.8%) | Teenager (32%) | Middle Age (37.6%) |  |  | Teenager (28.1%) |
| Location | 15 (100%) | 15 (39.1%) | 25 (56%) | 3 (52%) | 3 (40.7%) | 15 (25.4%) |
|  |  | 3 (30.7%) | 9 (44%) | 26 (47.4%) | 26 (32.9%) | 3 (7.6%) |
|  |  | 26 (27%) |  |  | 25 (14.4%) | 26 (16.8%) |
|  |  |  |  |  | 9 (12%) | 25 (26.9%) |
|  |  |  |  |  |  | 9 (23.3%) |

It is clear to see that Cluster 1, the readers lived in Location 15, their ages are Youg or Middle Age, has a max size of the number of vital readers. We can further probe what patterns of collections they borrowed. For the vital readers of Cluster 1, their average number of borrowings is about 121 times per person.

Language and literature books (800 Class) are accounted for the most borrowed, about 46.96% of their borrowings. Among them, the borrowing volumes of the novel (Classification Number 857), Chinese Children's Literature (Classification Number 859), and Japanese Literature (Classification Number 861) are the most. For the relationship between the books they borrowed, we applied association rule algorithms to conduct data mining and obtained the rules listed in Table 3 below.

**Table 3.** Association rules of borrowing books

| Rule | Importance | Confidence |
|---|---|---|
| 887, 875→874 | 0.230 | 0.980 |
| 863, 876→861 | 0.201 | 0.944 |
| 523, 859→872 | 0.749 | 0.684 |
| 496, 177→494 | 0.543 | 0.839 |

For Cluster 1 vital readers, those who borrowed books of Oceania literature (Classification Number 887) and German literature (Classification Number 875), have a 98% chance to borrow American literature (Classification Number 874). Also, those who borrowed Taiwanese literature (Classification Number 863) and French literature (Classification Number 876), have a 94.4% chance to borrow Japan literature (Classification Number 861).

## 4.2   The Useful Many Collections

75% circulation volume was provided by 25% vital readers of the case public library. These 75% circulations are corresponding to collection items, organized by the New Classification Scheme for Chinese Libraries, which belong to 800 Linguistics and Literature are borrowed most frequently, followed by 400 Applied Sciences and then 900 Arts. Since Language and literature books (800 Class) are accounted for the most borrowed, the number of such books borrowed exceeding 1.11 million, we separately analyze the useful many collections by Classification Number 800 and non-800.

For useful many collections of Classification Number 800, we cluster them according to the second layer Classification Number and material type. There are nine clusters as shown in the following Table 4.

**Table 4.** Clustering analysis of the useful collection of classification number 800

|  | Cluster 1 | Cluster 2 | Cluster 3 | Cluster 4 | Cluster 5 | Cluster 6 | Cluster 7 | Cluster 8 | Cluster 9 |
|---|---|---|---|---|---|---|---|---|---|
| Size | 284338 | 193818 | 129450 | 123707 | 97152 | 92904 | 92210 | 82435 | 22765 |
| Classification number | 850 (100%) | 850 (100%) | 870 (100%) | 870 (100%) | 860 (66%) | 850 (28.2%) | 860 (92.1%) | 800 (64.3%) | 810 (90%) |
|  |  |  |  |  | 800 (19.4%) | 870 (14.3%) | 810 (7.4%) | 880 (22%) | 830 (8.6%) |
|  |  |  |  |  | 880 (14.4%) | 800 (53.3%) |  | 830 (6.9%) |  |
| Material type | Chinese Book (100%) | Children Book (99.7%) | Chinese Book (99.9%) | Children Book (96.4%) | Children Book (97.5%) | Attachment (89.6%) | Chinese Book (98.6%) | Chinese Book (96.8%) | Children Book (93.5%) |

From Table 4, it is known that Chinese books and children's book account for the most; while the subject takes the most in Classification Number 850 and 870, which are divided into Cluster 1 to Cluster 4. In addition, Cluster 6 contains collections of linguistics and literature (800 categories) and their attachments.

Similarly, for useful many collections of Classification Number non-800, we cluster them according to the first layer Classification Number and material type to result in ten clusters. There are seven clusters their sizes are greater than 0.1 million as depicted in Table 5.

**Table 5.** Clustering analysis of the useful collection of classification number non-800

|  | Cluster 1 | Cluster 2 | Cluster 3 | Cluster 4 | Cluster 5 | Cluster 6 | Cluster 7 |
|---|---|---|---|---|---|---|---|
| Size | 285821 | 201215 | 197333 | 142955 | 141069 | 141738 | 110093 |
| Classification number | 400 (100%) | 000 (10.7%) 600 (16.7%) Others (72.4%) | 700 (51%) 100 (49%) | 500 (100%) | 300 (64.9%) 200 (34.2%) | 900 (100%) | 900 (56.1%) 700 (27.2%) 000 (14.7%) |
| Material type | Chinese Book (100%) | Chinese Book (23.4%) Journal (71.2%) | Chinese Book (100%) | Chinese Book (100%) | Chinese Book (79.4%) Attachment (15.6%) | Chinese Book (63.3%) Comic Book (32.5%) | Children Book (55.4%) Attachment (15.1%) Comic Book (6.7%) Video (22.1%) |

Books of Applied sciences (400 Class) account the most, self-forming Cluster 1; Cluster 3 has History and Geography books (700 Class) and Philosophy books (100 Class); Cluster 6 has Arts books (900 Class) and Comic books.

### 4.3    Vital Few Readers and Useful Many Collections

The amount of borrowing contributed by vital few readers accounts for a large portion of the two-years circulation (about 75%), with a total of more than 13.65 million. Considering the significant differences between various clusters, we discover the top five regions (15, 3, 26, 25 and 9) with the largest borrowing volume for further clustering analysis. There are about 7.72 million borrowing data in these five districts; we result in 10 clusters through clustering analysis according to the first layer Classification Number, age, and circulation district(location); as listed in Table 6.

Cluster 2 shows that at Location 15, there are 98.6% books of Classification Number 800 borrowed by 97.50% vital readers who are young people, children, teenagers, and middle-aged people. Also, Cluster 7 shows that at Location 25, there are 72.9% and 13.8% useful books of Classification Number 800 and Classification Number 400 borrowed by 98.50% vital readers who are young people, children, and old people.

**Table 6.** Clustering analysis of the vital readers and useful collections

| | Cluster 1 | Cluster 2 | Cluster 3 | Cluster 4 | Cluster 5 | Cluster 6 | Cluster 7 | Cluster 8 | Cluster 9 | Cluster 10 |
|---|---|---|---|---|---|---|---|---|---|---|
| Size | 1599902 | 1232384 | 1137236 | 859996 | 762382 | 615549 | 484426 | 303568 | 396332 | 326450 |
| Age | Youth (41.5%) | Youth (36.8%) | Youth (44.4%) | Youth (35.6%) | Youth (45.8%) | Youth (28%) | Youth (63.9%) | Middle Age (80.8%) | Youth (45.7%) | Youth (28%) |
| | Middle Age (25.3%) | Middle Age (31.3%) | Middle Age (30.8%) | Middle Age (28.3%) | Middle Age (32.6%) | Middle Age (30.7%) | Kid (26.5%) | Teenager (19.2%) | Middle Age (49.1%) | Middle Age (45.7%) |
| | Kid (21.6%) | Kid (20%) | Kid (16.3%) | Kid (20.6%) | Kid (14.1%) | Kid (26.7%) | Old (8.1%) | | | Kid (17.8%) |
| | Teenager (10.1%) | Teenager (9.4%) | Teenager (6.2%) | Teenager (13%) | Teenager (5.6%) | Teenager (12.2%) | | | | Teenager (6.7%) |
| Classification number | 800 (49.5%) | 800 (98.6%) | 900 (22.6%) | 800 (98.3%) | 900 (14.8%) | 800 (20%) | 800 (72.9%) | 900 (18.9%) | 800 (99%) | 800 (90.3%) |
| | 900 (10.8%) | | 400 (19.2%) | | 400 (23.3%) | 900 (27.7%) | 400 (13.8%) | 400 (16.1%) | | |
| | 400 (10.6%) | | 500 (15.5%) | | 500 (13%) | 400 (12%) | | 500 (20%) | | |
| | 500 (7.6%) | | 300 (11.8%) | | 300 (11.2%) | 500 (8.3%) | | 300 (14%) | | |
| | | | | | 700 (9.8%) | 300 (6.8%) | | 700 (8.3%) | | |
| Location | 26 (100%) | 15 (100%) | 15 (100%) | 3 (100%) | 3 (100%) | 9 (100%) | 25 (100%) | 25 (100%) | 9 (100%) | 25 (100%) |

# 5 Conclusion

In this study, we focus on data mining applied to the case public library whose two-years circulation data has uncovered the existence of the Pareto Principle. For 25% vital few readers and 75% useful many collections, we employ clustering analysis and association rule techniques to explore the unidentified valuable information patterns. The results of this research can be provided for library managers as references for marketing and management. For example, the vital readers of Location 15, young or middle-aged people, who borrowed books of Oceania literature (Classification Number 887) and German literature (Classification Number 875), have a 98% chance to borrow American literature (Classification Number 874). Such information is much useful for library readers' target marketing. For the useful many collections, there about 284338 books whose Classification Number are 850, are Chinese books. Also, there about 193818 books whose Classification Number are 850, are possible Children books in 99.7% probability. The information about books borrowed by vital readers can be provided for library managers to help in the decision-making of library collection development.

Another problem that libraries have always been being the use of collections; new books or popular books have a high frequency of borrowing, but other collections? In the case of limited library space, how do you decide which books to deselection? Does data mining technology help solve this problem? In the future, there are still many things to do in the application of data mining in library topics!

# References

1. Nicholson, S.: The basis for bibliomining: frameworks for bringing together usage-based data mining and bibliometrics through data warehousing in digital library services. Inf. Process. Manag. **42**(3), 785–804 (2006)
2. Shieh, J.C.: Bibliomining. Mandarin Library & Information Service, Taipei (2009)
3. Xiang, Z., Hao, Z.: Personalized requirements oriented data mining and implementation for college libraries. Comput. Modell. New Technol. **18**(2B), 293–300 (2014)
4. Hajek, P., Stejskal, J.: Library usage mining in the context of alternative costs: the case of the municipal library of Prague. Libr. Hi Tech **35**(4), 565–583 (2017)
5. Han, J., Kamber, M., Pei, J.: Data mining: concepts and techniques. Morgan Kaufmann, Waltham (2011)
6. Bajpai, J., Metkewar, P.S.: Data quality issues and current approaches to data cleaning process in data warehousing. Glob. Res. Dev. J. Eng. **1**(10), 14–18 (2016)
7. Siguenza-Guzman, L., Saquicela, V., Avila-Ordóñez, E., Vandewalle, J., Cattrysse, D.: Literature review of data mining applications in academic libraries. J. Acad. Libr. **41**(4), 499–510 (2015)
8. Al-Daihani, S.M., Abrahams, A.: A text mining analysis of academic libraries' tweets. J. Acad. Libr. **42**(2), 135–143 (2016)
9. Wu, F., Hu, Y.H., Wang, P.R.: Developing a novel recommender network-based ranking mechanism for library book acquisition. Electron. Libr. **35**(1), 50–68 (2017)
10. Renaud, J., Britton, S., Wang, D., Ogihara, M.: Mining library and university data to understand library use patterns. Electron. Libr. **33**(3), 355–372 (2015)
11. Goodall, D., Pattern, D.: Academic library non/low use and undergraduate student achievement: a preliminary report of research in progress. Libr. Manag. **32**(3), 159–170 (2011)
12. Yang, Y.T., Shieh, J.C.: Target marketing public libraries' vital readers: before. In: Wilimowska, Z., Borzemski, L., Świątek, J. (eds.) Information Systems Architecture and Technology: Proceedings of 39th International Conference on Information Systems Architecture and Technology – ISAT 2018. ISAT 2018. Advances in Intelligent Systems and Computing, vol. 854. Springer, Cham (2019)
13. Koch, R.: The 80/20 Principle: The Secret of Achieving More with Less. Currency Doubleday, New York (2011)
14. Mesbahi, M.R., Rahmani, A.M., Hosseinzadeh, M.: Highly reliable architecture using the 80/20 rule in cloud computing data centers. Fut. Gener. Comput. Syst. **77**, 77–86 (2017)
15. Trueswell, R.L.: Some behavioral patterns of library readers: the 80/20 rule. Wilson Libr. Bull. **43**(5), 458–461 (1969)
16. Nash, J.L.: Richard trueswell's contribution to collection evaluation and management: a review. Evid. Based Libr. Inf. Pract. **11**(3), 118–124 (2016)
17. Hardesty, L.: Use of library materials at a small liberal arts college. Libr. Res. **3**(3), 261–282 (1981)
18. Singson, M., Hangsing, P.: Implication of 80/20 rule in electronic journal usage of UGC-infonet consortia. J. Acad. Libr. **41**(2), 207–219 (2015)
19. Burrell, Q.L.: The 80/20 rule: library lore or statistical law? J. Doc. **41**(1), 24–39 (1985)
20. Nisonger, T.E.: The "80/20 rule" and core journals. Serials Libr. **55**(1–2), 62–84 (2008)
21. Schneier, B.: Applied Cryptography: Protocols, Algorithms and Source Code in C, 20th Anniversary edn. Wiley, New York (2015)

# Information Systems Architecture and Governance of Enterprise Information Technology at a Regulatory Institution

Leila Goosen[1]([⊠])● and Hlole A. Soga[2]

[1] University of South Africa, Pretoria 0003, South Africa
GooseL@unisa.ac.za
[2] University of South Africa, Johannesburg 1710, South Africa

**Abstract.** In contexts where enterprises do not recognize associated value, problems remain of ensuring that Information Technology (IT) performs the role of reaching enterprise strategies. Literature agree that the problems IT managers face include partnering with enterprises to meet their needs by providing quality IT services, resulting in calls to focus on Governance of Enterprise IT (GEIT). The IT department at a regulatory institution faced problems, including growth and reliance on IT. The institution had not implemented GEIT, failed audits and faced risks, as infrastructure was outdated, service delivery poor, and customers were complaining that IT was not adding value. The methodology included semi-structured interviews with executive committee members and survey questionnaires completed by heads of departments and IT staff. The research established whether introducing GEIT at the institution assisted the IT department in solving problems and improving their way of doing business. The value of the research lies in providing guidance to other IT departments on implementing GEIT to improve the way they operate and showing how implementation increased value.

**Keywords:** Computer systems security ·
Modeling of financial and investment managerial decisions ·
Risk assessment and management

## 1 Introduction

This paper is set against the background of enterprises becoming increasingly critically dependent on Information Technology (IT). Organizations "around the world have been investing in" information systems architecture and technology towards improving "their operational efficiency, the effectiveness of their" modeling of financial and investment managerial decisions, "and their strategic positions in the market over the last few decades" [1, p. 2]. In an article on translating enterprise goals into supporting IT goals in the financial sector, Van Grembergen, Van Brempt and Haes [2] pointed out that even in contexts where many of these enterprises do recognize the associated value, the problem remains to assure that IT effectively performs this essential role in order to reach these enterprises' strategies. Cartlidge, et al. [3] added that the problems that IT managers are continually being faced with include coordinating and partnering

© Springer Nature Switzerland AG 2020
Z. Wilimowska et al. (Eds.): ISAT 2019, AISC 1052, pp. 141–150, 2020.
https://doi.org/10.1007/978-3-030-30443-0_13

with the enterprise in order to meet enterprise needs by providing high quality IT services. These problems resulted in calls for a focus on the Governance of Enterprise IT (GEIT).

Similar to the scenarios as described by Doughty and Grieco [4], the problems that the IT department from the particular regulatory institution focused on in this paper was confronted with included that not only had they expanded, but the level of dependency that they had on IT had also increased significantly over recent years. The institution had not previously implemented any form of governance of enterprise IT, and in the absence of any such structures, it failed audits and faced significantly increased financial and operational risks. Infrastructure was outdated, and customers were always complaining about poor service delivery decisions and investments relating to whether IT should be improved or replaced being made, as IT plans were not aligned with the institution's strategies [4]. It is important that these problems be solved, and this paper will show how the institution is going about in working towards a solution for addressing these problems.

The research process reported on in this paper was thus firstly carried out in order to develop a credible semblance of the reality as it existed at this regulatory institution. The methodological approach that was used in doing the research for the paper included semi-structured interviews with members of the executive committee, together with survey questionnaires completed by heads of departments and staff members from the IT department. All three of the research tools used contained similar research questions, with this paper specifically focusing on the following ones:

- Which IT-related issues had the regulatory institution experienced in the past 12 months, with regard to e.g. IT security or privacy incidents?
- Which major IT-related initiatives were being planned for the next 12 months, including ones related to data or information?
- What were the drivers of GEIT activities at this regulatory institution?
- What are the outcomes of GEIT at this regulatory institution, including e.g. those related to improved management of GEIT-related risks?
- What is the state of awareness and uptake of IT-related certifications at the regulatory institution, including e.g. for Certified Information Security Managers?

In terms of IT strategies being aligned to business strategies, according to Wiedenhoft, Luciano and Magnagnagno [5], strategic alignment seeks to ensure the link between business and IT plans by defining, maintaining and validating the IT value proposition, thereby aligning IT operations with the organizational operations.

With regard to challenges faced by the ICT Department, according to Ivanoski [6], the use of regulatory technology there is a significant strategic opportunity for organizations to better connect their IT and automation strategy with solid compliance and risk management requirements and controls.

## 2   Literature Review

"Although there is no single, universal definition of" Information and Communication Technology (ICT), "the term is generally accepted to mean all devices, networking components, applications and systems that combined allow people and" organizations "(i.e., businesses, non-profit agencies, governments and criminal enterprises) to interact in the digital world" [7, p. 1].

"Governance concerns the structure, functions, processes, and" organizational "traditions that have been put in place within the context of a program's" authorizing "environment to ensure that the [program] is run in such a way that it achieves its objective and transparent manner" [8, p. 1].

There are three kinds of governance, which should be considered in corporate environments: corporate governance, ICT governance and Information Security (IS) governance". The paper by Kumsuprom, Corbitt and Pittayachawan [9, p. 514] focused on "the available standards and other frameworks by focusing on ICT governance, Information Security" governance and IS management. "IS governance specifically is used to align with the ICT governance framework as an integrated strategy in order to achieve effective corporate governance". IS "governance focuses on the leadership, organizational structures, and processes in order to help the organization provide superior relevant processes to safeguard information … . Significantly, its benefits lead to "(a) increased predictability and reduced uncertainty of business operation by lowering information security-related risk to a definable and acceptable level, (b) assurance of effective information security policy and policy compliance, and (c) a firm foundation for efficient and effective risk management, process improvement, and rapid incident response related to securing information".

"Due to a direct link between corporate … and IT governance, many corporate governance mechanisms are translated into the IT governance domain" [10, p. 2].

Caporarello [11, p. 3] indicated that the objective of IT governance "is to define structures, processes, and mechanisms", as well as "decision making rights and responsibility about main IT issues, to control and monitor the effectiveness of such decisions, and to mitigate IT-related risks".

The glossary of terms in the King III Code of Principles [12, p. 53] defined information security as "the protection of information from a wide range of threats in order to ensure business continuity," minimize business risk and maximize "return on investments and business opportunities".

According to Tshinu, Botha and Herselman [13, p. 40], "the use of different best practices presented in frameworks such as" the Information Technology Infrastructure Library (ITIL) … and International Organization for Standardization (ISO) 17799." The IT Governance Institute mentioned "that as all frameworks complement each other, users can benefit from using … with more detailed standards such as ITIL for service delivery … and ISO 17799 for information security."

## 3   Research Methodology

The research process reported on in this paper was firstly carried out in order to develop a credible semblance of the reality as it existed at this regulatory institution. The methodological approach that was used in doing the research for the paper included semi-structured interviews with members from the executive committee, together with survey questionnaires completed by heads of departments and staff members from the

IT department. All three of the research tools used contained similar research questions, with this paper specifically focusing on the objectives as detailed in the introduction.

The research was also used to establish whether introducing GEIT at this regulatory institution assisted the IT department in solving the problems as described in the introductory section of the paper and improving the way in which they do business.

### 3.1    Research Design

A mixed methods study was decided upon, with a triangulation design being followed, combining both qualitative and quantitative modes of inquiry or approaches to research for collecting data "at about the same time" [14, p. 25].

### 3.2    Quantitative Research

Given the need for "interaction with practitioners in the" IT department, and those in "the management field, and the possibility of" obtaining first-hand information on what the perceptions of different" Heads of Department (HODs) were in relation to GEIT, a mainly quantitative "research method was selected for" this study [13, p. 43].

According to McMillan and Schumacher [14, p. 21], quantitative research designs emphasize objectivity to identify, measure and describe phenomenal characteristics.

Within the sub-classification of quantitative research, this non-experimental research used a descriptive design to provide "a summary of an existing phenomenon by using numbers to characterize" it, in a form of a survey questionnaire [14, p. 22].

"In a survey research design," investigators select a sample to acquire information about participants' opinions, attitudes, beliefs and information, by asking them certain questions [14, p. 22]. "A sample can be defined as a group of (a) relatively smaller number of people selected from a population for investigation" purposes [15, p. 11].

Permission for this research was received from the Chief Information Officer (CIO).

### 3.3    Qualitative Research

An interactive qualitative research design was also used, in the form of a phenomenological study [14, p. 23], which attempted to describe participants' perceptions, perspectives and understandings. As also described by Kumsuprom, et al. [9, p. 516], data were collected in the form of semi-structured interviews with members of the Executive Committee (Exco). A "digital voice recorder was used with" participants' prior consent, in order to ensure accurate transcription. During these interviews, structured questions were used. Short notes were also taken. With regard to "the advantage of using semi-structured interviews", Tshinu, Botha and Herselman [13, p. 44] indicated that "the interviewer is allowed to use probes with a view to clearing up" any misunderstandings that could occur, as well as "vague responses or to ask for elaboration of incomplete answers." Respondents may be encouraged "to proceed. It provides the opportunity to discuss with practitioners face-to-face, with the possibility of … reordering questions to facilitate the flow of" the discussion.

## 3.4   Data Collection Tools

Regarding data collection, the technique selected for primary data collection from Exco members "to discuss the research topic" was semi-structured interviews [13, p. 44].

The questionnaires were broadly based on the Global Status Report on the Governance of Enterprise IT [16] - these were distributed to 22 HODs at the regulatory institution, as well as to all IT department employees, as experts in IT at the organization.

In the structured "questions (also called limited response" questions) used, participants were provided with a suitable list of options [14, p. 206].

# 4   Discussion of Results

## 4.1   Demographics of Respondents

Of the eight Exco members, six agreed to be interviewed as part of this research project, for a response rate of 75%. The Chief Information Officer is the main IT decision maker or Head of IT working at an operational level, with the remaining Exco participants being non-IT-related executives working at a strategic level, such as the Chief Executive Officer, Chief Financial Officer, Chief Operations Officer or Managing Director. Eighty-three percent indicated their main area of responsibility as general management, while 33% specified other regulatory functions.

For the HODs that the questionnaire was emailed to, eight responded, for a 36% response. Of these HODs, the main area of responsibility was operations (37%), while 63% indicated other, specifying regulatory functions, senior management and risk.

There are 30 staff members in the IT department, of which 18 responded to the questionnaire - this represents a 60% participation rate. Only 28% of these staff members indicated that they formed part of the senior management team.

## 4.2   Results

Problems "with external IT service providers" [16, p. 18] were the highest-scoring IT-related issues experienced by IT respondents in the past twelve months (see Table 1), while insufficient IT skills and an insufficient number of IT staff were jointly second highest "at the top of the list for" IT respondents [16, p. 19]. Other prevalent issues experienced by HODs in the past twelve months also included insufficient IT skills and an insufficient number of IT staff, together with problems with regard to the implementation of new IT systems being the joint highest-scoring IT-related issues.

In terms of cross-analysis, although increasing IT costs and IT disaster recovery or business continuity issues were most frequently indicated by Exco members, other participants rated these significantly lower. Other issues that generated different responses were "the insufficient number of IT staff, which was mentioned by" 50% of Exco and HODs and 56% of IT respondents [16, p. 19].

The most prominent IT-related initiatives planned for the next twelve months included major IT system implementations or upgrades for all three groups of

**Table 1.** IT-related issues experienced in the past 12 months.

|  | EXCO | HODs | IT |
|---|---|---|---|
| Insufficient IT skills | 67% | **50%** | 56% |
| Problems with external IT service providers | 67% | 38% | **61%** |
| Insufficient number of IT staff | 50% | **50%** | 56% |
| Problems implementing new IT systems | 50% | **50%** | 39% |
| Increasing IT costs | **83%** | 25% | 33% |
| IT disaster recovery or business continuity issues | **83%** | 38% | 22% |
| Serious operational IT incidents | 33% | 38% | 33% |
| IT security or privacy incidents | 33% | 13% | 11% |
| Return on IT investment not as expected | 0% | 0% | 6% |
| None of the above | 0% | 0% | 6% |

participants (see Table 2). In terms of cross-analysis, the views of Exco members, HODs and IT respondents were quite similar, with all of them including "major IT system implementations or upgrades" "and major IT infrastructure initiatives" (50% for Exco members and HODs, vs. 67% for IT respondents) [16, p. 18].

**Table 2.** Major IT-related initiatives planned for next 12 months.

|  | EXCO | HODs | IT |
|---|---|---|---|
| **Major IT system implementations** or upgrades | **83%** | **75%** | **94%** |
| Major IT infrastructure initiatives | 50% | 50% | 67% |
| IT-supported regulatory compliance initiatives | 67% | **75%** | 50% |
| IT risk management initiatives | 50% | **75%** | 50% |
| Data or information initiatives | **83%** | 38% | 28% |
| Outsourcing IT services | 17% | 0% | 67% |
| IT cost reduction initiatives | 17% | 25% | 44% |
| Green IT/sustainability initiatives | **83%** | 13% | 6% |
| Changing internal IT costing arrangements such as implementing chargeback of IT costs to departments | 0% | 13% | 22% |
| None of the above | 0% | 0% | 6% |

Ensuring that current IT functionality is aligned with current business needs was the most important driver of this institution's GEIT activities, while the least frequently mentioned driver was managing costs (Table 3). In terms of cross-analysis, the views across Exco members, HODs and IT department staff were fairly consistent for the two items indicated most frequently. In light of the glossary of terms in the King III Code of Principles [12, p. 53] defining information security as "the protection of information from a wide range of threats in order to" minimize business *risk* and maximize *return* on investments, the fact that so few respondents indicated achieving better balance between innovation and *risk* avoidance to improve *return* as a driver of IT governance activities is of some concern.

**Table 3.** Drivers of IT governance activities.

| | EXCO | HODs | IT |
|---|---|---|---|
| Ensuring that current IT functionality is aligned with current business needs | 67% | 50% | 83% |
| Complying with industry and/or governmental regulations | 33% | 13% | 22% |
| Increasing agility to support future changes in the business | 0% | 13% | 17% |
| Achieving better balance between innovation and risk avoidance to improve return | 0% | 13% | 11% |
| Avoiding negative incidents | 0% | 13% | 11% |
| Managing costs | 0% | 0% | 6% |

The most commonly experienced outcome of GEIT practices is improved communication and relationships between business and IT, followed closely by improvements in the management of IT-related risk (Table 4). Enhanced transparency of IT and its activities, tracking and monitoring of IT performance and IT delivery of business objectives were also experienced. Improved return on IT investments and IT innovation, together with lower IT costs, were seldom selected. The outcomes identified revealed both "more intangible and longer-term benefits, such as improved" communication and relationships between business and IT, management of IT-related risk and transparency of IT and its activities, as well as "tangible and shorter-term aspects, such as lower IT costs," and improved return on IT investments and IT innovation - however, the former grouping seems to be favored [16, p. 32]. Please note that although an item regarding improved business competitiveness had also been available, none of the respondents selected this.

**Table 4.** Outcomes of IT governance.

| | EXCO | HODs | IT |
|---|---|---|---|
| Improved communication and relationships between business and IT | **50%** | **50%** | **44%** |
| Improved management of IT-related risk | **50%** | 38% | **44%** |
| Improved transparency of IT and its activities | 17% | 25% | 39% |
| Improved tracking and monitoring of IT performance | 33% | 13% | 33% |
| Improved IT delivery of business objectives | 17% | 25% | 22% |
| Improved return on IT investments | 0% | 13% | 11% |
| Lower IT costs | 0% | 13% | 0% |
| Improved IT innovation | 0% | 0% | 6% |

In terms of awareness and the uptake of IT-related certifications at this institution, it is clear from Table 5 that the overwhelming majority of employees in the IT department have completed the ITIL Foundations certification. It also seems possible that almost a quarter of employees have the follow-up ITIL Service Manager certification, with the PRINCE2 Foundations certification being seemingly just about equally popular - however, that particular follow-up certification, PRINCE2 Practitioner, lies significantly lower on the table. Fifteen percent of respondents indicated that a significant number and/or at least some of this IT department's employees have the Project Management Professionals or Certified Associate in Project Management certifications respectively.

**Table 5.** Awareness and uptake of IT-related certifications.

| | A significant number of our IT employees have this certification | Some of our IT employees have this certification | Aware but no one in our organization is certified | Not aware of the certification |
|---|---|---|---|---|
| ITIL Foundations | 78% | 11% | 0% | 11% |
| ITIL Service Manager | 6% | 33% | 22% | 11% |
| PRINCE2 Foundations | 0% | 44% | 11% | 11% |
| Project Management Professionals (PMP) | 6% | 17% | 28% | 17% |
| Certified Associate in Project Management | 0% | 28% | 11% | 28% |
| PRINCE2 Practitioner | 0% | 17% | 22% | 22% |
| Certified Information Systems Security Professionals | 0% | 17% | 17% | 28% |
| Certified Information Systems Auditor (CISA) | 0% | 6% | 22% | 28% |
| Certified Information Security Manager (CISM) | 0% | 6% | 17% | 28% |
| Certified in the Governance of Enterprise IT | 0% | 6% | 17% | 39% |

For the other certifications mentioned, Certified Information Systems Security Professionals, Information Systems Auditors, Information Security Managers and those in the Governance of Enterprise IT, much lower numbers in terms of awareness and uptake are reported, with increasingly larger numbers of respondents not being aware of these certifications at all.

Admittedly, all of this work in Sect. 4.2 on a relatively small number of people being involved in the quantitative surveys. In this type of work, however, qualitative research could raise some doubts, and as a rule, such studies are often used in the initial stages of research to build assumptions and a basic plan of the research, due to the difficulty in assessing the reliability of the information obtained. This study could be continued, with repetition after some time allowing for the verification of doubts.

# 5   Conclusion

The purpose of the research reported on in this paper was firstly to develop a credible semblance of the reality as it existed at this regulatory institution, as well as to establish whether introducing GEIT at the institution assisted the IT department in solving the problems as described in the introductory sections of the paper and improving the way in which they do business. In this section, a short summary of the results as reported in this paper will be provided in answer to the research questions:

- With regard to IT-related issues that the regulatory institution had experienced in the past 12 months, insufficient IT skills "and problems with external IT service providers" topped the list [16, p. 18], while IT security or privacy incidents were only indicated by five participants.
- The most prominent IT-related initiatives planned by all three groups of participants in the next twelve months relate to major IT system implementations or upgrades. Although data or information initiatives were also most frequently indicated by Exco members, this item placed around the middle of these particular results.
- Ensuring that current IT functionality is aligned with current business needs was the most important driver of this institution's GEIT activities, while the least frequently mentioned driver was managing costs.
- The most commonly experienced outcome of GEIT practices is improved communication and relationships between business and IT, followed closely by improvements in the management of IT-related risk.
- In terms of awareness and the uptake of IT-related certifications at this institution, an overwhelming three-quarters of employees in the IT department have completed the ITIL Foundations certification. Other certifications, such as e.g. Certified Information Security Managers, however, were rated significantly lower.

The information in this paper not only holds benefit for the reader and the community in providing guidance to other IT departments (at regulatory institutions) on implementing GEIT, in order to improve the way that they operate daily. An additional benefit of having done this research could lie in showing how implementing these same procedures could possibly lead to increased value being added. Firms would be able to benefit from using insights from Chen, Wang, Nevo, Benitez and Kou [17], who showed that firms that use IT to support core competencies experience improved strategic flexibility, possibly enhancing performance.

# References

1. Matta, M., Cavusoglu, H., Benbasat, I.: Understanding the board's involvement in information technology governance: theory, review and research agenda (2016). https://ssrn.com/abstract=2778811. Accessed 06 June 2017
2. Van Grembergen, W., Van Brempt, H., De Haes, S.: Translating business goals into supporting IT goals in the financial sector. Review of Business and Economics **53**(1), 56–68 (2008)

3. Cartlidge, A., Hanna, H., Rudd, C., Macfarlane, I., Windebank, J., Rance, S.: An introductory overview of ITIL V3. In: Cartlidge, A., Lillycrop, M. (eds.) IT Service Management Forum, London (2007)

4. Doughty, K., Grieco, F.: IT governance: pass or fail? Inf. Syst. Control J. **2** (2005)

5. Wiedenhoft, G.C., Luciano, E.M., Magnagnagno, O.A.: Information technology governance in public organisations: identifying mechanisms that meet its goals while respecting principles. J. Inf. Syst. Technol. Manag. **14**, 69–87 (2017)

6. Ivanoski, J.: Regulatory technology: innovating compliance and your business (2017). https://www.cio.com/article/3237147/regulation/regulatory-technology-innovating-compliance-and-your-business.html. Accessed 16 July 2018

7. Rouse, M.: Definition: ICT (information and communications technology, or technologies) (2017). http://searchcio.techtarget.com/definition/ICT-information-and-communications-technology-or-technologies. Accessed 13 June 2017

8. World Bank: Governance and Management (2016). http://www.cugh.org/sites/default/files/World%20Bank%20Governance%20and%20Management.pdf. Accessed 16 May 2016

9. Kumsuprom, S., Corbitt, B., Pittayachawan, S.: ICT risk management in organizations: case studies in Thai business. In: 19th Australasian Conference on Information Systems, Christchurch (2008)

10. Huygh, T., De Haes, S., Joshi, A., Van Grembergen, W., Gui, D.: Exploring the influence of Belgian and South-African corporate governance codes on IT governance transparency. In: Proceedings of the 50th Hawaii International Conference on System Sciences (2017)

11. Caporarello, L.: IT Governance: A Framework Proposal and an Empirical Study. LUISS University, Rome (2008)

12. King Committee on Governance: King III Code of Governance Principles, Institute of Directors in Southern Africa (2009)

13. Tshinu, S.M., Botha, G., Herselman, M.: An integrated ICT management framework for commercial banking organisations in South Africa. Interdiscip. J. Inf. Knowl. Manag. **3**, 39–53 (2008)

14. McMillan, J.H., Schumacher, S.: Research in Education: Evidence-Based Inquiry. Pearson, Boston (2010)

15. Alvi, M.: A manual for selecting sampling techniques in research. munich personal RePEc Archive. University of Karachi (2016). https://mpra.ub.uni-muenchen.de/70218/1/MPRA_paper_70218.pdf. Accessed 23 July 2018

16. IT Governance Institute (ITGI): Global Status Report on the Governance of Enterprise IT. ITGI, Rolling Meadows, IL (2011)

17. Chen, Y., Wang, Y., Nevo, S., Benitez, S., Kou, G.: Improving strategic flexibility with information technologies: Insights for firm performance in an emerging economy. J. Inf. Technol. **32**(1), 10–25 (2017)

# Models of Production Management

# Comparison of Exact and Metaheuristic Approaches in Polynomial Algorithms for Scheduling of Processes

Mieczysław Drabowski[✉]

Department of Theoretical Electrical Engineering and Computing Science,
Cracow University of Technology, Warszawska 24, 31-115 Kraków, Poland
drabowski@pk.edu.pl

**Abstract.** The paper presents analysis of results obtained by implemented meta-heuristic algorithms for processes scheduling, which were discussed and detailed presented within series proceedings on 37th, 38th and 39th International Conferences on Information Systems Architecture and Technology ISAT, respectively. This analysis is based primarily of comparisons of the results obtained by these algorithms with the results obtained by with polynomial algorithms, known with subject literature.

**Keywords:** Complex system of processes · Scheduling · Allocation ·
Optimization · NP-complete problems · Algorithms: simulated annealing ·
Genetic · Ant Colony Optimization · Greedy algorithm ·
Exact, polynomial algorithm

## 1  Introduction. Model of Scheduling of Processes in Embedded Systems

Since 2016, I had the honor to present problems related to the design of embedded systems on a high level of abstraction. All these researches were presented in conference proceedings of the International Conferences on Information Systems Architecture and Technology organized by Wrocław University of Science and Technology. These proceedings were published by Springer's "Advances in Intelligent Systems and Computing" series in years 2017–2019.

The new system model was presented, which was based on a modified computer system model for tasks scheduling [1]. In this model, among others, the following were changed: goals of optimization, in addition to the best scheduling of tasks, also the best resource configurations taking into account, along with the time optimization criteria, also criterions of the cost and power consumption and degree of dependability [2, 3].

The objective of computer aided design of complex embedded computer systems (i.e. that contain sets of processors and additional resources, that perform great number of programs) is to find the optimal solution, in accordance with the requirements and constraints imposed by of the stated specification of the system. The specification is most often given in the form of sets of dependent processes. The speed of the designed embedded system depends primarily on how the processes are scheduled, the length of

© Springer Nature Switzerland AG 2020
Z. Wilimowska et al. (Eds.): ISAT 2019, AISC 1052, pp. 153–163, 2020.
https://doi.org/10.1007/978-3-030-30443-0_14

this schedule and the allocation of processes to resources. So, therefore, processes scheduling algorithms for the criterion of speed optimization (i.e. of time of execution) are of great importance when concerns automatically aided embedded system design.

The following meta-heuristic approaches were examined in 2016-2018, which suboptimally solved the problems of scheduling of processes and allocation of processes and resources. Approaches based the Ant Colony Optimization and Branch & Bound algorithm [4] and hybrid approach i.e. genetic with simulated annealing with Boltzmann tournaments [5] were presented in [1, 2], respectively.

Further research should take into account the following questions: what meta-heuristics are needed for calculations in the design of these systems? Which ones are the best and which ones give weaker solutions? Do the results obtained in the results of the implementation of these heuristics give exact and optimal results for polynomial scheduling problems? What is the distance between suboptimal and optimal solutions? Can this distance be estimated? Does obtaining good (optimal or suboptimal) results for polynomial problems guarantee for the same algorithms achieving equally good results for NP-complete problems?

Attempts to answer some of these questions are presented in this paper.

## 2   Results of Comparison Computations

### 2.1   The Comparison with Exact Polynomial Algorithms

To show convergence of heuristic algorithms towards optimum, one can compare their results with optimal results of existing, exact, polynomial algorithms for certain exemplary problems of processes scheduling. If a heuristic algorithm finds an optimal solution to polynomial problems, it is very probable that solutions found for NP-complete problems will also be optimal or at least approximated to optimal [6]. Heuristic algorithms described herein were tested with known polynomial algorithms and all of them achieved optimal solutions for those problems. The comparisons utilized polynomial exact algorithms: Coffman – Graham Algorithm [7] for identical, parallel processors and Baer Algorithm [8] for uniform parallel processors. The schedule length (makespan) is the optimality criterion for these algorithms.

Comparisons of metaheuristic algorithms (presented in previous papers: ACO algorithm and hybrid algorithm - genetic with simulated annealing) solutions with selected exact polynomial algorithms will be presented in following examples.

### Comparison with results of Coffman and Graham Algorithm
Scheduling of dependent processes which constitute a digraph with unit executions times (UET processes) on two identical processors in order to minimize schedule length – i.e. $C_{max}$. Computation complexity of the algorithm is $O(n^2)$.

*Test problem 1:*
2 identical processors (in example 1a), 3 identical processor (in example 2a), 15 processes with unit executions times, digraph with processes – Fig. 1.

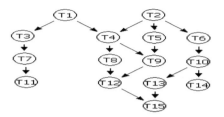

**Fig. 1.** Digraph of processes used for the comparison of metaheuristic algorithms with Coffman and Graham algorithm (test problem 1)

Solutions of optimal scheduling for two processors obtained as a result of Coffman and Graham algorithm use (example 1a) is show in Gantt chart, Fig. 2. The schedule with the same length was found by meta-heuristic algorithms, that were tested, also Fig. 2.

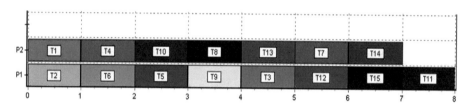

**Fig. 2.** Optimal schedule, for two processors, obtained Coffman and Graham algorithm and also meta-heuristic algorithms (example 1a)

Solutions of no-optimal scheduling for three processors obtained as a result of Coffman and Graham algorithm use (example 1b) is show in Gantt chart Fig. 3. The Coffman-Graham algorithm finds an optimal solution for only two processors. Solutions of optimal scheduling for three processors obtained as a result of metaheuristic algorithm use (example 1b) is show in Gantt chart Fig. 4.

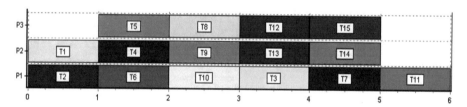

**Fig. 3.** Scheduling for 3 processors by Coffman and Graham algorithm (example 1b)

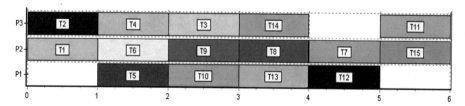

**Fig. 4.** Optimal scheduling for 3 processors by metaheuristic algorithms (example 1b)

For two processors (example 1a) metaheuristic algorithms similarly to Coffman and Graham algorithm obtained optimal schedule. It was the same in the case of three processors (example 1b) – both algorithms obtained the same scheduling. Coffman and Graham algorithm is optimal only for two identical processors. For task graph under research it also found optimal scheduling for 3 identical processors i.e. minimum schedule length.

Another test problem is shown by the non-optimality of Coffman and Graham algorithm for processor number greater than 2.

*Test Problem 2:*

2 identical processors (in example 2a), 3 identical processors (in example 2b), 12 processes with unit executions times, digraph of processes is show Fig. 5.

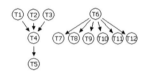

**Fig. 5.** Graph of processes used for the comparison of metaheuristic algorithms with Coffman and Graham algorithm (test problem 2)

Solutions of optimal scheduling for two processors obtained as a result of Coffman and Graham algorithm use (example 2a) is show in Gantt chart Fig. 6. The schedule with the same length was found by meta-heuristic algorithms, that were tested, also Fig. 6. Solutions of no-optimal scheduling for three processors obtained as a result of Coffman and Graham algorithm use (example 2b) is show in Gantt chart Fig. 7. Solutions of optimal scheduling for three processors obtained as a result of meta-heuristic algorithms use (example 2b) is show in Gantt chart Fig. 8.

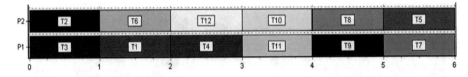

**Fig. 6.** Optimal schedule for 2 processors obtained Coffman and Graham algorithm and meta-heuristic algorithms (example 2a)

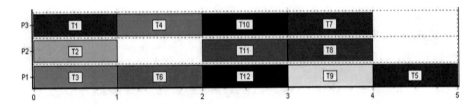

**Fig. 7.** Non-optimal schedule for 3 processors in Coffman - Graham algorithm (example 2b)

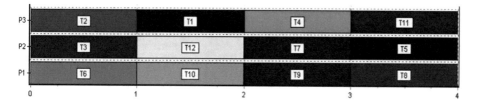

**Fig. 8.** Optimal schedule for 3 processors in metaheuristic algorithms (example 2b)

For the problem of two processors (example 2a) both algorithms obtained optimal schedule. In the case of three processors (example 2b) the Coffman and Graham algorithm did not find optimal schedule, whereas the heuristic algorithm did find it without any difficulty.

In another test example both algorithms were compared for the problem of processes scheduling on two identical processors with unit and different executions times. *Test Problem 3:*

2 identical processors, 5 processes with unit executions times (example 3a), 5 processes with different executions times (example 3b), digraph of processes Fig. 9.

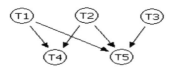

**Fig. 9.** Graph of processes, used for the comparison of metaheuristic algorithm with Coffman and Graham algorithm (test problem 3)

Solutions of optimal scheduling for processes with unit executions times obtained as a result of Coffman and Graham algorithm use (test problem 3a) is show in Gantt chart Fig. 10.

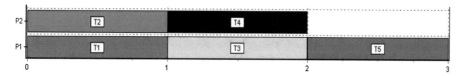

**Fig. 10.** Optimal schedule for processes with unit executions times by Coffman and Graham algorithm (example 3a)

Solutions of optimal schedule for processes with unit executions times obtained as a result of heuristic algorithm use (example 3a) is show in Gantt chart Fig. 11.

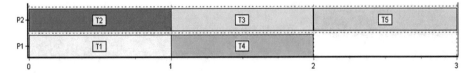

**Fig. 11.** Optimal problem schedule for processes with unit executions times, obtained by heuristic algorithm (example 3a)

Solutions of no-optimal scheduling for dependent processes with different executions times, obtained as a result of Coffman and Graham algorithm use (example 3b) is show in Gantt chart Fig. 12.

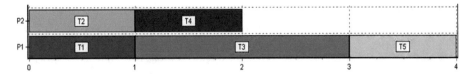

**Fig. 12.** Non-optimal schedule for dependent processes with different executions times, obtained by Coffman and Graham algorithm (example 3b)

Solutions of optimal schedule for dependent processes with different executions times, obtained as a result of heuristic algorithm use (example 3b) is show in Gantt chart Fig. 13.

**Fig. 13.** Optimal schedule for dependent processes with different executions times, obtained by metaheuristic algorithms (example 3b)

Both compared approaches (exact and metaheuristic) obtain optimal schedule for the problem with processes with unit executions times (example 3a). For processes with different executions times (example 3b) the Coffman and Graham algorithm does not obtain optimal schedule, whereas the metaheuristic approach does obtain.

**Comparison with results of Baer Algorithm**

Scheduling of non-preemptive processes with unit executions times, which create a digraph of anti-tree type on two uniform processors in order to minimize schedule length $C_{max}$. Computation complexity of the algorithm is $O(n^2)$.

*Test Problem:*

2 uniform processors with speed coefficients b1 = 2, b2 = 1 and 11 processes with unit executions times, digraph of dependent processes (anti-tree) in Fig. 14.

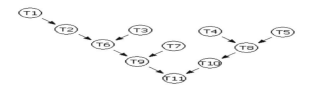

**Fig. 14.** Graph of processes used for the comparison of Baer and metaheuristic algorithms

Solutions on optimal scheduling for the problem solved with Baer algorithm and obtained as a result of metaheuristic algorithm use is show in Gantt chart Fig. 15.

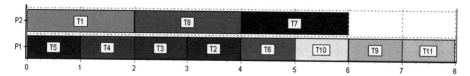

**Fig. 15.** Optimal scheduling for the problem solved with Baer algorithm and obtained as a result of metaheuristic algorithm use

For the problem optimized with Baer algorithm and the metaheuristic algorithm also obtains optimal solution. The makespan computed by these algorithms is the same.

## 2.2  Comparison of Suboptimal Algorithms for Non-polynomial Problems of Processes Scheduling

### NP-complete Scheduling Problem

Because in general case the problem of scheduling for non-preemptable processes is strong NP-complete, in some applications one can use polynomial approximate algorithms. For example such algorithms are list algorithms [7].

They are usually considered five types of list scheduling rules were compared: HLFET (Highest Levels First with Estimated Times), HLFNET (Highest Levels First with No Estimated Times), RANDOM, SCFET (Smallest Co-levels First with Estimated Times), SCFNET (Smallest Co-levels First with No Estimated Times).

The number of cases, in which the solution differs less than 5% from optimal solution, is accepted as an evaluation criterion for the priority allocation rule. If for 90% of examined examples the sub-optimal solution fit in the above range, the rule would be

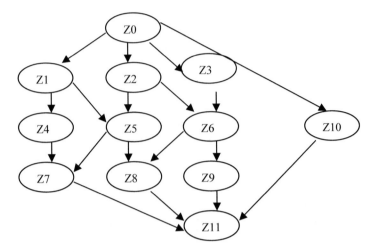

**Fig. 16.** The graph of tasks used for the comparison of metaheuristic and list algorithms

described as "almost optimal". This requirement is met only by HLFET rule, which gives results varying from optimum by 4,4% on average [7, 8].

Example:

2 identical processors, 12 processes with different performance times: (Z0,1), (Z1,1), (Z2,7), (Z3,3), (Z4,1), (Z5,1), (Z6,3), (Z7,2), (Z8,2), (Z9,1), (Z10,3), (Z11,1), digraph of processes is show in Fig. 16.

Scheduling obtained as a result of metaheuristic algorithm operation is show in Fig. 17.

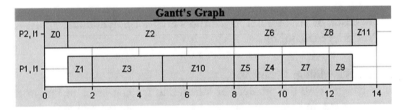

**Fig. 17.** Scheduling obtained with metaheuristic algorithm

The schedule length obtained by metaheuristic algorithms is compliant with the scheduling which was obtained by means of the best list scheduling available for this case and which is HLFET ("almost optimal").

## Comparison with PDF/HIS Algorithm for type STG Digraphs and for Strong NP-complete Problems

For research purposes a set of graphs was utilized from the website below: http://www. kasahara.elec.waseda.ac.jp/schedule/index.html. Graphs of processes made available therein were divided into groups because of the number of processes. Minimum scheduling length was calculated by means of PDF/HIS algorithm (i.e. *Parallelized Depth First/Implicit Heuristic Search*) for every processes graph [9].

STG graphs are vectored, a-cyclic digraphs of processes. Different process performance times, discretionary sequence constraints as well as random number of processors and there characteristic cause with STG scheduling problems to be strong NP-complete problems. Out of all solved problems heuristic algorithms under research did not find an optimal solution (assuming this is the solution obtained with PDF/IHS algorithm) only for three of them. However, results obtained are satisfactory, because the deviation from optimum varies from 0,36% to 4,63% (results is show in Table 1).

**Table 1.** Comparison ACO algorithm and algorithm of genetic with simulated annealing with PDF/IHS algorithm.

| STG | Number of processes | Number of processors | PDF/IHS $C_{max}$ | Ant colony $C_{max}$ | Number of iterations | Difference [%] | Genetic & Simulated annealing $C_{max}$ | Number of iteratons | Difference [%] |
|---|---|---|---|---|---|---|---|---|---|
| rand0008 | 50 | 2 | 281 | 281 | 117 | 0 | 281 | 80 | 0 |
| rand0107 | 50 | 8 | 155 | 155 | 389 | 0 | 155 | 411 | 0 |
| rand0174 | 50 | 16 | 131 | 131 | 180 | 0 | 131 | 190 | 0 |
| rand0017 | 100 | 2 | 569 | 569 | 171 | 0 | 569 | 92 | 0 |
| rand0066 | 100 | 4 | 253 | 253 | 4736 | 0 | 257 | 3644 | 1,58 |
| rand0106 | 100 | 8 | 205 | 205 | 861 | 0 | 205 | 927 | 0 |
| rand0174 | 100 | 16 | 162 | 162 | 265 | 0 | 162 | 216 | 0 |
| rand0020 | 300 | 2 | 827 | 846 | 5130 | 2,30 | 830 | 4840 | 0,36 |
| rand0095 | 300 | 8 | 382 | 394 | 5787 | 3,14 | 384 | 5253 | 0,52 |
| rand0136 | 300 | 16 | 324 | 339 | 2620 | 4,63 | 324 | 3067 | 0 |

Algorithms were also investigated by scheduling processes represented with the same graph (50 STG processes) on a different number of processors. The results are demonstrated in Table 2 (for *STG rand0008*) and Table 3 (for *STG rand0107*) [9].

**Table 2.** Minimization of $C_{max}$ of dependent processes (*STG rand0008.stg*)

| Number of processes | Number of processors | PDF/IHS $C_{max}$ | Ant colony $C_{max}$ | Number of iterations | Genetic $C_{max}$ | Number of iterations |
|---|---|---|---|---|---|---|
| 50 | 2 | 228 | 228 | 132 | 228 | 92 |
| 50 | 4 | 114 | 114 | 1401 | 114 | 925 |
| 50 | 8 | 57 | 61 | 4318 | 58 | 4442 |
| 50 | 16 | 48 | 48 | 58 | 48 | 33 |

**Table 3.** Minimization of $C_{max}$ of dependent processes (*STG rand0107.stg*)

| Number of processes | Number of processors | PDF/IHS $C_{max}$ | Ant colony $C_{max}$ | Number of iterations | Genetic $C_{max}$ | Number of iterations |
|---|---|---|---|---|---|---|
| 50 | 2 | 267 | 267 | 388 | 267 | 412 |
| 50 | 4 | 155 | 160 | 4487 | 160 | 3339 |
| 50 | 8 | 155 | 155 | 89 | 155 | 112 |
| 50 | 16 | 155 | 155 | 10 | 155 | 8 |

In all researched problems algorithms under comparison found optimal solution. The only difference can be observed in the number of iterations needed to find an optimal solution. Genetic algorithm needed less iterations than ACO one to find the solution.

## 3 Conclusions

Conducted research shows that presented metaheuristic algorithms for processes scheduling obtain good solutions, irrespectively of investigated problem complexity. These results are considered optimal or sub-optimal, for whose deviation from optimum does not exceed 5% [10]. Metaheuristic algorithms proposed for task scheduling problems, especially presented genetic with simulated annealing and ACO algorithm, should be so a good tool for computer aided synthesis of embedded systems, in particular in optimizing the speed of the system, because the speed of the system depends of the optimal scheduling of its processes, i.e. of the minimum length of the schedule [11].

These studies also indicate ways to assessing heuristic algorithms applied for various applications and therefore may have some universal importance.

## References

1. Drabowski, M.: Modification of concurrent design of hardware and software for embedded systems – a synergistic approach. In: Grzech, A., Świątek, J., Wilimowska, Z., Borzemski, L. (eds.) Information Systems Architecture and Technology: Proceedings of 37th International Conference on Information Systems Architecture and Technology – ISAT 2016, vol. 522, pp. 3–13. Springer, Heidelberg (2017)
2. Drabowski, M., Kiełkowicz, K.: A hybrid genetic algorithm for hardware–software synthesis of heterogeneous parallel embedded systems. In: Świątek, J., Borzemski, L., Wilimowska, Z. (eds.) Information Systems Architecture and Technology: Proceedings of 38th International Conference on Information Systems Architecture and Technology – ISAT 2017, vol. 656, pp. 331–343. Springer, Heidelberg (2018)
3. Drabowski, M.: Concurrent, coherent design of hardware and software embedded systems with higher degree of reliability and fault tolerant. In: Borzemski, L., Świątek, J., Wilimowska, Z. (eds.) Information Systems Architecture and Technology: Proceedings of

39th International Conference on Information Systems Architecture and Technology – ISAT 2018, vol. 852, pp. 7–18. Springer, Heidelberg (2019)

4. Montgomery, J., Fayad, C., Petrovic, S.: Solution representation for job shop scheduling problems in ant colony optimization. LNCS, vol. 4150, pp. 484–491 (2006)

5. Goldberg, D.E.: Genetic Algorithms in Search, Optimization and Machine Learning. Addison-Wesley, Reading (1989)

6. Błażewicz, J., Drabowski, M., Węglarz, J.: Scheduling multiprocessor tasks to minimize schedule length. IEEE Trans. Comput. **35**(5), 389–393 (1986)

7. Coffman Jr., E.G.: Computer and Job-Shop Scheduling Theory. Wiley, New York (1976)

8. Garey, M.R., Johnson, D.S.: Computers and Intractability: A Guide to the Theory of NP-Completeness. Freeman, San Francisco (1979)

9. Graphs STG. http://www.kasahara.elec.waseda.ac.jp/schedule/

10. Pricopi, M., Mitra, T.: Task scheduling on adaptive multi-core. IEEE Trans. Comput. **C-59**, 167–173 (2014)

11. Agraval, T.K., Sahu, A., Ghose, M., Sharma, R.: Scheduling chained multiprocessor tasks onto large multiprocessor system. Computing **99**(10), 1007–1028 (2017)

# Implementation of Scrum Retrospective in the Process of Improving Logistics Organization

Paweł Rola[(⊠)] and Dorota Kuchta

Faculty of Computer Science and Management, Wrocław University of Science and Technology, ul. Ignacego Łukasiewicza 5, 50-371 Wrocław, Poland
pawel.rola@pwr.edu.pl

**Abstract.** Agile approach, and especially Srum, has gained huge importance in project management and numerous organisations acknowledge its utility for improving their processes. However, agile approach has been mainly applied to IT organisations. The main goal of this paper is to consider possible advantages of its introduction into the transport sector. It has to be underlined that the decision and the process to implement agile approach in an organization are extremely difficult. The first step of agile implementation is conducting a trial implementation in accordance with the principles of one or more of agile methods or practices in a relatively small area. In the present paper the case of a forwarding company is discussed in which the agile practice called retrospective was implemented. The short term goal of this implementation was to support process of collaborative problem identification and solving, the result of which should be improving own internal process of forwarding agency. The implementation and the advantages identified so far are presented and analysed.

**Keywords:** Scrum · Retrospective · Logistics · Forwarding agencies · Case study

## 1 Introduction

The main aim of road transport is transporting passengers and goods by road. It is the most developed area of transport. The advantage of road transport is the direct transport from the sender to the recipient, faster, easier transportation resulting from technology, availability and operability [2]. Road freight transport in the EU 28 member countries is the main freight transport activity. Total goods transport activities in the EU-28 has reached 3524 billion t-km (not including transport between EU and rest of the world). Road transport accounted for 49% of this, with maritime transport (domestic and intra EU operations) as the second most important mode (31.8%), rail for 11.7%, inland waterways for 4.3%, oil pipelines for 3.2% and air transport just 0.1% [3] (Fig. 1).

The road transport system, as main freight transport activity, is based on national and international economic prosperity. It is a connection between locations of production and consumption for materials and goods. Modern, minimum-inventory supply chains, for an efficient execution of flow of goods must rely on road transport. A supply chain is defined as a process with a complete set of activities wherein raw materials are

© Springer Nature Switzerland AG 2020
Z. Wilimowska et al. (Eds.): ISAT 2019, AISC 1052, pp. 164–175, 2020.
https://doi.org/10.1007/978-3-030-30443-0_15

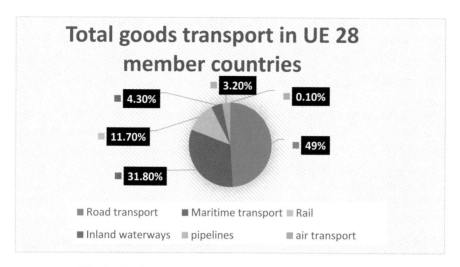

**Fig. 1.** Total goods transport in UE 28 members countries [3].

transformed into final products, then delivered to customers by distribution, logistics, and retail [4]. Efficiently organized transport and logistics has become a competitive advantage for organizations. The biggest part of logistics costs can be considered as costs related with transportation and inventory carrying [5]. For an even better performance of activities in the field of logistics and transport, organizations decide to implement these processes through outsourcing them to dedicated forwarding agent or agency.

Forwarding agent is a kind of coordinator and the manager of the transport process. The organization of the transport process is a process which requires various activities. The effect of these activities should be delivered with accepted quality of service and at the lowest possible cost level [6]. Company providing services in rail freight transport (forwarding agency) must have drawn up a very good strategy to stay on the transport market in the long term. This requires preparing high-quality strategy plans, their transformation to operational plans, monitoring actual processes and comparison of actual processes with the plan [7]. Many forwarding agencies are currently looking for the possibility of obtaining the same as before or even better results, thanks to the usage of less resources - especially time and money. One possibility to achieve this goal is to improve own internal process. Improving a process is understood as modifying it in a manner that allows to achieve economic effects in demand and adapt the organization to the changing business environment. Currently, it is considered that 70% to 80% of all costs of client's demand satisfaction are associated to administrative processes, so it becomes fundamental to recognize the importance of these administrative areas [8]. The article presents an attempt to use the Scrum method elements for improving a process in forwarding agency, in order to reflect on the possible usefulness of implementing the agile approach in other sectors than IT.

The rest of this paper is structured as follows. In Sect. 2, authors present the identification and solving process. In Sects. 3 and 4 respectively, the agile manifesto and Scrum framework are presented. Section 5 scope is Sprint Retrospective. Research methodology and documents are presented in Sect. 6. Section 7 contains organization description. In Sect. 8 retrospection framework used in study is presented. In Sect. 9 authors present a discussion of case study results. Conclusions can be found in Sect. 10.

## 2   Problems Identifying and Solving

Improving processes should be preceded by the identification of potential problems. Identification of problems in the field of management can be described as four stages approach [9]:

- searching for the most important problems;
- formulating problems;
- collecting information about the problems;
- investigating the causes of the problems.

This definition sets the problem of identification as an issue involving a diagnostic, description, registration and assessment of the significance of the identified problems. Problem recognition is important and much more difficult than the solution itself [10]. However, both identification and potential solution plan should be developed on the basis of collaboration of group of workers conducting internal process in the forwarding agency. Collaborative solving problem is considered as one of the core competencies of the 21st century [11]. Collaborative problem solving involves a complex process whereby two or more agents attempt to solve a problem by sharing the understanding and effort required to come to a solution and pooling their knowledge, skills and efforts to reach that solution [12]. Good collaboration implies balanced and equal participation in which knowledge is co-constructed and all members contribute different pieces of information or build upon each other's explanations to co-create a complete solution [13].

Agile methods have been widely used in software engineering over the last decade [14]. Due to their success, demand for the extension of agile principles to other domains has risen [15]. Research on agile software development, such as studies of particular practices, like pair programming, retrospection, daily scrum etc., might be described as being at an intermediate, or even a mature state [16]. Therefore, relatively little attention has been devoted to investigate the possibility of adopting agile techniques in areas not related to software development (the only identified examples describing positive effects of the implementation of Scrum or its parts in areas not related to software development are [46–49]).

In this context it is worthwhile to consider implementing agile techniques for group problems identifying and solving in area of improving certain internal processes in logistic companies, which is the main objective of the paper.

## 3   Agile Manifesto

Project management is a branch of management whose acceptance and recognition has continued to grow. It is widely used in logistics and production systems, but not in the agile version: in the logistics and production sector usually the so called traditional project management is applied [17]. Project management concepts evolve along with the acquisition of experience in the field of project implementation by enterprises. These changes contributed to the creation of agile project management approach, which indicated the main principles developed in the 2001 and contained in the Agility Manifesto [18]:

- People and interaction over processes and tools;
- Working software over comprehensive documentation;
- Customer collaboration over formal arrangements;
- Responding to change over following of the plan.

This means that items listed on the right hand side of the above phrases are valuable, but more valuable items are listed on the left hand side. One of the most frequently applied methodology from the Agile family is the Scrum approach (e.g., [19, 20] and many others [16, 21–23]).

## 4   Scrum Framework

The Scrum method is embedded in the theory of empirical process control. Empiricism is a theory where the origin of all knowledge is sense experience. Scrum uses iterative and incremental approaches to achieve better predictability and risk control [24].

Scrum assumes five events (Sprint Planning Meeting, Daily Scrum, Sprint, Sprint Review, Sprint Retrospective). Scrum Team consists of three scrum roles (Scrum Master, Product Owner, Development Team) and of tree artefacts (Product Backlog, Sprint Backlog, Increment).

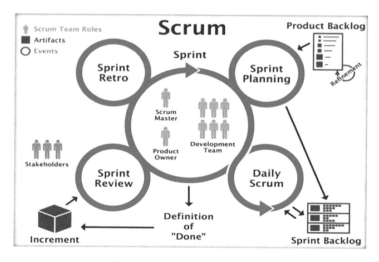

**Fig. 2.**  Scrum framework [25]

The Scrum framework is simple as a concept. Figure 2 shows the diagram of the Scrum framework. In this section, the authors present only a brief description of the Scrum, because the framework is well known and described in many publications ([23, 24, 26–28] and others). Product Owner is the person representing the knowledge and understanding of the needs of the business realized through product design. She or He is responsible for the Product Backlog Items list, called Product Backlog. Product Backlog has Backlog Items arranged in the order of significance that represents functional and non-functional requirements of the product.

During the Sprint Planning, the Development Team selects Product Backlog Items to be implemented in the next Sprint. The Development Team is composed of members with different competences and skills. They are called Developers, regardless of their competences and skills (developers, testers, business analysts, etc.) The Development Team is a self-organizing entity, with no assigned team leader. The development team manages itself in the given frame and is responsible for developing Increment. Product Backlog Items is decomposed to less complicated tasks which are stored in the Sprint Backlog whose owner is Development Team. The Sprint is a time-box event during which the Scrum Team performs work on the Increment. Each Sprint can be considered as small project and has own goal describing what is to be implemented.

Every day a short meeting, called Daily Scrum, is organized. It is a meeting during which each team member answers three questions:

- What has been done since the last meeting to achieve the goal of the sprint?
- What will be done before the next meeting to achieve the goal of the sprint?
- What obstacles stand in the way of achieving the goal of the sprint?

The result of each sprint should be an Increment, which is the sum of all the Product Backlog items completed during the Sprint and the value of the previous increments.

Increment presentation takes place during the Sprint Review, at this meeting the Product Owner introduces the increment to stakeholders. The event which ends Sprint is Sprint Retrospective. It is a meeting whose aims is to check what happened in the last Sprint, taking into account people, dependencies, processes and tools. During this meeting, the team determines the practices that have proven in action and those that need to be improved. The result of such a meeting may be a list of impediments or improvements. As the retrospective event will be considered as Scrum event implemented in process of improvement of the forwarding agency, it will be defined in next section.

## 5   Scrum Retrospective

During the Sprint Retrospective, the important question is: What did work well or wrong in the past Sprint? Feedback obtained from the team shows its strengths, which should be nurtured and developed. Sprint Retrospective is the opportunity for the team to discuss what works and what does not work, and to determine changes. Retrospectives are typically conducted in face-to-face meetings, in which the team members first identify problems that have occurred. Subsequently, they conduct lightweight root cause analysis, collaboratively creating a cause-effect diagram visualizing the causes of problems [29]. Retrospective is one of the central practices of Scrum framework and it has been found that it is employed very frequently [30]. During the Retrospective the team identifies problems with the execution of tasks and duties. Mutual, sincere and substantial exchange of information on the process not only builds the unity of the whole team, but also helps to improve the comfort and increase the productivity of Scrum Team. Although the Retrospection is treated as mandatory, this event itself is not strictly defined. The Scrum team can use any techniques or tools to support the Retrospection.

## 6   Research Methodology

For purpose of this research a case study was applied as a research method. This method is often used especially in the field of management. Case study presents the best solutions or practices and provides proposals of practical solutions concerning management problems to organizations or managers that can be very valuable. Precisely the confrontation of theoretical and empirical results contributes to the development of any field of science [31]. The presented case study is single case study according to the definition in [32]. In the research process carried out using the case study, various methods, techniques and tools may be used. Useful methods include: documentation analysis, observation, method of interview, test implementation or the survey method [33]. Case studies are valuable in situations when the analysed problem is often encountered in practice, but there is no access to statistical data due to the novelty of a problem.

Two methods of collecting data and information were applied in this case study. The first one was direct observation conducted of one of the authors. The author was an Agile coach supporting the team as leader of the Retrospection. The second method were individual interviews with the team members concerning their feelings and perceptions regarding the use of Retrospection (as an agile practice) in their company for improving internal processes.

## 7   Organization Description

The object of the research was a freight forwarding organization. It was a company providing services in European Union. It has mainly experience in inland and maritime transport and is a medium-sized enterprise. Medium-sized enterprises are enterprises which employ less than 250 employees and whose annual one turnover does not exceed 50 million euro and/or the total annual balance sheet does not exceed 43 million euro [34].

## 8   Retrospection Framework

The team performing Retrospection consisted of senior managers of the organization. Team members were managers of operating divisions (financial and administrative ones), branches managers and members of the board of directors. In total in the retrospective meeting twelve members of the team and one leader of retrospection took part. No one of the team members had never participated in a retrospective meeting. The leader of the Retrospection was the Agile coach. The main purpose of the Retrospection was to improve the cooperation of one of the departments with the rest of the organization. The goal was set by the chief executive officer of the freight forwarding company. The period that covered by the Retrospection was the last year of the organization's activity. The Retrospection was carried out during the annual reporting meeting. Retrospective meeting lasted four hours and had a workshop character. Before the Retrospection, the team members received information about the purpose of the meeting and the planned workshop character. The Retrospection was divided into three separate phases:

1. Time line – its main goal was to stimulate memories of what happened during the previous period of time;
2. Sail boat – its main goal was to identify what slows the team down and what actually helps them;
3. Dot voting – its main goal was to choose a course of action that can be undertaken quickly and easily as part of a wider range of changes or solutions to a problem.

Such phasing has been proposed by the authors of this paper. According to their expert knowledge and experience, it is an effective way to carry out the Retrospection. All of the tree above-mentioned phases (in chronological order) of Retrospection are defined in more detail below.

## 8.1  Time Line

This part was based on a technique that is a modification of the technique presented by [35]. The key goal of this part of Retrospection was to stimulate memories of what happened during the previous period of time (past year) which was under examination, create a picture of the work from many perspectives, examine assumptions about who did what and when. Team members write on cards significant events during the considered period of time (past year) and then post them in chronological order on a formed timeline. The Retrospective leader encouraged the team to discuss the events and to understand facts and feelings formulated by individual team members. After identifying all relevant events and determination of the moment they happened, each participant was asked to draw a trend line presenting his or her individual satisfaction. The vertical axis represents team members satisfaction, where the very top of the axis is the highest satisfaction and the very bottom of the axis is the lowest satisfaction.

## 8.2  Sailboat

This part was based on technique presented by [36]. This Retrospective technique helps to identify what slows the team down and what actually helps the team to achieve their objectives (defined by the chief executive officer). The Retrospective leader uses a sailboat as a metaphor for the team. The team identifies anchors (impediments) and wind (positive forces) and chooses an area to improve. The Retrospective leader prepares a graphical interpretation of sail boat on a white board and leads the team through the retrospection. Promised land it is the goal of the retrospective. However, there are "anchors" that are holding the team back, preventing the team from making progress towards the promised land. But there are also "opportune winds", things that help to sail to the promised land. After introduction, the team starts a brainstorming session during which it identifies the factors, events, activities, etc. which can be treated as an anchor or a wind. In next stage team chooses the most important issue that stops them or helps to sail.

### 8.3   Dot Voting

This part was based on technique presented by [37]. Dot voting is a technique for public semi-anonymous voting and decision making process. Participants are given identical sets of one or more coloured dots, one dot per vote, which they stick onto paper sheets, each listing one of the available choices. In the case of the Retrospection presented here, authors decided to modify this technique. Each of the team members had a pool of dots in two distinct colours. Each colour defined distinct aspects: the first colour described how important the factor was (the more important it was, the more dots it should be assigned). The second colour described how easy it was to make changes in the area (the easier it was, the more dots it should be assigned). With such a modification, after the conclusion of the voting process the team could identify possible areas for changes that are, potentially, most important and easiest to implement. At the end the team defines actions that had to be undertaken in accordance with the obtained results.

## 9   The Results Discussion

Due to the fact that the management of the organization did not allow to reveal its identity (because of concerns about the disclosure of confidential data to market competition), the results of this case study presented here do not include data making such an identification possible. As it was mentioned earlier, case study involves a number of constraints: little representativeness performance, intuitiveness and subjectivity courts etc. [38]. The interviews with team members started after the conducting Retrospection. Team members were asked to express their opinions about the Retrospection. Due to lack of formal hierarchical relations between the team members and the authors of paper, authors believe that they have obtained sincere comments. Below we present subjective opinions of individual team members.

> TM # 1 "… For me the most efficient element was aspect of graphical visualization. It's amazing how the simple picture of sail boat can improve the work. I have never encountered such tool …"

The sailboat as a metaphor used by the leader of retrospection can be treated as gamification. Gamification seeks for improvement of the team engagement, motivation, and performance while carrying out certain tasks; it does so by incorporating game mechanics and elements, thus making those tasks more attractive [39]. Since its inception around 2010, gamification has become one of the top technology and software trends. However, gamification has also been regarded as one of the most challenging areas of software engineering [40] as is it seems to be clear, from the opinion quoted above, that gamification can be successfully used as an element of retrospection also in other industries not related to the software development.

> **TM # 2** "… A good idea was the first part of the meeting. We have recalled events from the past year and then marked the level of satisfaction according to these memories. I realized how important this information is …"

Human actions and activities take place on some timescale [41]. As the time passes by, people gradually forget events which occurred and were related to a specific topic, including what the most important event was, how it started and what its turning point and consequences were [42]. The dimension of time seems to have an enormous impact on the importance we attach to various experiences [43]. The use of graphic representation of the time line helps with chronological ordering of all events that occurred in the analysed period of time. Further comparison of the events mentioned by various members of the team builds a less subjective (containing more points of view) picture of reality and events. As people reflect on memories of past experiences, and report them to others, these memories have the power to guide the future people behavior [44].

After building a common chronology of events in the organization, each participant could express its subjective evaluation of individual events by means of a trend line of satisfaction. The relatively long period of time (one year) for which the Retrospection was conducted, in the authors opinion has brought additional benefits. Long term user experience measurements provide complementary insights to momentary measurements, for instance by yielding an understanding of changes in user experiences and allowing the prediction of long term team behaviour [45]. Long term analysis allows to determine how individual events in future may affect the team members. In the examined case, the satisfaction trend line dropped significantly (for most of team members) in the months in which the organization implemented new market procedures. It was related to the enormous amount of additional work that the team had to do before implementation. Also, an upward trend in the morale in the company was noticed soon after one of main managers had left the company. In informal conversations team members admitted that it was a person who elicited a lot of stressful situations in the organization.

> **TM # 3** "… At the output of this meeting we had a list of issues that were the most important and we knew which of them will be relatively cheap for us"

The proposed modified dot voting is a method that supports capturing "low hanging fruits" (i.e. easy benefits with respect to cost). Determining the importance of the problem and in parallel determining the difficulty of implementing changes in the area is widely included in the agile principle. The Pareto principle (or the "80–20 rule") states than in many events, 80% of the effects come from 20% of the causes. Agile management methods recommend to focus on the requirements that bring the highest value to the customer. In the Retrospection carried out during this study the low-hanging fruits were identified. The low-hanging fruit are simply targets and goals that are easily achievable.

## 10    Conclusions

Each agile approach, Scrum included, is a conglomerate of numerous roles, events, artefacts, and rules and the long term objective of any organisation which has made the decision to implement it should be to implement all of it elements [24]. However, it is a difficult and time consuming process. On the other hand, even a partial implementation of Scrum or any other agile approach can be valuable for a team or an organization. This is the first important conclusion following from the present paper, which presents the case of implementing one of Scrum practices for the improvement of internal processes of a forwarding company. The quoted opinions of the team members confirm the validity and valence of this agile practice for this goal.

According to best knowledge of the authors, this study is one of the first studies to contribute to our understanding of how Agile practices can be used to improve internal processes of organizations operating in the logistics sector. This is another important conclusion: non only the IT sector, but also the transport sector and many other ones might profit from the agile approach. Of course, the agile approach cannot be used for numerous typically logistics projects, which have to be managed by means of traditional project management, but even apparently "non-agile" organisations may draw important benefits from applying agility to their selected areas (as e.g. here – improving internal processes).

The very introductory character of the presented study and the necessity of continuation are obvious to the authors. Moreover, they are conscious that another crucial area not covered by the present paper is the Scrum implementation (as a whole framework) in logistics projects. It should be investigated in future studies.

## References

1. Sanchez, L.M., Nagi, R.: A review of agile manufacturing systems. Int. J. Prod. Res. **39**, 3561–3600 (2001)
2. Potkány, M., Krajčírová, L.: Cost reporting of the transport company and its use in decision-making. Procedia Eng. **192**, 731–736 (2017). https://doi.org/10.1016/j.proeng.2017.06.126
3. European Commission: EU Transport in figures, statistical pocketbook 2016 (2018)
4. Yazdani, M., Zarate, P., Coulibaly, A., Zavadskas, E.K.: A group decision making support system in logistics and supply chain management. Expert Syst. Appl. **88**, 376–392 (2017). https://doi.org/10.1016/j.eswa.2017.07.014
5. Bazaras, D., Palšaitis, R.: Logistics situation in Lithuania – changes during 10 years. Procedia Eng. **187**, 726–732 (2017). https://doi.org/10.1016/j.proeng.2017.04.447
6. Starkowski, D., Bieńczak, K., Zwierzycki, W.: Samochodowy transport krajowy i międzynarodowy. Kompendium wiedzy praktycznej, Przepisy w transporcie drogowym TOM IV. SYSTHERM D. Gazińska sp. j., Poland, Poznań (2011)
7. Dolinayová, A., Ľoch, M.: Controlling instruments used for monitoring and evaluation processes in the rail freight companies. Procedia Econ. Financ. **34**, 113–120 (2015). https://doi.org/10.1016/S2212-5671(15)01608-1
8. Monteiro, J., Alves, A.C., do Sameiro Carvalho, M.: Processes improvement applying Lean Office tools in a logistic department of a car multimedia components company. Procedia Manuf. **13**, 995–1002 (2017). https://doi.org/10.1016/j.promfg.2017.09.097

9. Pilcer, H.: Identyfikacja problemów w organizacji 754 (1997)
10. Kozielecki, J.: Rozwiązywanie problemów. Państwowe Zakłady Wydawnictw Szkolnych, Warszawa (1969)
11. Griffin, P., Care, E.: Assessment and Teaching of 21st Century Skills: Methods and Approach. Springer, Heidelberg (2014)
12. OECD: Organisation for Economic Co-operation and Development: PISA 2015: Draft collaborative problem solving framework (2015). https://www.oecd.org/pisa/pisaproducts/Draft%20PISA%202015%20Collaborative%20Problem%20Solving%20Framework%20.pdf. Accessed 01 May 2018
13. Sampson, V.D., Clark, D.B.: Comparison of more and less successful groups in collaboration **41**, 63–97 (2011)
14. Lindsjørn, Y., Sjøberg, D.I.K., Dingsøyr, T., Bergersen, G.R., Dybå, T.: Teamwork quality and project success in software development: a survey of agile development teams. J. Syst. Softw. **122**, 274–286 (2016)
15. Vlaanderen, K., Jansen, S., Brinkkemper, S., Jaspers, E.: The agile requirements refinery: applying SCRUM principles to software product management. Inf. Softw. Technol. **53**, 58–70 (2011). https://doi.org/10.1016/j.infsof.2010.08.004
16. Dybå, T., Dingsøyr, T.: Empirical studies of agile software development: a systematic review. Inf. Softw. Technol. **50**, 833–859 (2008)
17. Project Management Institute: A guide to the project management body of knowledge (PMBOK guide) (2013)
18. agilemanifesto.org: Manifesto for Agile Software Development. Kluwer Academic Publishers, February 2001. http://agilemanifesto.org/. Accessed 24 Apr 2014
19. Hirotaka, T., Nonaka, I.: The New New Product Development Game, January–February, pp. 137–146 (1986)
20. Coplien, J.O.: Borland software craftsmanship: a new look at process, quality, and productivity (1994)
21. Rising, L., Janoff, N.S.: The Scrum software development process for small teams. IEEE Softw. **17**, 26–32 (2000)
22. Holmström, H., Fitzgerald, B., Ågerfalk, P., Conchúir, E.: Agile practices reduce distance in global software development. Inf. Syst. Manag. **23**, 7–18 (2006)
23. Moe, N., Dingsøyr, T., Dybå, T.: A teamwork model for understanding an agile team: a case study of a Scrum project. Inf. Softw. Technol. **52**, 480–491 (2010)
24. Schwaber, K., Sutherland, J.: The Scrum Guide the Definitive Guide to Scrum: The Rules of the Game, 1991–2013 edn. www.scrum.org. Accessed April 2014
25. ArBo_HaCkEr: Los 12 principios del manifiesto ágil (2016). http://arbo.com.ve/los-12-principios-del-manifiesto-agil/. Accessed 25 Apr 2018
26. Dingsøyr, T., Hanssen, G.K., Dybå, T., Anker, G., Nygaard, J.O.: Developing software with Scrum in a small cross-organizational project. In: Richardson, I., Runeson, P., Messnarz, R. (eds.) Software Process Improvement. LNCS, vol. 4257, pp. 5–15. Springer, Heidelberg (2006)
27. Kuchta, D., Rola, P.: Estymacja czasu trwania projektów zarządzanych metodą Scrum - propozycja zastosowania metody delfickiej. In: Anonymous, pp. 51–63 (2015)
28. Eloranta, V., Koskimies, K., Mikkonen, T.: Exploring ScrumBut—an empirical study of Scrum anti-patterns. Inf. Softw. Technol. **74**, 194–203 (2016)
29. Bjørnson, F.O., Wang, A.I., Arisholm, E.: Improving the effectiveness of root cause analysis in post mortem analysis: a controlled experiment. Inf. Softw. Technol. **51**, 150–161 (2009)
30. Vallon, R., da Silva Estácio, B.J., Prikladnicki, R., Grechenig, T.: Systematic literature review on agile practices in global software development. Inf. Softw. Technol. **96**, 161–180 (2018). https://doi.org/10.1016/j.infsof.2017.12.004

31. Sudoł, S.: Nauki o zarządzaniu. Węzłowe problemy i kontrowersje. Wydawnictwo "Dom Organizatora", Toruń (2007)
32. Dyer, W., Wilkins, A.: Better stories, not better constructs, to generate better theory: a rejoinder to Eisenhardt. Acad. Manag. Rev. **16**, 613–619 (1991)
33. Apanowicz, A.: Metodologia nauk. Wydawnictwo Dom Organizatora, Toruń (2003)
34. Komisja Europejska: Zalecenie Komisji 2003/361/WE, Dzienniku Urzędowym Unii Europejskiej L 124 opublikowana r.: p. 36, 20 May 2003
35. Derby, E., Larsen, D.: Agile Retrospectives: Making Good Teams Great. Pragmatic Bookshelf, Raleigh (2006)
36. Hohmann, L.: Innovation Games: Creating Breakthrough Products Through Collaborative Play. Addison-Wesley, Upper Saddle River (2007)
37. Segar, A.: The Power of Participation: Creating Conferences That Deliver Learning, Connection, Engagement, and Action. CreateSpace Independent Publishing Platform, Scotts Valley (2015)
38. Lachiewicz, S., Matejun, M.: Studia przypadków karier menedżerskich absolwentów Politechniki Łódzkiej. In: Anonymous Kształcenie menedżerów na uczelni technicznej, red. I. Staniec edn. Wydawnictwo PŁ, Łódź, p. 88 (2010)
39. García, F., Pedreira, O., Piattini, M., Cerdeira-Pena, A., Penabad, M.: A framework for gamification in software engineering. J. Syst. Softw. **132**, 21–40 (2017). https://doi.org/10. 1016/j.jss.2017.06.021
40. Morschheuser, B., Hassan, L., Werder, K., Hamari, J.: How to design gamification? A method for engineering gamified software. Inf. Softw. Technol. **95**, 219–237 (2018). https:// doi.org/10.1016/j.infsof.2017.10.015
41. Lemke, J.L.: Across the scales of time: artifacts, activities, and meanings in ecosocial systems. Mind Cult. Act. **7**, 273–290 (2000)
42. Lin, F., Liang, C.: Storyline-based summarization for news topic retrospection. Decis. Support Syst. **45**, 473–490 (2008). https://doi.org/10.1016/j.dss.2007.06.009
43. Karapanos, E., Martens, J., Hassenzahl, M.: On the retrospective assessment of users' experiences over time: memory or actuality?, pp. 4075–4080 (2010)
44. Norman, D.A.: Memory is more important than actuality. Interactions **16**, 24–26 (2009)
45. Karapanos, E., Zimmerman, J., Forlizzi, J., Martens, J.: User experience over time: an initial framework, pp. 729–738 (2009)
46. Streule, T., Miserini, N., Bartlomé, O., Klippel, M., de Soto, B.G.: Implementation of Scrum in the construction industry. Procedia Eng. **164**, 269–276 (2016). https://doi.org/10.1016/j. proeng.2016.11.619
47. Edin Grimheden, M.: Can agile methods enhance mechatronics design education? Mechatronics **23**, 967–973 (2013). https://doi.org/10.1016/j.mechatronics.2013.01.003
48. Mergel, I.: Agile innovation management in government: a research agenda. Gov. Inf. Q. **33**, 516–523 (2016). https://doi.org/10.1016/j.giq.2016.07.004
49. Tomek, R., Kalinichuk, S.: Agile PM and BIM: a hybrid scheduling approach for a technological construction project. Procedia Eng. **123**, 557–564 (2015). https://doi.org/10. 1016/j.proeng.2015.10.108

# Accuracy Assessment of Artificial Intelligence-Based Hybrid Models for Spare Parts Demand Forecasting in Mining Industry

Maria Rosienkiewicz[(✉)] [iD]

Faculty of Mechanical Engineering, Wroclaw University of Science
and Technology, Wrocław, Poland
maria.rosienkiewicz@pwr.edu.pl

**Abstract.** The paper addresses the problem of spare parts demand forecasting in mining industry. The paper proposes new hybrid models combining traditional forecasting techniques based on time series (ARIMA, SES, Holt's model, TES, SMA, EMA, WMA, ZLEMA, SBA) with artificial intelligence-based methods. Three new approaches are developed - (1) hybrid forecasting econometric model, (2) hybrid forecasting artificial neural network model and (3) hybrid forecasting support vector machine model. The assessment of the proposed hybrid models is conducted by a comparison with traditional methods and is based on relative forecast error *ex post* and coefficient of determination. Empirical verification of the proposed models is built upon real data from an underground copper mine. The forecasts according to 9 traditional techniques and 3 hybrid models are computed for 10 cases.

**Keywords:** Spare parts · Demand forecasting · Artificial neural network · Support vector machine · Hybrid model · Lumpy demand

## 1 Introduction

The management of spare parts is considered as one of the most neglected areas of management whereas its importance cannot be overemphasized [11, 28]. A wide range of important research topics is addressed by spares management – i.a. inventory control, maintenance and reliability, supply chain management and spare parts demand forecasting [24]. In many industrial sectors, especially those which are of key importance to the national economy (aerospace, automotive, mining and railway), the issue of recurrent failures causes a need of implementation well thought-out parts management policies. One of the important aspects of such policies is spares demand forecasting.

When analyzing various sectors of economy, it can be noticed that in particular enterprises from underground mining are characterized by very high failure rates of machines [21]. It is mainly caused by the very specific working environment of mining equipment. Machines are almost constantly in motion and very often break down due to difficult working environment characterized by high temperatures, high humidity and poor road conditions. It also is worth emphasizing that the complexity of these

© Springer Nature Switzerland AG 2020
Z. Wilimowska et al. (Eds.): ISAT 2019, AISC 1052, pp. 176–187, 2020.
https://doi.org/10.1007/978-3-030-30443-0_16

machines and the high loads to which they are subjected impose very strict requirements on their reliability and maintenance [6, 15]. In consequence the issue of accurate
spare parts demand forecasting is very important because it directly influences the
availability of machines and their maintenance processes.

From the manufacturing companies and mines point of view, the practical application of forecasting models is not an easy task. As most companies in Poland are
currently at the stage of the third industrial revolution, the process of implementation
Industry 4.0 technologies and transformation towards smart factories or intelligent
mines is still ahead [12]. This means that in the most companies in Poland, obtaining
relevant data to build effective and accurate forecasting models can still be a big
challenge – in particular, when data characterizing explanatory variables are desirable.
The lack of available, reliable and at the same time complete and comparable statistical
data makes it difficult to build effective forecasting models. Hence there is a need to
develop a solution that would allow constructing a set of potential explanatory variables when access to data is limited.

The need of development new ways enhancing companies in transformation
towards smart factories and the need of solving problem of limited data leads to a
conclusion that research on innovative solution addressing this issues is crucial. To
answer these needs the Author proposes new hybrid spare parts forecasting models
dedicated to mining industry. The main aim of the study presented in this paper is to
assess accuracy of the proposed artificial intelligence-based hybrid models for spare
parts demand forecasting. To verify the forecasting accuracy, data coming from a
copper mine will be used.

## 2 Literature Review on Hybrid Demand Forecasting Approaches

Spare parts demand is very hard to predict due to the fact that it is burdened with a large
degree of uncertainty and is often characterized by unpredictable fluctuations. It is
lumpy and intermittent. A detailed and up-to-date review of research on the development of spare parts demand forecasting methods can be found i.a. in [3, 9, 27, 30].

Literature analysis shows that research on the application of artificial intelligence to
forecasting problems is currently developing very dynamically. At the same time, a
rapid increase of publications devoted to hybrid forecasting approaches can be
observed. According to [22] hybrid models are constructed in order to increase the
accuracy of forecasts obtained through the use of individual models. Moreover, [22]
emphasizes that in the case of hybrid methods it is assumed that the forecasts based on
many techniques are simply more accurate in comparison to individual ones.

Analysis of hybrid forecasting methods described in the literature leads to a conclusion that researchers combine forecasting models and methods in many different
ways. For example Więcek proposed a hybrid method, which combines auto-regression
models (ARIMA), artificial neural networks and spectral analysis techniques in order to
forecast demand based on time series [32]. Amin-Naseri and Rostami Tabar have
developed a hybrid forecasting approach combining a multi-layered perceptron neural
network and a traditional recursive method for forecasting future demands [1]. This

method is based on the approach that an artificial neural network dedicated to classification issues forecasts occurrences of non-zero demands, and next, a traditional recursive method estimates the value of non-zero demand. According to their assumptions, if the output from the network is 1, then the demand value should be calculated using conventional methods, and if it is 0, it means that there will be no demand in the forecasted period. Grzeszczyk proposed an integrated forecasting method, which includes a quantitative analysis based on the ANN model (subsystem 1), a qualitative analysis based on rough set theory (subsystem 2) and a subsystem integrating results from the listed subsystems 1 and 2 [14]. In this approach quantitative forecasts and qualitative parameters are the input to the ANN model. Another approach uses a vector of seasonality indicators and a trend vector as the entry to an artificial neural network model [19]. Hua et al. have considered spare parts demand forecasting in an enterprise from the petrochemical industry. They have developed a mechanism integrating "the demand autocorrelated process and the relationship between explanatory variables and the nonzero demand of spare parts during forecasting occurrences of nonzero demands over lead times" [17]. Next hybrid approach, developed by Hua and Zhang, uses support vector machine model in order to forecast occurrences of nonzero demand of spare parts and then integrates the forecast being an output from the SVM and the relationship of occurrence of nonzero demand with explanatory variables [16]. Rosienkiewicz et al. have used outputs of traditional forecasting methods (moving average and exponential smoothing) as inputs to an artificial neural network model. This hybrid approach has been implemented in a furniture factory as a tool supporting demand forecasting [29]. Bounou et al. have created a hybrid of Syntetos-Boylan method (being a modification Croston's method) and exponential smoothing [4]. Wan et al. have developed a hybrid, two-stage approach to forecast intervals of market clearing prices (MCPs). In this method first extreme learning machine (ELM) is used in order to estimate point forecasts of MCPs, and subsequently the maximum likelihood method is applied to estimate the noise variance [31]. Another hybrid approach combining artificial neural network model and traditional forecasting is proposed by Omar et al. In their model historical sales data, popularity of article titles, and the prediction result of a time series based on ARIMA are inputs to backpropagation neural network (BPNN) [25]. Marques and Gomes proposed a hybrid method combining moving average and Elman neural network model [23]. Another approach being a hybrid of ARIMA model and ANN model have been developed by Areekul et al. [2].

Based on the literature review, the three main groups of hybrid demand forecasting approaches can be distinguished:

1. Hybrid approach composed of an artificial intelligence-based method and a traditional quantitative forecasting technique [1, 2, 16, 19, 23, 25, 29, 31, 32],
2. Hybrid approach composed of an artificial intelligence-based method, a traditional quantitative forecasting technique and a traditional qualitative forecasting technique [4, 14],
3. Hybrid approach composed of a few traditional quantitative forecasting techniques [17].

Application areas of the above-mentioned hybrid approaches include forecasting of electricity prices, sales forecasting in the publishing industry, short-term price forecasting in deregulated market, sales forecasting and spare parts demand forecasting in the petrochemical industry. Hybrid models combining ANN and simulation modeling have also been developed for the purpose of ensuring stability of a production system [5]. Results of the literature analysis show that although increasing number of scientific papers is focusing on hybrid approaches, little research has been done on spare parts demand forecasting based on hybrid models. Therefore this paper aims to fill in this gap.

# 3 Research Methodology

Based on the conclusions from the literature review and experiences gained thanks to cooperation with industrial companies, a new hybrid models have been developed, which aim to obtain more accurate forecasts in comparison to traditional forecasting methods. Moreover, the proposed approach is trying to solve a common problem in industrial practice – insufficient data or the lack of available data. In the paper three hybrid models are proposed:

- a hybrid forecasting econometric model,
- a hybrid forecasting artificial neural network (ANN) model,
- a hybrid forecasting support vector machine (SVM) model.

The study discussed in this paper is a continuation of research on spare parts demand forecasting dedicated to mining industry performed by Rosienkiewicz et al. [28]. Figure 1 presents a research methodology composed of 4 main steps: (1) initial analysis, (2) forecasts computation based on traditional forecasting methods, (3) hybrid forecasting models development and (4) accuracy assessment of estimated forecasts.

According to the presented research methodology, in the initial phase, spare parts should be selected for which forecasts are to be created, and next adequate data should be gathered. Afterwards, 9 forecasting methods ($F_1$-$F_9$) will be applied:

- $F_1$: autoregressive-integrated moving average (abbr.: ARIMA),
- $F_2$: simple exponential smoothing (abbr.: SES),
- $F_3$: Holt's model (abbr.: Holt),
- $F_4$: trigonometric exponential smoothing (abbr.: TES),
- $F_5$: simple moving average (abbr.: SMA),
- $F_6$: exponential moving average (abbr.: EMA),
- $F_7$: weighted moving average (abbr.: WMA),
- $F_8$: zero-lag exponential moving average (abbr.: ZLEMA),
- $F_9$: Syntetos-Boylan method (abbr.: SBA).

Formulas describing each of the method are given in Table 1. To check components of all formulas please refer to sources given in the last column ($\hat{y}_t$ is a forecasted value of variable $y$ in the $t$ period).

Subsequently, after the forecasts according to traditional techniques are computed, hybrid models are constructed – one based on econometric modeling and two based on

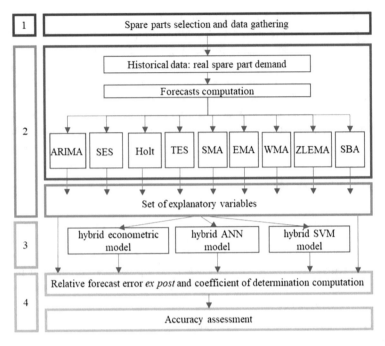

**Fig. 1.** Research methodology diagram

artificial intelligence – hybrid ANN model and hybrid SVM model. Explanatory variables set (EXS) composed of forecasts coming from 9 traditional methods ($F_1$-$F_9$) is an input to each of the hybrid model. Hybrid econometric model is given by the following formula:

$$\hat{y} = \hat{\alpha}_0 + \hat{\alpha}_1 F_1 + \hat{\alpha}_2 F_2 + \cdots + \hat{\alpha}_m F_m. \tag{10}$$

Bayesian Schwarz information criterion (BIC) is used to select appropriate subset of explanatory variables from the EXS. Hybrid ANN model is developed on the basis of neurons. The basic model of the neuron can be described by the following expression [20]:

$$y = f\left(\sum_{p=1}^{P} w_p u_p + u_0\right), \tag{11}$$

where $u_p$, $p = 1,2,\ldots, P$ – inputs ($F_1$-$F_9$ for hybrid model), $u_0$ - bias, $w_p$ – weights, $f(\cdot)$ – activation function. A mathematical function converts the weighted sum of the signals to form the output of the neuron. In case of hybrid regression SVM model the functional dependence of the dependent variable $y$ on a set of explanatory variables $x$ (in this case x = $F_1$, $F_2$, ..., $F_9$) will be estimated. The relationship between the explanatory and dependent variables is given by a deterministic function $f$ and the addition of some noise [34]:

**Table 1.** Forecasting methods formulas

| Method | Formula | No./ Source | |
|--------|---------|-------------|---|
| ARIMA | $\hat{y}_t = c + \Phi_1 y\prime_{t-1} + \cdots + \Phi_p y\prime_{t-p} + \Theta_0 \varepsilon_t + \Theta_1 \varepsilon_{t-1} + \cdots + \Theta_q \varepsilon_{t-q}$ | (1) | [18] |
| SES | $\hat{y}_t = \alpha y_{t-1} + (1 - \alpha)\hat{y}_{t-1}$ | (2) | [13] |
| Holt | $\hat{y}_t = F_{t-1} + S_{t-1}, \; F_t = \alpha y_t + (1 - \alpha)(F_{t-1} + S_{t-1}),$ $S_t = \beta(F_t - F_{t-1}) + (1 - \beta)S_{t-1}$ | (3) | [13] |
| TES | $\hat{y}_t = l_{t-1} + \varphi b_{t-1} + \sum_{i=1}^{T} s_{t-m_i}^{(i)} + d_t, \; l_t = l_{t-1} + \varphi b_{t-1} + \alpha d_t,$ $b_t = (1 - \varphi)b + \varphi b_{t-1} + \beta d_t, \; s_t^{(i)} = s_{t-m_i}^{(i)} + \gamma_i d_t,)$ $d_t = \sum_{i=1}^{p} \rho_i d_{t-1} + \sum_{i=1}^{q} \theta_i \varepsilon_{t-1} + \varepsilon_t$ | (4) | [10] |
| SMA | $\hat{y}_t = \frac{1}{m} \sum_{j=-k}^{k} y_{t+j}$ | (5) | [18] |
| EMA | $\hat{y}_t = \frac{y_{t-1} + (1-\alpha)y_{t-2} + (1-\alpha)^2 y_{t-3} + \ldots + (1-\alpha)^n y_{t-(n+1)}}{1 + (1-\alpha) + (1-\alpha)^2 + (1-\alpha)^3 + \ldots + (1-\alpha)^n}$ | (6) | [26] |
| WMA | $\hat{y}_t = \sum_{j=-k}^{k} a_j y_{t+j}$ | (7) | [18] |
| ZLEMA | $\hat{y}_t = \frac{2}{(n+1)}\left(2y_{t-1} - y_{lag}\right) + \left(1 - \frac{2}{(n+1)}\right) \times \hat{y}_{t-1}$ | (8) | [33] |
| SBA | $\hat{y}_t = \left(1 - \frac{\alpha}{2}\right)\frac{Z_{t-1}}{P_{t-1}}, Z_t = \alpha X_t + (1 - \alpha)Z_{t-1}$ $P_t = \alpha G_t + (1 - \alpha)P_{t-1}$ | (9) | [8] |

$$y = f(x) + noise. \tag{12}$$

The functional form for $f$ which can correctly predict new cases can be achieved by training the SVM model on a sample set – a process involving the sequential optimization of an error function (for details see [34]). Radial basis function (RBF) will be the kernel type $K$ used in the hybrid SVM model [34]:

$$K(X_i, X_j) = \exp(-\gamma |X_i - X_j|^2), \tag{13}$$

where $K(X_i, X_j) = \phi(X_i) \cdot \phi(X_j)$, $\phi$ – transformation.

After the hybrid models are developed, the last step of the research methodology can be applied – accuracy assessment of estimated forecasts for each analyzed method. The comparison of accuracy of the new hybrid models with traditional forecasting methods will be based on two indicators - relative forecast error *ex post I* given by the following formula:

$$I = \sqrt{\frac{\sum_{t=1}^{m}(y_t - \hat{y}_t)^2}{\sum_{t=1}^{m} y_t^2}}, \tag{14}$$

and coefficient of determination $R^2$ defined as follows:

$$R^2 = \frac{\sum_{i=1}^{n}(\hat{y}_t - \bar{y})^2}{\sum_{i=1}^{n}(y_t - \bar{y})^2}. \tag{15}$$

The most accurate model will be the one with the lowest value of $I$ and the highest value of $R^2$. In case the model with the lowest $I$ will not be the one with the highest $R^2$, as the most accurate will be considered the one characterized with the lowest $I$.

## 4    Accuracy Analysis of the Hybrid Models

To empirically verify and assess the proposed hybrid models, real data from an underground mine were used. The data were gathered within the research project "Adaptation and Implementation of Lean Methodology in Copper Mines" co-financed by the Polish National Centre for Research and Development. According to the proposed research methodology, at first spare parts to be analyzed had to be selected – hence 10 types characterized by the highest usage were chosen – 5 representatives of spare parts and 5 representatives of consumable materials. Table 2 presents the name of each spare part (consumable material), attached symbol (letters A-J) and the sample size $n$ (representing number of months).

**Table 2.** Spare parts and consumables – legend.

| Symbol | Spare part/Consumable material | n |
|--------|-------------------------------|-----|
| A | Hydraulic oil | 58 |
| B | Brake fluid | 30 |
| C | Engine coolant | 26 |
| D | Bearing grease | 35 |
| E | Engine oil | 25 |
| F | Tyre type A | 34 |
| G | Air filter element | 72 |
| H | Brake pump | 34 |
| I | Tyre type B | 38 |
| J | Inner tube | 70 |

Initial data analysis showed that daily forecasts are inaccurate and characterized by very high errors. Therefore, due to the fact that the demand was lumpy and intermittent, the daily data was aggregated to monthly values. Following the research methodology proposed in the previous chapter, after initial analysis, the forecasts were computed according to 9 traditional techniques – ARIMA, SES, Holt, TES, SMA, EMA, WMA, ZLEMA and SBA – for each analyzed spare part/ consumable material (A-J). Next, received forecasts were transformed into sets of explanatory variables (10 sets – one per investigated case). Each set was composed of 9 variables ($F_1$-$F_9$) representing traditional techniques. These sets formed inputs to each of the hybrid model. Subsequently, three hybrid models were created – hybrid econometric model (hybrid_ECO), hybrid artificial neural network model (hybrid_ANN) and hybrid support vector machine model (hybrid_SVM). Computation on traditional forecasting and hybrid econometric model were performed in *R language* version 3.5.3, whereas hybrid ANN model and hybrid SVN model were developed in *Statistica* version 13.1. After all models were developed and forecasts calculated, accuracy assessments was performed. For each case and each investigated method/ model relative forecast error *ex post I* and coefficients of determination $R^2$ were computed. Obtained results are presented in Tables 3 and 4. The most accurate outcomes are marked in bold (the lowest $I$ in Table 3 and the highest $R^2$ Table 4).

**Table 3.** Relative forecast errors *ex post*.

|            | A      | B      | C      | D      | E      | F      | G      | H      | I      | J      |
|------------|--------|--------|--------|--------|--------|--------|--------|--------|--------|--------|
| ARIMA      | 0,294  | 0,214  | 0,364  | 0,377  | 0,415  | 0,489  | 0,503  | 0,461  | 0,647  | 0,432  |
| SES        | 0,287  | 0,208  | 0,339  | 0,506  | 0,395  | 0,573  | 0,568  | 0,885  | 0,881  | 0,499  |
| Holt       | 0,296  | 0,211  | 0,316  | 0,514  | 0,413  | 0,528  | 0,558  | 0,830  | 0,801  | 0,463  |
| TES        | 0,287  | 0,188  | 0,256  | 0,377  | 0,389  | 0,420  | 0,516  | 0,719  | 0,645  | 0,430  |
| SMA        | 0,292  | 0,218  | 0,340  | 0,567  | 0,396  | 0,535  | 0,565  | 0,861  | 0,828  | 0,459  |
| EMA        | 0,286  | 0,209  | 0,324  | 0,503  | 0,391  | 0,545  | 0,549  | 0,844  | 0,831  | 0,475  |
| WMA        | 0,288  | 0,211  | 0,338  | 0,535  | 0,390  | 0,551  | 0,561  | 0,866  | 0,848  | 0,474  |
| ZLEMA      | 0,324  | 0,257  | 0,450  | 0,674  | 0,482  | 0,746  | 0,726  | 1,158  | 1,167  | 0,646  |
| SBA        | 0,493  | 0,226  | 0,302  | 0,500  | 0,529  | 0,598  | 0,527  | 0,745  | 0,660  | 0,632  |
| hybrid_ECO | 0,257  | **0,143** | 0,207 | 0,374 | 0,323 | **0,344** | 0,503 | 0,227 | 0,614 | 0,424 |
| hybrid_ANN | **0,256** | 0,179 | **0,196** | 0,415 | **0,286** | 0,353 | **0,339** | 0,230 | **0,042** | **0,245** |
| hybrid_SVM | 0,283  | 0,159  | 0,205  | **0,365** | 0,396 | 0,374 | 0,490 | **0,209** | 0,090 | 0,411 |

The computed $I$ and $R^2$ show that the proposed hybrid models are definitely more accurate than traditional forecasting methods (ARIMA, SES, Holt, TES, SMA, EMA, WMA, ZLEMA and SBA). For all 10 investigated cases (*A-J*) the most accurate forecasting methods (according to $I$ criterion) were hybrid models – 2 out of 10 hybrid_ECO, 6 out of 10 hybrid_ANN and 2 out of 10 hybrid_SVM.

In order to graphically visualize obtained results, an average relative forecast *error ex post I* and an average coefficient of determination $R^2$ were computed and presented in figures below (Figs. 2 and 3).

An analysis of $I$ and $R^2$ shows that the most accurate forecasts can be obtained from the proposed in the paper hybrid models. The most accurate are artificial intelligence-based models – ANN hybrid model ($I = 25\%$, $R^2 = 72\%$) and SVM hybrid model ($I = 30\%$, $R^2 = 64\%$). Moreover, it can be noticed that all traditional forecasting methods are characterized by high errors ($I$ 44–66%) and low fit ($R^2$ 22–48%).

**Table 4.** Coefficients of determination.

|  | A | B | C | D | E | F | G | H | I | J |
|---|---|---|---|---|---|---|---|---|---|---|
| ARIMA | 0,765 | 0,608 | 0,017 | 0,387 | 0,450 | 0,420 | 0,170 | 0,797 | 0,586 | 0,581 |
| SES | 0,772 | 0,628 | 0,013 | 0,139 | 0,471 | 0,288 | 0,158 | 0,002 | 0,017 | 0,488 |
| Holt | 0,771 | 0,647 | 0,008 | 0,082 | 0,447 | 0,348 | 0,140 | 0,001 | 0,010 | 0,551 |
| TES | 0,771 | 0,636 | 0,316 | 0,398 | 0,468 | 0,537 | 0,129 | 0,001 | 0,097 | 0,586 |
| SMA | 0,763 | 0,608 | 0,000 | 0,033 | 0,461 | 0,330 | 0,140 | 0,000 | 0,017 | 0,547 |
| EMA | 0,771 | 0,636 | 0,013 | 0,110 | 0,470 | 0,313 | 0,158 | 0,002 | 0,018 | 0,519 |
| WMA | 0,770 | 0,621 | 0,004 | 0,083 | 0,476 | 0,312 | 0,156 | 0,000 | 0,019 | 0,524 |
| ZLEMA | 0,737 | 0,508 | 0,005 | 0,152 | 0,431 | 0,171 | 0,138 | 0,000 | 0,020 | 0,341 |
| SBA | 0,580 | 0,618 | 0,107 | 0,006 | 0,197 | 0,151 | 0,091 | 0,006 | 0,068 | 0,419 |
| hybrid_ECO | 0,810 | **0,780** | 0,477 | 0,398 | 0,602 | 0,687 | 0,138 | 0,894 | 0,097 | 0,586 |
| hybrid_ANN | **0,815** | 0,663 | **0,634** | 0,292 | **0,688** | **0,766** | **0,618** | 0,893 | **0,996** | **0,864** |
| hybrid_SVM | 0,774 | 0,730 | 0,527 | **0,464** | 0,497 | 0,667 | 0,200 | **0,913** | 0,992 | 0,612 |

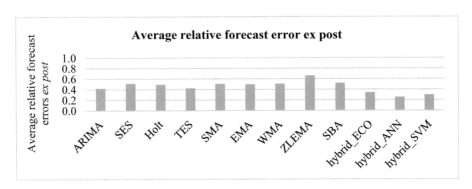

**Fig. 2.** Average relative forecast error *ex post*.

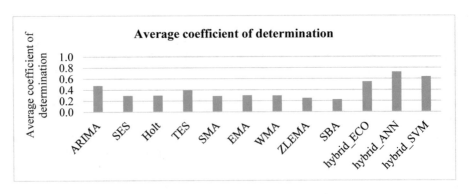

**Fig. 3.** Average coefficient of determination.

# 5   Conclusions

The issue of accurate spare parts demand forecasting is important in many sectors – especially in mining, aerospace, automotive and railway industries. Nowadays, an increasing number of companies implementing Industry 4.0 technologies will shift their maintenance strategies towards predictive maintenance. As it will result in a much bigger access to data, it is important to develop accurate forecasting models which will be able to efficiently take advantage of this possibility.

The proper use of the new hybrid spare parts forecasting models can benefit its user in a number of ways. Accurate spares demand forecasts can directly and indirectly enhance maintenance processes and production organization due to i.a.: shortening repair time, increase in operational availability of machines, lowering risk that production plan will not be executed and the risk of mechanics and operators' work disorganization. Moreover, implementation of hybrid forecasting models can bring benefits including elimination or significant reduction of costs caused by excessive or insufficient inventories of spare parts, elimination of space problems, depreciation problems and minimizing bullwhip effect. They should also positively influence indicators like Overall Equipment Effectiveness (OEE), Mean Time To Repair (MTTR) and Mean Time Between Failures (MTBF). The new models can help the mines to make progress in following the global trends towards the intelligent mine [28].

The future research will focus on the further investigation on ability of other artificial intelligence-based models and methods (i.a. ELM, K-Nearest Neighbors (KNN) and genetic algorithms [7]) to forecast spare parts demand forecasting. The models proposed in the paper will also be implemented to other case studies in order to better asses their accuracy and efficiency.

# References

1. Amin-Naseri, M.R., Tabar, B.R.: Neural network approach to lumpy demand forecasting for spare parts in process industries. In: 2008 International Conference on Computer and Communication Engineering, pp. 1378–1382. IEEE, Kuala Lumpur (2008)
2. Areekul, P., et al.: Notice of violation of IEEE publication principles - a hybrid ARIMA and neural network model for short-term price forecasting in deregulated market. IEEE Trans. Power Syst. **25**(1), 524–530 (2010)
3. Bacchetti, A., Saccani, N.: Spare parts classification and demand forecasting for stock control: investigating the gap between research and practice. Omega **40**(6), 722–737 (2012)
4. Bounou, O., et al.: Parametric approaches for spare parts demand. Int. J. Supply Chain. Manag. **7**(4), 432–439 (2018)
5. Burduk, A., et al.: Hybrid model of an expert system for assessing the stability of a production system. Logforum **14**(4), 507–518 (2018)
6. Burduk, A., Jagodziński, M.: Assessment of production system stability with the use of the FMEA analysis and simulation models. In: Jackowski, K., et al. (eds.) Intelligent Data Engineering and Automated Learning – IDEAL 2015, pp. 216–223. Springer, Cham (2015)
7. Burduk, A., Musiał, K.: Optimization of chosen transport task by using generic algorithms. In: Saeed, K., Homenda, W. (eds.) Computer Information Systems and Industrial Management, pp. 197–205. Springer, Cham (2016)

8. Croston, J.D.: Forecasting and stock control for intermittent demands. J. Oper. Res. Soc. **23** (3), 289–303 (1972)
9. De Gooijer, J.G., Hyndman, R.J.: 25 years of time series forecasting. Int. J. Forecast. **22**(3), 443–473 (2006)
10. De Livera, A.M., et al.: Forecasting time series with complex seasonal patterns using exponential smoothing. J. Am. Stat. Assoc. **106**(496), 1513–1527 (2011)
11. Gopalakrishnan, P., Banerji, A.K.: Maintenance and Spare Parts Management. PHI Learning - Private Limited, New Delhi (2011)
12. Gracel, J., et al.: Inżynierowie Przemysłu 4.0: [nie]gotowi do zmian? Astor Whitepaper, Kraków (2017)
13. Gruszczyński, M., et al.: Ekonometria i badania operacyjne. Wydawnictwo Naukowe PWN, Warszawa (2009)
14. Grzeszczyk, T.A.: Zintegrowana metoda prognozowania w zarządzaniu przedsiębiorstwem. Zastosowania metod statystycznych w badaniach naukowych II, 8 (2003)
15. Hadi Hoseinie, S., et al.: Reliability analysis of drum shearer machine at mechanized longwall mines. J. Qual. Maint. Eng. **18**(1), 98–119 (2012)
16. Hua, Z., Zhang, B.: A hybrid support vector machines and logistic regression approach for forecasting intermittent demand of spare parts. Appl. Math. Comput. **181**(2), 1035–1048 (2006)
17. Hua, Z.S., et al.: A new approach of forecasting intermittent demand for spare parts inventories in the process industries. J. Oper. Res. Soc. **58**(1), 52–61 (2007)
18. Hyndman, R.J., Athanasopoulos, G.: Forecasting: principles and practice ; [a comprehensive introduction to the latest forecasting methods using R ; learn to improve your forecast accuracy using dozenss of real data examples (2018)
19. Jurczyk, K., Kutyba, A.: Prognozowanie wielkości sprzedaży z wykorzystaniem sztucznych sieci neuronowych na przykładzie przedsiębiorstwa branży kwiatowej. Logistyka, vol. 2, CD 1, pp. 323–333 (2015)
20. Korbicz, J., et al.: Diagnostyka procesów. Modele. Metody sztucznej inteligencji. Zastosowania. Wydawnictwa Naukowo-Techniczne (2002)
21. Król, R., et al.: Analiza awaryjności układów hydraulicznych samojezdnych maszyn roboczych stosowanych w KGHM POLSKA MIEDŹ S.A. Prace Naukowe Instytutu Górnictwa Politechniki Wrocławskiej **128**(36), 127–139 (2009)
22. Maciąg, A., et al.: Prognozowanie i symulacja w przedsiębiorstwie. Polskie Wydawnictwo Ekonomiczne (2013)
23. Marques, N.C., Gomes, C.: Implementing an intelligent moving average with a neural network. In: Conference: ECAI 2010 - 19th European Conference on Artificial Intelligence, Lisbon, Portugal, 16–20 August 2010, Proceedings (2010)
24. Molenaers, A., et al.: Criticality classification of spare parts: a case study. Int. J. Prod. Econ. **140**(2), 570–578 (2012)
25. Omar, H., et al.: A hybrid neural network model for sales forecasting based on ARIMA and search popularity of article titles. Comput. Intell. Neurosci. **2016**, 1–9 (2016)
26. Praekhaow, P.: Determination of trading points using the moving average methods. In: Presented at the GMSTEC 2010: International Conference for a Sustainable Greater Mekong Subregion, Bangkok, Thailand (2010)
27. do Rego, J.R., de Mesquita, M.A.: Spare parts inventory control: a literature review. Production **21**(4), 645–666 (2011)
28. Rosienkiewicz, M., et al.: A hybrid spares demand forecasting method dedicated to mining industry. Appl. Math. Modell. **49**, 87–107 (2017)

29. Rosienkiewicz, M., et al.: Development of lean hybrid furniture production control system based on glenday sieve, artificial neural networks and simulation modeling. Drvna industrija **69**(2), 163–173 (2018)
30. Van Horenbeek, A., et al.: Joint maintenance and inventory optimization systems: a review. Int. J. Prod. Econ. **143**(2), 499–508 (2013)
31. Wan, C., et al.: A hybrid approach for probabilistic forecasting of electricity price. IEEE Trans. Smart Grid **5**(1), 463–470 (2014)
32. Więcek, P.: Hybrydowa metoda predykcji popytu w logistyce. Logistyka **4**, 9 (2015)
33. http://www.mesasoftware.com/papers/ZeroLag.pdf. Accessed 10 Aug 2018
34. http://documentation.statsoft.com/STATISTICAHelp.aspx?path=MachineLearning/MachineLearning/Overviews/SupportVectorMachinesIntroductoryOverview. Accessed 15 Feb 2019

# Single Machine Scheduling
# with Precedence Constrains, Release
# and Delivery Times

Natalia Grigoreva[✉]

St. Petersburg State University, Universitetskaja nab. 7/9,
199034 St. Petersburg, Russia
n.s.grig@gmail.com

**Abstract.** The goal of this paper is to propose algorithms for scheduling problem, where set of jobs performed on a single processor. Each job has a release time, when it becomes available for processing, a processing time and a delivery time. We study the case in which there exist precedence constrains among jobs and preemption is not allowed. The objective is to minimize the time, by which all jobs are delivered. The single machine scheduling problem is one of the classic NP-hard optimization problems, and it is useful in solving flowshop and jobshop scheduling problems. We develop branch and bound algorithm for the problem. We propose an approximation algorithm to find an upper bound, solve the preemptive version of the problem to provide a lower bound and use a binary branching rule, where at each branch node, a complete schedule is generated.

To illustrate the effectiveness of our algorithms we tested them on randomly generated set of jobs.

**Keywords:** Single processor · Branch and bound algorithm ·
Release and delivery time · Precedence constrains

## 1 Introduction and Related Work

This paper is concerned with a single machine scheduling problem of minimizing the makespan. Following the 3-field classification scheme proposed by Graham *et al.* [8], the problem under consideration is denoted by $1|r_j, q_j, prec|C_{\max}$. The problem relates to the scheduling problem [4], has many applications and it is $NP$-hard in the strong sense [13].

We consider a system of jobs $U = \{u_1, u_2, \ldots, u_n\}$. Each job is characterized by its execution time $t(u_i)$, its release time $r(u_i)$ and its delivery time $q(u_i)$. Precedence constructions between jobs are represented by a directed acyclic task graph $G = \langle U, E \rangle$. $E$ is a set of directed arcs, an arc $e = (u_i, u_j) \in E$ if and only if $u_i \prec u_j$. The expression $u_i \prec u_j$ means that the job $u_j$ may be initiated only after completion of the job $u_i$. If $u_i \prec u_j$ we call job $u_i$ a predecessor of job $u_j$ and job $u_j$ a successor of job $u_i$. Release time $r(u_i)$ is the time at which

© Springer Nature Switzerland AG 2020
Z. Wilimowska et al. (Eds.): ISAT 2019, AISC 1052, pp. 188–198, 2020.
https://doi.org/10.1007/978-3-030-30443-0_17

the job is ready to start processing, and its delivery begins immediately after processing has been completed. At most one job can be processed at a time, but all jobs may be simultaneously delivered. The set of jobs is performed on a single processor. Job preemption is not allowed. The schedule defines the start time $\tau(u_i)$ of each job $u_i \in U$. The makespan of the schedule $S$ is the quantity

$$C_{\max} = \max\{\tau(u_i) + t(u_i) + q(u_i)|u_i \in U\}.$$

The objective is to minimize $C_{\max}$, the time by which all jobs are delivered.

The problem is equivalent to model $1|r_j, prec, |L_{\max}$ with release times and due dates, rather than delivery times. In this problem the objective is to minimize the maximum lateness of jobs.

The $1|r_j, q_j, prec|C_{\max}$ is a key component of several more complex scheduling problems. Problem is useful in solving flowshop and jobshop scheduling problems [1,6] and plays a central role in some real industrial application [20].

Baker et al. [3] developed algorithm to solve the problem with preemption allowed. Solving the preemptive version of the problem provides an efficient lower bound to the problem without preemption and will be used in our branch and bound method.

The problem $1|r_j, q_j|C_{\max}$ has been studied by many researches. Potts [18] proposed an $O(n^2 \log n)$ iterated algorithm, Nowicki and Smutnicki [16] presented a more efficient $3/2$ approximation algorithm which runs in $O(n \log n)$. Hall and Shmoys improved Potts algorithm and modified it to solve the problem with precedence constraints too. The worst-case performance ratios of there algorithm is equal $4/3$.

The works of Baker and Su [3], McMahon and Florian [15], Carlier [5], Grabowski et al. [7], Pan and Shi [17] developed branch and bound algorithms to solve the problem without precedence constraints using different branching rules and bounding techniques. The most efficient algorithm is algorithm by Carlier, which optimally solves instances with up to thousand of jobs. This algorithm constructs a full solution in each node of the search tree.

The branch and bound algorithms proposed by Baker and Su [3], Grabowski et al. [7] and Chandra *et al.* [6] can apply to the problem with precedence constraints. Liu [14] modified Carlier algorithm to solve the problem with precedence constraints very efficiently.

In fact, most heuristic algorithms can construct a nondelay schedule, which was defined by Baker [3] as a feasible schedule, with no processor being kept idle at the time when it could begin processing a task. An inserted idle time schedule (IIT) was defined by Kanet and Sridharam in [12] as a feasible schedule in which a processor is kept idle at the time when it could begin processing a task. In [9] we considered scheduling with inserted idle time for $m$ parallel identical processors and proposed the branch and bound algorithm for the problem with precedence-constrained tasks.

In [10] we proposed IIT algorithm for $1|r_j, q_j|C_{\max}$ problem without precedence constraints, which runs in $O(n \log n)$ times. By combining this algorithm and branch and bound method we developed *BB* algorithm, which can find optimal solutions.

In this paper, we propose the approximation algorithm HIJR and, by combining the HIJR algorithm and $B\&B$ method, we develop $BBI$ algorithm, which can find optimal solutions for the problem with precedence constraints. We use a binary branching rule, where at each branch node a complete schedule is generated.

This paper is organized as follows. The approximation algorithm HIJR is presented in Sect. 2. In Sect. 3 we consider properties of schedules, constructed by HIJR algorithm, which help to recognize optimality of schedules and to obtain its lower bound. The branch and bound algorithm is described in details in Sect. 4. Computational studies are provided in Sect. 5. Section 6 contains a summary of this paper.

## 2   Approximation Algorithm HIJR

Let $S_k$ be the set of jobs that has been scheduled after $k$ iterations. Algorithm HIJR is a greedy algorithm in the sense that at each iteration it adds any ready job to $S_k$. Algorithm HIJR generates the schedule, in which processor is kept idle at the time when it could begin processing a job.

**Definition 1.** *A job $u \notin S_k$ is called the ready job at the level $k$, if all its predecessors are included in the partial solution $S_k$.*

Let *time* be the completion time of the last completed job in $S_k$.
Then $time := \max\{\tau(u) + t(u) \mid u \in S_k\}$. $time := 0; k := 0; S_0 := \emptyset$.

The approximation schedule $S$ is constructed by HIJR algorithm as follows.

Step 1. Renumber the jobs in topological order.

Step 2. If $u_i \prec u_j$ then we change $r(u_j) = \max\{r(u_j), r(u_i) + t(u_i)\}$ and $q(u_i) = \max\{q(u_i), q(u_j) + t(u_j)\}$. This replacement does not affect the feasibility of any schedule.

Step 3. Let $r_{\min} = \min\{r(u) \mid u \in U\}$, and $q_{\min} = \min\{q(u) \mid u \in U\}$.
Define the lower bound $LB$ of the optimal makespan [5]:
$LB = \max\{r_{\min} + \sum_{i=1}^{n} t(u_i) + q_{\min}, \max\{r(u_i) + t(u_i) + q(u_i) \mid u_i \in U\}\}$.

Step 4. If there is no job $u_i$, such as $r(u_i) \leq time$, then $time := \min\{r(u_i) \mid u_i \notin S_k\}$. Select the ready job $u$ with maximum delivery time $q(u) = \max\{q(u_i) \mid r(u_i) \leq time\}$.

Step 5. Select the ready job $u^*$ with maximum delivery time, such as $q(u^*) = \max\{q(u_i) \mid time < r(u_i) < time + t(u) \ \& \ q(u_i) > LB/3\}$.

Step 6. If there is no such job $u^*$ or $q(u) \geq q(u^*)$, then $v = u$. Goto 11.

Step 7. Define the idle time of the processor $idle(u^*) = r(u^*) - time$. If $q(u^*) - q(u) < idle(u^*)$, then we select job $v = u$. Goto 11.

Step 8. Select the ready job $u_1$ which can be executed during the idle time of the processor without increasing the start time of the job $u^*$, namely such as $q(u_1) = \max\{q(u_i) \mid time \geq r(u_i) \ \& \ t(u_i) \leq idle(u^*)\}$. If we find $u_1$, then we select job $v = u_1$. Go to 11.

Step 9. Select the ready job $u_2$ with maximum delivery time, such as $q(u_2) = \max\{q(u_i) \mid time < r(u_i) \ \& \ r(u_i) + t(u_i) \leq r(u^*)\}$.

If we find $u_2$, then we select job $v = u_2$. Go to 11.

Step 10. We select the job $v = u^*$.

Step 11. Define the start time of job $v$: $\tau(v) := \max\{time, r(v)\}$; $time := \tau(v) + t(v)$; $k := k + 1$. If $k < n$, then go to 4.

We construct the approximation schedule $S = S_n$ and find the objective function $C_{\max}(S) = \max\{\tau(u_i) + t(u_i) + q(u_i) \mid u_i \in U\}$.

The algorithm sets on the processor the job $u^*$ with the large delivery time $q(u^*)$. If this job is not ready, then the processor will be idle in the interval $[time, r(u^*)]$. In order to avoid too much idle of the processor the inequality $q(u^*) - q(u) \geq idle(u^*)$ is verified on step 7. If the inequality is hold, we choose job $u^*$. In order to use the idle time of the processor we select job $u_1$ or $u_2$ to perform in this interval. Job $u^*$ starts at $\tau(u^*) = r(u^*)$. The HIJR algorithm runs in $O(n^2)$ times.

*Example 1.* We apply this algorithm to the example with 4 jobs: $a, b, c, v, a \prec c$. $r(a) = 1; r(b) = r(c) = r(v) = 0; t(a) = t(c) = 1; t(b) = M - 1; t(v) = 1/2$; $q(a) = q(v) = 0; q(b) = 1; q(c) = M - 1$.

After steps 1 and 2 of HIJR algorithm we have new release time $r(c) = 1$ and new delivery time $q(a) = M - 2$.

Lower bound of the optimal makespan is equal to $LB = M + 3/2$. Algorithm HIJR generates schedule $S = (v, a, c, b)$. $C_{\max}(S) = M + 2$.

The problem can be solved using extended Jackson's rule (algorithm Schrage [19]): whenever the machine is free and one or more jobs are available for processing, schedule an available job with the greatest delivery time.

Algorithm Schrage generates schedule $S_r = (b, a, c, v)$. $C_{\max}(S_r) = 2M - 2$.

# 3    Lower Bounds of the Optimal Makespan

We consider the properties of the schedules, constructed by the algorithm HIJR. If we have the schedule $S$, then for each job $j$ we have the start time $\tau(j)$. The objective function is $C_{\max}(S) = \max\{\tau(j) + t(j) + q(j) \mid j \in U\}$.

**Definition 2.** *Critical job $j_c$ is the first processed job such as $C_{\max}(S) = \tau(j_c) + t(j_c) + q(j_c)$.*

The delivery of critical job is completed last. Obviously, to reduce the objective function, the critical job must to be started earlier.

**Definition 3.** *The sequence $J(S) = (j_a, j_{a+1}, \ldots, j_c)$ in schedule $S$, such as $j_c$ is the critical job and job $j_a$ is the earliest-scheduled job so that there is no idle time between the processing of jobs $j_a$ and $j_c$ is called the critical sequence.*

The job $j_a$ is the first job in the schedule or there is any idle time before $\tau(j_a)$.

**Definition 4.** *The job $j_u$ in a critical sequence $J(S)$ is called the interference job, if $q(j_u) < q(j_c)$ and $q(j_i) \geq q(j_c)$ for $i > u$, and there are no any successors of job $j_u$ in $J(S)$.*

Now we define the delayed job, which can be started before a critical sequence.

**Definition 5.** *The job $j_v$ in a critical sequence is called the delayed job, if there are no any predecessors of job $j_v$ in $J(S)$ and $r(j_v) = \min\{r(j_i)|r(j_i) < r(j_a)\}$.*

**Lemma 1.** *If there is no interference job and there are no delayed jobs, then the schedule $S$ is optimal.*

We can represent the critical sequence as $J(S) = (S_1, j_u, S_2)$, where $S_1$ sequence jobs before the interference job $j_u$, and $S_2$ sequence jobs after $j_u$.
Let $r_{min}(S_2) = \min\{r(j_i)|j_i \in S_2\}$

Let $LB(P)$ be the lower bound of the optimal makespan for problem $P$. The next Lemma bounds deviation of $C_{\max}(S)$ from the lower bound $LB(P)$.

**Lemma 2.** *If $j_u$ is the interference job in the critical sequence $J(S) = (S_1, j_u, S_2)$, then $LB(P) = C_{\max}(S) - t(j_u) + \delta(S)$, where $\delta(S) = r_{min}(S_2) - \tau(j_u)$.*

*Proof.* Let $T(J(S))$ be the total processing time of jobs from the critical sequence $T(J(S)) = \sum_{i=a}^{c} t(j_i)$, then $C_{max}(S) = r(j_a) + T(J(S)) + q(j_c)$.

If $q(j_u) < q(j_c)$ and there are no any successors of interference job $j_u \in S_2$, then all jobs from $S_2$ are not available for processing at time $\tau(j_u) = r(j_a) + T(S_1)$ and then $\tau(j_u) < r_{min}(S_2)$. For an optimal schedule it is true: $C_{opt} \geq r_{min}(S_2) + T(S_2) + q(j_c)$, because $q(j_i) \geq q(j_c)$ for $j_i \in S_2$. Then

$$C_{max}(S) - C_{opt} \leq r(j_a) + T(J(S)) + q(j_c) - r_{min}(S_2) - T(S_2) - q(j_c) =$$

$$= r(j_a) + T(S_1) + t(j_u) - r_{min}(S_2) = t(j_u) - \delta(S),$$

where $\delta(S) = r_{min}(S_2) - \tau(j_u)$. Then, we obtain $LB(P) = C_{max}(S) - t(j_u) + \delta(S)$.

This lemma shows how it is possible to improve the value of the objective function by removing the interference job from the critical sequence. Our next lemma allows us to clarify lower bounds of the optimal makespan.

**Lemma 3.** *1. If $j_u$ is the interference job, then we generate two new problem:*
*1.1. Problem $PL$: constraint that the critical job precedes the interference job is added to problem $P$ $j_c \prec j_u$. Then $LB(PL) = C_{\max}(S) + \delta(S) - q(j_c) + q(j_u)$.*
*1.2. Problem $PR$: constraint that the interference job precedes $S_2$ sequence is added $j_u \prec j_i$, where $j_i \in S_2$. Then $LB(PR) = C_{\max}(S) - \tau(j_u) + r(j_u)$.*
*2. If there is no interference job for the critical sequence, but there is the delayed job $j_v$, then we can generate two new problem:*
*2.1. Problem $PL$: constraint that the delayed job $j_v$ precedes the critical sequence is added $j_v \prec j_a$. Then $LB(PL) = C_{\max}(S) - r(j_a) + r(j_v)$.*
*2.2. Problem $PR$: constraint that the first job $j_a$ of the critical sequence precedes the delayed job $j_v$ is added $j_a \prec j_v$. Let $J_1 = \{j_i \in J(S) \mid r(j_i) \geq r(j_a)\}$ and $T(J_1) = \sum_{i \in J_1} t(j_i)$. Then $LB(PL) = C_{\max}(S) - T(J(S)) + t(j_v) + T(J_1)$.*

*Proof.* 1.1. If $j_u$ is the interference job for the critical sequence $J(S)$ and the constraint $j_c \prec j_u$ is added, then $C_{opt}(PL) \geq r_{min}(S_2) + T(S_2) + t(j_u) + q(j_u)$.

$$C_{max}(S) - C_{opt}(PL) \leq r(j_a) + T(J(S)) + q(j_c) - r_{min}(S_2) - T(S_2) - t(j_u) - q(j_u) =$$

$$= r(j_a) + T(S_1) - r_{min}(S_2) + q(j_c) - q(j_u) = -\delta(S) + q(j_c) - q(j_u).$$

$LB(PL) = C_{max}(S) + \delta(S) - q(j_c) + q(j_u)$.

1.2. If $j_u \prec j_i$, where $j_i \in S_2$, then $C_{opt}(PR) \geq r(j_u) + t(j_u) + T(S_2) + q(j_c)$.

$$C_{max}(S) - C_{opt}(PR) \leq r(j_a) + T(S_1) - r(j_u) = \tau(j_u) - r(j_u).$$

Then $LB(PR) = C_{max}(S) - \tau(j_u) + r(j_u)$.

2.1. If the constraint $j_v \prec j_a$ is added, then $C_{opt}(PL) \geq r(j_v) + T(J(S)) + q(j_c)$. Then

$$LB(PL) = C_{max}(S) - r(j_a) + r(j_v).$$

2.2. If the constraint $j_a \prec j_v$ is added, then $C_{opt}(PR) \geq r(j_a) + T(J_1) + t(j_v) + q(j_c)$.

$$C_{max}(S) - C_{opt}(PR) \leq T(J(S)) - T(J_1) - t(j_v).$$

Then $LB(PR) = C_{max}(S) - T(J(S)) + T(J_1) + t(j_v)$.

## 4    Branch and Bound Algorithm

**Search Tree.** We construct a schedule $S$ in every node of the tree. The upper bound $UB$ is the value of the best solution known so far. In every node of the search tree we apply the HIJR algorithm and construct the schedule $S$ with objective function $C_{max}(S)$. Then, we check the optimality condition and, if it is satisfied, the node is closed. Otherwise, we specify the lower bound $LB$ using lemma 2 and 3. If $LB \geq UB$, then the current node cannot improve $UB$, and it is eliminated. Else branch to a left and a right node.

**Lower Bounds.** In Sect. 2 we define two simple lower bounds for the makespan of the optimal schedule $S_{opt}$. We can compute the best lower bound by a preemptive algorithm. The preemptive problem $1|r_j, q_j, prmp|C_{max}$ can be solved in $O(n^2)$ time [2]. The makespan of an optimum preemptive schedule is used as a lower bound for the problem by many researches [3,5,6]. We apply the Baker algorithm [3] for computing the lower bound of the optimal solution. For most randomly generated instances tested in our paper the makespan of an optimum preemptive schedule is very close to the makespan of an optimum schedule.

**Upper Bounds.** Two heuristics are used to find upper bounds for the branch and bound algorithm at each branch node: Schrage's heuristic and HJRI heuristic.

**Branch and Bound Algorithm** $BBI$.

Now we formally describe our branch and bound algorithm $BBI$, as follows:

We associate with each node of the search tree the problem $P$ and the lower bound $LB$. The upper bound $UB$ is the value of the best solution known so far.

1.1. Lower bound. Find a preemptive optimal schedule $S_p$ by the Baker algorithm. If no job is preempted, then the schedule $S_p$ is optimal for this node, this node is closed and $UB = \min(UB, C_{\max}(S_p))$. Else $LB = \max(LB, C_{\max}(S_p))$.

1.2 Find a schedule $S_r$ by Schrage's heuristic. If $UB > C_{\max}(S_r)$ then $UB = C_{\max}(S_r)$. If $LB < UB$, then find a schedule $S$ by HIJR algorithm. If $UB > C_{\max}(S)$ then $UB = C_{\max}(S)$. If $LB = UB$, then the schedule $S$ is optimal for this node and this node is closed.

1.3. Find critical sequence $J(S)$ and analyze it. If there are no interference and delayed jobs, then $S$ is optimal for this node and this node is closed.

1.3.1. If there is the interference job $j_u$ in $J(S)$, then we update $LB$ such as $LB = \max(LB, C_{\max}(S) - t(j_u) + \delta(S))$. If $LB \geq UB$, then the current node is eliminated, else branch to a left and a right node as follows:

**Left Node.** The precedence constraint $j_c \prec j_u$ is added. Update $LB$ such as $LB = \max(LB, C_{\max}(S) - \delta(S) + q(j_c) - q(j_u))$.

If $LB \geq UB$, then the left node is eliminated else set $r(j_u) = r(j_c) + t(j_c)$ and recalculate $r(j)$ for all successors of job $j_u$.

**Right Node.** The precedence constraint $j_u \prec S_2$ added. Update $LB$ such as $LB = \max(LB, C_{\max}(S) - r(j_a) - T(S_1) + r(j_u))$.

If $LB \geq UB$, then the right node is eliminated else set $q(j_u) = q(j_c) + T(S_2)$ and recalculate $q(j)$ for all predecessors of job $(j_u)$.

Step 1.3.2. If there is no an interference job, but there is the delayed job $j_v$ in $J(S)$, we update $LB$ such as $LB = \max(LB, C_{\max}(S) - r(j_v) + r(j_a))$.

If $LB < UB$, then branch to a left and a right node as follows:

**Left Node.** The precedence constraint $j_v \prec j_a$ is added. Update $LB$ such as $LB = \max(LB, C_{\max}(S) - r(j_a) + r(j_v))$.

If $LB \geq UB$, then the left node is eliminated else set $q(j_v) = q(j_a) + t(j_a)$ and recalculate q(j) for all predecessors of job $j_v$.

**Right node.** The precedence constrain $j_a \prec j_v$ added. Update $LB$ such as $LB = \max(LB, C_{\max}(S) - T(J(S)) + t(j_v) + T(J_1))$.

If $LB \geq UB$, then the right node is eliminated else set $r(j_v) = r(j_a) + t(j_a)$ and recalculate $r(j)$ for all successors of job $j_v$.

The algorithm works as long as there are unclosed nodes.

## 5    Computational Results

To illustrate the efficiency of our approach we tested it on randomly generated instances.

We conduct computational studies over two sets of random instances and compare the $BBI$ algorithm with the Carlier algorithm [5] and $BBH$ algorithm by Chandra et al. [6]. These three algorithms use different approximation algorithms to generate a schedule at each vertex of the search tree. Carlier algorithm uses Schrage's heuristic, $BBI$ algorithm uses HIJR algorithm and $BBH$ algorithm uses H heuristic [6]. All algorithms use a binary branching technique and a preemptive schedule to define a lower bound. The program for the $BBI$ algorithm is coded in Object Pascal and compiled with Delphi 7. All the computational experiments were carried out on a laptop computer 1.9 GHz speed and 4 GB memory. At first, we consider instances without precedence constraints. The number of jobs considered is from $n = 50$ to $n = 5000$. Job processing time are generated with discrete uniform distributions between 1 and 50. Release dates and delivery times are generated with discrete uniform distributions between 1 and $Kn$, for $K = 20$. Parameter $K$ controls the range of heads and tails. Pan found [17] that most hard ones occur when $14 \leq K \leq 25$ and we chose $K = 20$.

For each $n$ and $K = 20$ we generate 100 instances. The computational results are summarized in Table 1. The first column of this table contains the number of jobs $n$. Columns V, VM, T contain mean number of the search tree nodes, maximum number of the search tree nodes and mean computing time (in sec), for $BBI$ algorithm, respectively. Columns VC, VMC, TC contain the same values for the Carlier algorithm and columns VH, VMH, TH contain the same values for $BBH$ algorithm.

Table 1 shows the performance of this three algorithms according to the variation of the number of jobs. Out of the 700 instances, 491, 235 and 405 are solved at root node by $BBI$ algorithm, the Carlier algorithm and $BBH$ algorithm , respectively. From this table we observe that average computing time and mean number of the search tree nodes increases with $n$ for the Carlier algorithm. The $BBI$ algorithm performs better then the Carlier algorithm in finding an optimal schedule with fewer branch nodes for instances. Mean number of the nodes does not increase with $n$ and average computing time increases with $n$ at a slow speed for algorithm BBI.

**Table 1.** Performance of algorithms according to the variation of $n$.

| $n$ | V | VC | VH | VM | VMC | VMH | T | TC | TH |
|---|---|---|---|---|---|---|---|---|---|
| 50 | 1.05 | 10.7 | 1.5 | 2 | 17 | 24 | 0.000 | 0.000 | 0.000 |
| 100 | 1.00 | 15.8 | 1.4 | 1 | 28 | 3 | 0.000 | 0.005 | 0.004 |
| 300 | 1.06 | 28.4 | 1.3 | 2 | 188 | 59 | 0.000 | 0.077 | 0.010 |
| 500 | 1.15 | 46.2 | 1.5 | 2 | 115 | 19 | 0.012 | 0.353 | 0.048 |
| 1000 | 1.05 | 52.5 | 1.5 | 2 | 220 | 4 | 0.047 | 1.792 | 0.413 |
| 2000 | 1.10 | 34.0 | 1.5 | 2 | 77 | 3 | 0.204 | 5.149 | 0.847 |
| 5000 | 1.05 | 59.4 | 1.5 | 2 | 146 | 3 | 1.381 | 72.843 | 3.235 |

Mean number of the search tree nodes for the $BBH$ algorithm is slightly different from mean number of the search tree nodes for algorithm $BBI$, but average computing time is greater than this time for algorithm $BBI$.

**Table 2.** Performance of algorithms for instances with precedence constrains.

| $n$ | V | VH | VM | VMH | T | TH |
|---|---|---|---|---|---|---|
| 50 | 1.06 | 2.2 | 2 | 14 | 0.000 | 0.000 |
| 100 | 1.01 | 2.7 | 2 | 71 | 0.002 | 0.004 |
| 300 | 1.03 | 2.9 | 4 | 38 | 0.008 | 0.016 |
| 500 | 1.11 | 2.2 | 4 | 45 | 0.027 | 0.043 |
| 1000 | 1.10 | 2.1 | 5 | 51 | 0.086 | 0.201 |
| 2000 | 1.04 | 2.0 | 3 | 19 | 0.484 | 0.904 |
| 5000 | 1.07 | 2.0 | 3 | 11 | 1.274 | 3.143 |

We investigate instances with precedence constraints. The instance generation scheme is as follows [11]: for $1 \leq i < j \leq n$ a precedence constraint $i \prec j$ is generated when $\gamma < P_{ij}$, where $\gamma$ is generated from the uniform distribution over interval $[0,1]$, and $D = 0, 0.1, 0.3, \ldots, 0.9$.

$$P_{ij} = \frac{D(1-D)^{(j-i-1)}}{1 - D(1 - (1-D)^{(j-i-1)})}.$$

Table 2 presents the results of the algorithm $BBI$ and the algorithm $BBH$ for instances with precedence constrains. Parameter $D$ is equal 0.3. For instances with precedence constrains the algorithm $BBI$ generates on average fewer nodes of the search tree then the algorithm $BBH$. The average computational time of the algorithm $BBI$ is less than the average computational time of the $BBH$ algorithm.

Table 3 presents the results of the algorithm $BBI$ and the algorithm $BBH$ for instances with precedence constrains according to the variation of $D$. the The number of jobs $n = 500$. We can note that as $D$ increases, the instance becomes easier to solve. It is evident that precedence constraints decrease the number of feasible solutions. Our computational studies show that algorithm $BBI$ reduces equally the number of search nodes and the computational time.

**Table 3.** Performance of algorithms according to the variation of $D$.

| $D$ | V | VH | $VM$ | VMH | $T$ | TH |
|---|---|---|---|---|---|---|
| 0.1 | 1.4 | 3.7 | 20 | 61 | 0.022 | 0.038 |
| 0.3 | 1.3 | 2.2 | 23 | 38 | 0.019 | 0.036 |
| 0.5 | 1.1 | 2.0 | 14 | 35 | 0.017 | 0.033 |
| 0.7 | 1.1 | 1.5 | 11 | 12 | 0.014 | 0.029 |
| 0.9 | 1.04 | 1.2 | 9 | 11 | 0.012 | 0.024 |

# 6   Conclusions

In this paper we investigated algorithms for $1|r_j, q_j, prec|C_{\max}$ problem. We proposed the approximation algorithm HIJR and the branch and bound algorithm, which produces an optimal solution. The $BBI$ algorithm is based on the HIJR algorithm, uses a binary branching technique and a preemptive schedule to define a lower bound. Algorithm $BBI$ finds optimal solutions for all instances tested with up to 5000 jobs, which are randomly generated.

# References

1. Artigues, C., Feillet, D.: A branch and bound method for the job-shop problem with sequence-dependent setup times. Ann. Oper. Res. **159**, 135–159 (2008)
2. Baker, K., Lawner, E., Lenstra, J., Rinnooy Kan, A.: Preemptive scheduling of a single machine to minimize maximum cost subject to release dates and precedence constrains. Oper. Res. **31**, 381–386 (1983)
3. Baker, K., Su, Z.: Sequensing with due-dates and early start times to minimize maximum tardiness. Naval Res. Logist. Q. **21**, 171–176 (1974)
4. Brucker, P.: Scheduling Algorithms, 5th edn. Springer, Berlin (2007)
5. Carlier, J.: The one machine sequencing problem. Eur. J. Oper. Res. **11**, 42–47 (1982)
6. Chandra, C., Liu, Z., He, J., Ruohonen, T.: A binary branch and bound algorithm to minimize maximum scheduling cost. Omega **42**, 9–15 (2014)
7. Grabowski, J., Nowicki, E., Zdrzalka, S.: A block approach for single-mashine scheduling with release dates and due dates. Eur. J. Oper. Res. **26**, 278–285 (1986)
8. Graham, R., Lawner, E., Rinnoy Kan, A.: Optimization and approximation in deterministic sequencing and scheduling. A survey. Ann. Discret. Math. **5**(10), 287–326 (1979)
9. Grigoreva, N.: Branch and bound method for scheduling precedence constrained tasks on parallel identical processors. In: Proceedings of The World Congress on Engineering 2014, WCE 2014. Lecture Notes in Engineering and Computer Science, pp. 832–836, London, U.K. (2014)
10. Grigoreva, N.: Branch and bound algorithm for the single machine scheduling problem with release and delivery times. In: 2018 IX International Conference on Optimization and Applications (OPTIMA2018) Supplementory Volume, Petrovac, Montenegro, 1–5 October 2018 (2018)
11. Hall, L., Posner, M.: Generating experimental data forcomputation testing with machine scheduling applications. Oper. Res. **49**, 854–865 (2001)
12. Kanet, J., Sridharan, V.: Scheduling with inserted idle time:problem taxonomy and literature review. Oper. Res. **48**(1), 99–110 (2000)
13. Lenstra, J., Rinnooy Kan, A., Brucker, P.: Complexity of machine scheduling problems. Ann. Discret. Math. **1**, 343–362 (1977)
14. Liu, Z.: Single machine scheduling to minimize maximum lateness subject to release dates and precedence constraints. Comput. Oper. Res. **37**, 1537–1543 (2010)
15. McMahon, G., Florian, N.: On scheduling with ready times and due dates to minimize maximum lateness. Oper. Res. **23**(3), 475–482 (1975)
16. Nowicki, E., Smutnicki, C.: An approximation algorithm for a single-machine scheduling problem with release times and delivery times. Discret. Appl. Math. **48**, 69–79 (1994)

17. Pan, Y., Shi, L.: Branch and bound algorithm for solving hard instances of the one-mashine sequencing problem. Eur. J. Oper. Res. **168**, 1030–1039 (2006)
18. Potts, C.: Analysis of a heuristic for one machine sequencing with release dates and delivery times. Oper. Res. **28**(6), 445–462 (1980)
19. Schage, L.: Obtaining optimal solutionto resourse constrained network scheduling problems. Unpublished manuscript (1971)
20. Sourirajan, K., Uzsoy, R.: Hybrid decomposition heuristics for solving large-scale scheduling problems in semiconductor wafer fabrication. J. Sched. **10**, 41–65 (2007)

# Operational and Strategic Administration of Engineering Services Based on Information Models of Economic Life Cycle

Yuri Lipuntsov[(✉)] [iD]

Lomonosov Moscow State University, Moscow, Russia
lipuntsov@econ.msu.ru

**Abstract.** The development of information and communications technologies allow solving various classes of tasks, ranging from the optimization of individual transactions and ending with the construction of systems for implementing strategies at the level of the corporation, the sector or the economy of country as a whole. By increasing the scale of the task, the part of factors outside the control zone grows. However, with the growth of the informational maturity of the economy, the number of zones of inaccessibility are becoming smaller. The article considers the model of engineering services management. For the Russian economic reality, in addition to the micro-level models (a customer-oriented approach and the capabilities model), meso-level and macro-level models are actual. As a result of the transition from the planned system of economic activity to market mechanisms, the product chains, uniting a large number of different enterprises, were disrupted. Previously related companies were in various industrial holdings. The existing disunity is aggravated by the existing legal regulation of economic activity. With the object of creating a competitive environment, it does not allow to form the long contracts. These provisions also do not contribute to the formation of longtime strategies for both individual enterprises and the sector. The article shows how the use of information models, being indicative of the life cycle of an engineering project can help solve the accumulated problems.

**Keywords:** Engendering services · Life cycle · Capability model · Product chain

## 1 Introduction

A significant part of economic activity is associated with design-and-engineering projects. The life cycle of engineering projects is estimated for decades. Most of this period accounts for the utilize, and this stage of the life cycle is accompanied by the development and design of separate components of the underlying asset that fail, and sometimes it may require a complete replacement of the underlying asset. To perform current and thorough overhaul of the underlying asset, production, and supply of components is required. During a long service life, production conditions, standards, component suppliers may change significantly. These changes have a significant impact on the effectiveness of the services provided. The economic model based on the

© Springer Nature Switzerland AG 2020
Z. Wilimowska et al. (Eds.): ISAT 2019, AISC 1052, pp. 199–208, 2020.
https://doi.org/10.1007/978-3-030-30443-0_18

life cycle implies, on the one hand, the unification of participants at all stages of the life cycle, and on the other, the formation of a sustainable group of these participants, each of which clearly sees its own perspectives. The main idea of the economic model of the life cycle is a single view of the activity throughout the entire period of using the main asset. In addition to describing the main asset, a description of supported assets is described, which allows for the realization of various economic models, ranging from value flow and capability models to a sharing economy model. All this diversity of management approaches is based on active information support, which is created by information modeling methods.

## 2  Life Cycle Models

Many sectors of the economy, such as nuclear, hydroelectric engineering, various types of transport (air, motor, rail, sea), manufacturing, construction, are based on engineering products, systems or super-systems with a common service life, measured in decades. The main part of the costs in such projects is associated with support in the process of utilization: these are the costs of operating expenses for the provision of services, the costs of repair and overhaul of the main and supported equipment.

Various categories of participants are involved in providing services to the main asset, including developers of the main asset, its component's manufacturers, suppliers, organizations operating the main asset, or accompanying the work of the main asset.

During the period of utilization of the main and supported facilities, significant amounts of data are generated, the use of which can increase the value of the services provided to the client. To maximize value, there is a need to consider data coming from different stages of the life cycle from the perspective of assessing how customers use the services of the underlying asset.

Engineering viewpoint is mainly focused on the development of functions to achieve the parameters stated in the requirements. The use of scientific knowledge and experience is focused on the analysis of functional requirements. But this view does not always imply an assessment from the perspective of clients. In part, this is due to the lack of attention to the client-oriented approach, and from this point of view, considering the services of the main asset during its life cycle. To the cost reduction, the total cost of ownership of the main asset must be well analyzed.

All problems arising during the life cycle of the main asset can be divided into two categories: apparent from the engineering viewpoint, such as system failure, long delays, reduced overall equipment efficiency, etc. and non-obvious engineering issues that may arise as a result of obsolescence of the system or degradation of individual components. Apparent problems are obvious to customers, and to those who provide the service.

Traditional management methods are focused on solving well-defined problems related to the quality of the services provided, performance or similar indicators of services, cost-effectiveness. World-class companies are developing innovative business models to provide value-added solutions to their customers [1].

## 2.1 Information Model of the Life Cycle in Construction

Construction is one of the most developed domain areas in the field of information modeling based on the life cycle model. On average, the life cycle of a building is about 40 years. From this period, on average, 2 years are spent on project development, 5 years on construction, 30 years on the operation, and about a year on retire and dismantling. The main part of information modeling and filling occurs at the project and construction stage. Data collection, that is directly related to a particular object is not only the customer and the contractor but is distributed among many participants, located in various systems: architect, construction management, owner, communications manager, government authorities and utilities, construction engineer and other participants. Changing data or errors from one participant has an impact on other parts of the project. As a solution, [2] centralized information management is used through the information model, which describes certain objects, such as construction objects, buildings, structures, technological installations, etc.

The international community has developed a technology for information modeling, Building Information Modeling (BIM), designed to aggregate data from various construction participants. This modeling ideology has reached the level of international standards. Today, more than 30 countries participate in buildingSMART International (bSI) [3, 4]. The bSI has developed and produced the Industry Foundation Classes (IFC) Specification. The current IFC 4 version, released in March 2013, has been adopted as an international standard [3].

The implementation of this approach has made it possible to achieve a high degree of compliance with construction plans and legislative acts. By improving the accuracy of planning, adjustments to the budget have been reduced by 40%. Budgeting time reduced by 80%. Investors understood that using the system they get a cost saving due to collision detection: saving 10% of the contract value. Commissioning is 7% faster, budget accuracy is high - the error is about 3% [5].

Successful experience in the use of information modeling in the construction sector of Singapore has allowed expanding the application of methods in other areas of life: smart city, smart logistics and warehousing, smart farming and agriculture, smart waste management and other areas.

## 2.2 Information Model for Data Exchange in the Transport Industry

One of the types of engineering projects in which information exchange is actively used for operational management is short-term car rental - car sharing. This type of service is developing rapidly in Moscow, St. Petersburg and other major cities. The basis for the provision of such a service is the exchange of information between the operator, client, and vehicle. The main categories of data that circulate between the participants include information about the car (make, model), its technical state (gasoline residue, mileage, state of the central locking, etc.), the location of the car on a set of complete telemetry information. Information exchange allows for receiving operational information online. A feature of this service is the provision of a car for temporary use to a third party, so the service pays considerable attention to such a service as registration and identification of customers. In addition, the information on the client history of payments, rent, fines, bonuses, current balance etc. is accumulated. An essential part of the project is a payment system that provides convenience for the client.

Using a detailed information base for the utilize of the fleet allows getting a detailed cost estimate of the maintenance costs. In conjunction with the cost of purchasing cars (which can be carried out according to different schemes - leasing, buying on credit, rent), insurance is the value of the business model of a short-term lease. Based on these data, different rates are calculated. For this, activity-based costing and other information-advanced accounting methods can be used [6].

### 2.3 Life Cycle Model for Organizing the Interaction of Long-Term Engineering Projects

The basis for the organization of information interaction between the various participants of the engineering project is a model of the life cycle. As was shown, the life cycle of an engineering project in aggregated form can be represented as a combination of four stages: "Development", "Prepare", "Utilize" and "Retire". In the traditional wisdom of the life cycle of the main asset, the initial stages "Development" and "Prepare" are in the area of responsibility of the manufacturer of the asset. The "Utilize" and "Retire" steps are usually performing by the service provider. A traditional business model often militates against the efficient and cost-effective delivery of services during operation. To implement the economic model, in addition to the life cycle of the underlying asset, which is the basis of the engineering view, the life cycle of the supported assets is developed, which is the basis for the analysis of economic efficiency. Feedback from the "utilize" phase informs participants of the design and production stages of the next iteration or modernization of the underlying asset and supported assets.

Such a business model involves risk accounting and the provision of rewards for operating results between all interested parties. This encourages stakeholders to work together, solves questions about the required level of service, including the allocation of costs required to maintain the availability of the underlying asset, as well as in cases where the functionality of the underlying asset is not available. The higher the involvement of all participants in the distribution of income and expenses throughout the life cycle, the closer the cooperation will be and the more effective the business model.

Life cycle stages from an engineering viewpoint that can be added by economic models. One of the variants of such an extension is presented in the new British standard PAS 280: 2018, Through-life engineering services - Adding business value through a common framework, TES [7].

# 3   A Set of Models for Management in the Sector of Navigation Services

Currently, the active creation and development of global navigation satellite systems are underway. Areas of their use are expanding, requirements of consumers of navigation services are rising.

Outer space in the modern economy is perceived as a new space for entrepreneurship. Space is becoming a commercial space. In the sphere of space, the private space flight industry is developing. This is largely due to commercial motives, in contrast to political or other intentions. Space is at a stage in which a growing number of diverse participants around the world, including private companies, scientific communities, industry, and citizens. This is facilitated by the economy digitalization.

In terms of digitalization, Space is one of the main data providers. The creation and exchange of data have always been an important factor in the social and economic development of the state. Data has become an important resource for economic growth, job creation and the development of society. A significant part of the information exchange is associated with the geographic information services supplied by the Global Navigation Satellite Systems (GNSS). The growing demand for accurate location information combined with the evolution of GNSS technology creates the conditions for the development of this market. A smartphone user just needs to look at the screen to find out the exact time, location or get information about how to gain destination using an easy route.

To solve problems and actively develop various divisions in the space sector can become globally competitive only through full integration into society and the economy. Do this requires the space sector to be closely linked to the structure of society and the economy.

Therefore, in addition to a large number of engineering models, on the basis of which the stages of development and production of basic components for satellite navigation are performed, it is necessary to develop economic models. The basis of the economic model is the customer value chain, which intersects the "end-to-end" information service, which reflects the main activities and processes.

## 3.1   Engineering View on Navigation Services

The Global Navigation Satellite System (GNSS) is an infrastructure that allows users with a compatible device to determine their location, speed and time. GNSS signals are provided by various satellite positioning systems, including global and regional systems and satellite expansion systems. GNSS technology is used for many types of applications, including applications that are critical to safety.

Currently, in a regular mode, there are two global systems: GPS (USA), GLONASS (Russia). Two more systems are being finalized: Galileo (EU), BeiDou (PRC). To improve the quality of navigation services, regional systems and Satellite Based Augmentation System (SBAS) are used.

GNSS consists of three segments:

- Space segment
- Control segment
- User segment

The main space segment consists of satellites rotating in several orbital planes at an altitude of about 20,000 km. The constellation of satellites can be updated by adding new satellites and deactivating the old ones.

The task of the control segment is to continuously monitor and tuning the satellites, as well as monitor and adjust the content of the navigation messages sent by the satellites. It consists of a network of ground objects:

- Master Control Stations and backup stations.
- Ground-Station Communications Networks
- Monitoring stations

The user segment consists of a large number of different receivers that receive signals from satellites or other devices and processing it, calculate the position and synchronize the clock.

Delivery of navigation data from GNSS spacecraft is free of charge. The positioning accuracy, in this case, can vary from 2 to 10 m, depending on the type of user's navigation device, methods of processing the incoming signal, degree of calibration of user's navigation device, etc.

Improving the accuracy of positioning involves the use of additional systems that increase the accuracy of navigation. If at the beginning of building up the satellite navigation infrastructure the main financial burden was borne by the state budget, now the question of commercializing navigation services provided in addition to the main GNSS signal is on the agenda.

To improve the accuracy of positioning, there are several technologies that provide an amendment to the base signal. To calculate the correction, either ground stations or spacecraft are used. When using ground stations, the most common technique is the Real Time Kinematics (RTK), based on the use of code and phase measurements carrying primary GNSS signals. The technique allows achieving an accuracy of a centimeter. The main disadvantage of the RTK method is the need for a relatively close location of the reference station to the user, which allows minimizing the differential ionospheric delay [8]. To overcome such inconvenience, RTK networks are created that use a set of reference stations, which allows increasing the distance to 15 km.

The second technology - precise-point-positioning (PPP) - involves the use of additional spacecraft in precise baseline orbits. This method can provide positioning from a few centimeters to decimeters [9].

### 3.2    Economic Models of Navigation Services

The basis of the economic model is the value stream mapping to customer service. Delivery of the service begins with the Global Navigation Satellite System (GNSS) signal, then the signal is received by a hardware solution - a receiver, or a software solution - by a microcircuit. Signals are processed by devices, such as smartphones.

The processed signal is issued to applications - maps, navigators, and then the finished service is provided to end users. Figure 1 presents a general view of the model.

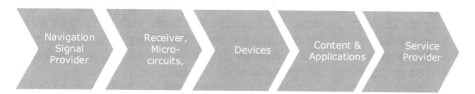

**Fig. 1.** The value stream mapping to the supply of GNSS data. Source: [10]

Further, the general model is detailed for individual industries, reflecting the specifics of the use of navigation services within the industry. Consider industry-specific features on the example of motor transport. The model for the use of navigation services in motor transport is currently not significantly different from the general model—it can include embedded devices, car manufacturers, suppliers of end-to-end navigation services, and a secondary market for navigation and monitoring devices.

Much more differences from the standard model appear when using navigation services for autonomous driving. Autonomous driving is a real prospect by 2030. Using navigation in a stand-alone vehicle implies an increase in positioning accuracy from 3 to 5 cm. To improve positioning accuracy, RTK and PPP technologies can be used. Traffic infrastructure of the new highway should be equipped with RTK stations, road condition informers, electronic road signs and other information providers that are integrated into the road system by integrators. Value chain for autonomous driving is shown in Fig. 2.

**Fig. 2.** Value chain of navigation services for autonomous driving Source: author's development

One of the main economic effects of the introduction of autonomous driving is to increase the density of traffic, reducing the distance between cars while maintaining vehicle speed The interACT project data indicate the three-time increase in the traffic: traffic capability of a three-lane road is 1700 cars with normal driving or 5500 cars with autonomous driving per hour. This is due to the reduction of the response rate to surrounding objects and active information interaction with them [11].

Road transport economy is changing: from the road's expansion and construction of interchanges, there is a transition to building an informational road infrastructure. To equip the road infrastructure with devices, they need to be produced, there is a need for a corresponding element base, the production of sensors, etc.

The use of a model of the life cycle of engineering projects involves the selection of the main modules of the engineering asset, a description of the delivery packages. Activity models are a tool to schematically reflect the processes performed with the necessary degree of detail, which allows to quickly understand the essence of the activity and identify intersectoral linkages. Activity models, value chains built on this basis and assessments of the capabilities of individual activity stages allow making process management transparent, identifying bottlenecks, and coordinating the efforts of various participants to achieve common goals [12].

Using such models, it is possible to obtain aggregated information to reproduce the business model of navigation services, including the cost of developing and creating satellites, the cost of placing satellites in orbit, the lifetime of the satellite in orbit and the service lives of individual components of spacecraft, the cost of creating maintaining control segment, etc.

### 3.3   Information Model of Navigation Services

An information model based on the life-cycle model of an engineering service implies a reflection of the main components of the underlying asset and supported assets. Taking into account the fact that commercialization of navigation services is supposed only in the high-precision positioning sector, the engineering system includes the GNSS space segment, the GNSS control segment, the positioning accuracy improvement system using ground-based RTK technology and their control center, and PPP technology. devices and control center. In addition, a user segment is presented, which includes navigation equipment equipped with a signal receiver, as well as applications using receiver signals.

The presented objects are involved in the process of providing navigation signals to end users. Since there are several suppliers of GNSS signals on the market of navigation services, in particular, GPS and GLONASS, the equipment must use GLONASS signals to provide accurate positioning.

Paid services based on the signal reception imply payment based on the quality of the supplied signal. The peculiarity of navigation services is that there is a one-way transmission of information - from the satellites to the client receiver. To build customer-oriented management, the operator must have information about the level of service provided. This information can be obtained either through the customer feedback or by transmitting data received by the customer to some operator, as is the case with vehicle dispatching.

From an information viewpoint, to build an information exchange system, it is necessary to form a single semantic space. To organize the collection and aggregation of data within the sector of activity, administrative work is necessary to separate the functions and responsibilities of data collection. The head offices of the corporation have an important role to play in this matter, since the development and implementation of a policy of navigation services are in their interests, including agreeing on the principles of organizing the interaction of building diverse market participants and developing a codirectional interaction [13, 14].

### 3.4    Models of Operational and Strategic Management

Having a complete information base allows you to implement a large number of economic models, including value stream mapping models and capability models.

Value stream mapping models allow business to interpret a proposed service or product in terms of value received by a customer. The user evaluates the provided navigation service according to such criteria as availability, accuracy, continuity, integrity, time for the first correction, reliability, and authentication. These service parameters are redefined by the quality of work of space and control GNSS segment, as well as by the systems that improve the quality of the navigation signal. For a detailed analysis of improving the quality of service, made a projection of quality criteria on each module composition, made on the basis of engineering models.

The criteria for evaluating the user segment and navigation equipment are power consumption, impact resistance, connectivity, compatibility, and traceability. These parameters are also predetermined by the presence of individual elements of the navigation equipment and are subject to decomposition for analysis.

An capability model shed light on the unique organizational capabilities of the team operating the engineering system, which is used to achieve their goals. Organization capabilities can be thought of as organizational-level skills related to personnel management, processes, and technology. An capability model describes a set of such skills that an organization needs to perform tasks.

One of the obvious problems and capabilities Russian navigation services sector is the restoration of product chains from the development and production of components to the creation of an operating system. The conclusion of long-term contracts involving the supply of components throughout the entire life cycle of the engineering project will allow restoring the product chains of the manufacturing sector and forming a national industrial outline of the navigation services sector in terms of sanctions. Import substitution, which is necessary to improve the quality of navigation services and reduce the cost of performing certain stages, implies a long-term development strategy not only for the sector as a whole but also for individual enterprises that need to feel confident in the medium and long-term perspective. Currently, some of them carry out one-time orders for the development of prototypes of individual components. Next comes the stage of creating an industrial design, an order for which a completely different manufacturer can receive, and the order for the production of an industrial batch - the third enterprise. At the same time, each of the manufacturers needs to keep in line, expand their skills. At the same time, the prospect of using these skills in the future is not always visible.

## 4    Conclusion

One of the main problems in the Russian engineering sector is disconnected production, formed as a result of the transition from a controlled management system to market mechanisms. It was assumed that market mechanisms will contribute to building a sound profile of production. In fact, the destruction of established industrial relations occurred, various enterprises of the product chain get in different industrial groups, and individual enterprises ceased to exist. Currently, there is a complex administrative task of rebuilding the production chains. When changing the management mechanism in the organization, the instability increases and the degree of control decreases. However, under current conditions, we have a fairly large potential in the field of information technology, which can significantly reduce this uncertainty and increase the amount of inspection of the upcoming changes. One of the important tools in improving manageability is the life cycle model of engineering projects, with which you can reproduce long production chains and define a national industrial body of the navigation services sector.

The transition to providing services with feedback from customers based on managing the information environment will allow implementing a commercialization policy, as well as developing a long-term strategy for developers and component manufacturers, as well as creating comfortable conditions for their self-organization and joint activities of agents and their communities in the information environment.

# References

1. Redding, L.: Through-life engineering services: a perspective from the literature. In: Through-life Engineering Services Motivation, Theory, and Practice, pp. 13–21. Springer, Heidelberg (2015)
2. Ton, T.: How information model in current digital economy can benefit investment. In: BIM Conference, Moscow (2017)
3. ISO 16739: 2013, Industry Foundation Classes (IFC) for data sharing in the construction and facility management industries, ISO/TC 184/SC 4 Industrial data. https://www.iso.org/standard/51622.html. Accessed 12 Dec 2018
4. ISO 29481-1:2016, Building information models – Information delivery manual, ISO/TC 59/SC 13 Organization of information about construction works. https://www.iso.org/standard/60553.html. Accessed 12 Dec 2018
5. Center for Integrated Facility Engineering (CIFE), BIM Statistics: based on 32 major projects using BIM, Stanford University, Stanford (2017)
6. CODECS D2.3 Workshop Documentation – Platforms, Pilots, Progress. https://www.codecs-project.eu/index.php?id=6. Accessed 2 Mar 2019
7. PAS 280 Through-life engineering services. Adding business value through a common framework. https://shop.bsigroup.com/ProductDetail?pid=000000000030371030. Accessed 18 Dec 2018
8. Alves, P.: Real-Time Kinematic with Multiple Reference Stations, Inside GNSS, vol. 4, July/August 2008
9. Laínez_Samper, M.: Multisystem real time precise-point-positioning, Coordinates, vol. VII, no. 2 (2011)
10. European GNSS Agency, GNSS Market Report, Publications Office of the European Union, Luxembourg (2017)
11. inLane, D6.3 Marketing Materials. https://inlane.eu/library/. Accessed 08 Mar 2019
12. Sokolov, I., Misharin, A., Kupriyanovsky, V., Pokusaev, O., Lipuntsov, Y.: Digital transport projects with global navigation satellite systems - the road to building integrated digital transport systems. Int. J. Open Inf. Technol. 7(1), 49–77 (2019)
13. Lipuntsov, Y.: Formation of the information space of the digital economy. Bull. Inst. Econ. Russian Acad. Sci. 6, 46–52 (2018)
14. Lipuntsov, Y.: Identifier and namespaces as parts of semantics for e-government information environment. Commun. Comput. Inf. Sci. 858, 78–89 (2018)

# Models of Organization Management

# Agent Model for Evaluating Efficiency of Regional Human Resource Management

Alexander V. Mamatov[1] , Igor S. Konstantinov[1] ,
Aleksandra L. Mashkova[1,2,3(✉)] , and Olga A. Savina[1,2(✉)]

[1] Belgorod National Research University, Belgorod, Russian Federation
aleks.savina@gmail.com
[2] Orel State University named after I.S. Turgenev, Orel, Russian Federation
[3] Central Economics and Mathematics Institute, Russian Academy of Sciences,
Moscow, Russian Federation

**Abstract.** In this paper we study influence of regional programs on the population resettlement in the Russian Federation at the example of the Belgorod region, where the Program of regional human resource development is being implemented. We divide population into clusters, grouped by a place of residence, age, education and employment; events of the Program are aimed at different clusters. We present the agent-based model of regional human resource dynamics. The model reflects sex-age structure, composition of households and spatial distribution of population; production; educational and administrative institutions in the region providing realization of the Program. To simulate social activity of the population in the agent-based model we need information about their reaction on the Program events. Within the survey we collect and process information about satisfaction with financial situation, desire to change job or region of residence of respondents in different clusters. Simulation on the basis of the agent-based model would show how these factors affect individual decisions and, consequently, size and structure of the regional human resource.

**Keywords:** Regional human resource · Spatial development · Migration ·
Agent-based modeling · Survey · Questionnaire

## 1 Introduction

Managing the regional human resource is an urgent task in sustainable socio-economic development of the country, since market mechanisms for regulating structure of the working population do not provide the required balance with socio-economic dynamics of the regions. Improvement of the resettlement system, distribution of productive forces and communication systems at the federal, regional and local levels is planned within the Strategy of spatial development of the Russian Federation [1, 4]. In turn, the Strategy should be taken into account within strategies of socio-economic development of the regions [3].

The aim of our research is to study influence of control actions (events of the Strategy) on the population resettlement in the Russian Federation. Particularly, in this paper we would accent at assessing impact of the regional human resource

© Springer Nature Switzerland AG 2020
Z. Wilimowska et al. (Eds.): ISAT 2019, AISC 1052, pp. 211–220, 2020.
https://doi.org/10.1007/978-3-030-30443-0_19

development program (using the example of the Belgorod region), and this methodology would further be adjusted to the general strategy at the federal level.

## 2   Research Methodology

In our study we have chosen agent-based modeling as a main method combining it with sociological surveys and statistical data analysis. The concept of agent-based modeling was proposed in the 1990s [8] and since then has been widely implemented in the analysis of economic, financial, social and environmental processes [7, 9, 11, 15]. There has been proposed a set of tools for simulation of population resettlement and urban planning using agent-based modeling [5, 6, 10, 14].

Application of agent-based approach allows to analyze influence of macro-level administrative decisions on the behavior of micro-level objects. The agent approach has already been used by our team to assess the effectiveness of social policy. In the study of 2008–2010, the impact of the activities of the federal target program "Russian language" on the dynamics of Russian language skills in Russia and abroad was evaluated [13]. Within this research the population (considered as set of agents) is grouped in clusters according to their social and demographic characteristics. For each cluster we set distribution of proficiency in Russian language. Language dynamics is characterized by indicators of the Program that show efficiency of the events. Modeling was performed for the Russian Federation and the CIS countries.

Methodology for research of regional human resource dynamics includes the following steps:

1. Reconstructing current territorial and demographic structure of population, administrative and economic system of the region in the agent-based computer model on the basis of statistical data.
2. Simulating dynamics of the system through decision-making procedures and behavior of agents that are affected by actions of the regional administration.
3. Conducting a series of experiments, statistical processing and analysis of the results.

In this article the issues of modeling dynamics of the system, connected with individual decisions of residents and impact of the events of the regional Program for the human resource development are discussed. At this stage, an urgent task is to collect and process information about the parameters of social activity of various categories of the population, including their satisfaction with the financial situation, desire to change job or region of residence, which affect their exposure to the Program events and, as a result, influence on the size and structure of the human resource of the region. To achieve this goal a series of sociological surveys are conducted.

## 3   Management of the Regional Human Resource

Human resource in the qualitative aspect is determined by demographic and migration processes, its quantitative character depends on the educational system in general and the professional retraining programs in particular. Administrative influence on demographic processes is indirect; the key channels are social security system: maternity transfers, transfers for low-income families with children, maternity capital. Immigration can be regulated legislatively by quoting the number of migrants and execution of the established requirements [12]. The educational system is regulated by setting the number of budgetary educational places and their distribution by groups of specialties. There are also educational orders from large enterprises, which provide targeted training of specialists for the industry.

In this study we have chosen the Belgorod region, since there the development strategy is currently being implemented, which is aimed, among other socio-economic tasks, at developing the regional human resource [3]. The Program of human resource development in the Belgorod region includes the following tasks: coordinating structure of the labor force and the needs of the labor market in the region; strengthening the relationship of educational institutions and the regional enterprises; support of business; social infrastructure development; designing a system of information support for the regional human resource management. Participants of the Program are regional administration, enterprises and organizations of the region, educational institutions for training and retraining of personnel at all levels and the population of the region.

The impact of events on each resident depends on their socio-economic characteristics and personal preferences. The significance of these factors differs among various categories of the population; therefore, it is necessary to divide the population into homogeneous groups – clusters (Table 1).

**Table 1.** Grouping characteristics of the population clusters.

| Feature | 1 | 2 | 3 | 4 | 5 | 6 | 7 | 8 | ... | 17 | 18 | 19 | 20 | 21 |
|---|---|---|---|---|---|---|---|---|---|---|---|---|---|---|
| Schoolchildren | + | + | | | | | | | | | | | | |
| Students | | | + | | | | | | | | | | | |
| Employees | | | | | + | | + | | | + | | + | | |
| Unemployed | | | | | | + | | + | | | + | | + | |
| Self-employed | | | | + | | | | | | | | | | |
| Pensioners | | | | | | | | | | | | | | + |
| Age under 35 years | | | | | + | + | | | | + | + | | | |
| Age over 35 years | | | | | | | + | + | | | | + | + | |
| Unqualified | | | | | + | + | + | + | | | | | | |
| Qualified | | | | | | | | | | + | + | + | + | |
| Town | + | | | | + | + | + | + | | | | | | |
| Village | | + | | | | | | | | + | + | + | + | |

Clustering agents in the corresponding feature space is based on the assumption that for agents, who are closed in this space, Program events should be similar, and for agents represented by distant points - different. The criteria for grouping are: place of residence (city, village); age (up to 35 years, over 35 years); education (schoolchild, student, qualified worker or unqualified worker); employment (working, unemployed, self-employed). Population clusters formed on the basis of the described criteria are presented in Table 1.

Events of the Program of the regional human resource development are aimed at different clusters of the population, and for each group the events have different economic and social efficiency. Program events and indicators of their effectiveness are presented in Table 2.

**Table 2.** Events of the Program of the regional human resource development.

| Sphere of the event | Event | Efficiency indicator of the event | Target cluster |
|---|---|---|---|
| Education | Increasing number of specialties in educational institutions in the region | Number and average USE grade of the target groups of specialties students | 1, 2 |
| | Increasing number of target educational quota from regional enterprises | | |
| | Assigning students to the enterprises of the region at the time of study | Number of graduates employed at enterprises of the region | 3 |
| Economy | Creation of workplaces in towns and villages of the region | Percentage of unemployed in the region | 6, 8, 10, 11, 14, 16, 18, 20 |
| | | Outflow of residents from the region | |
| | Subsidies for regional business | Number of self-employed in the region | 4 |
| | | Number of new jobs in commercial organizations | |
| Social sphere | Providing housing for young professionals | Outflow of young specialists from the region | 5, 13 |
| | Increasing the number of kindergartens, clubs, sections | | 5, 9, 13, 17 |
| | Regional pension benefits | Outflow of experienced specialists from the region | 7, 11, 15, 19 |

## 4  Agent Model of Human Resource Dynamics

The impact of the Program events on the decisions of residents and the changes resulting from these decisions in the size and structure of the regional human resource are studied on the basis of the agent-based model of regional human resource dynamics. The model includes a number of interrelated modules: "Demography and Migration", "Education", "Economy", "Social environment" and "Regional Administration", each of those reflects significant factors affecting quantitative and qualitative composition of the labor force (Fig. 1). The geography of the model is set of territorial units corresponding to the districts of the region; regional center and other towns are set separately [12].

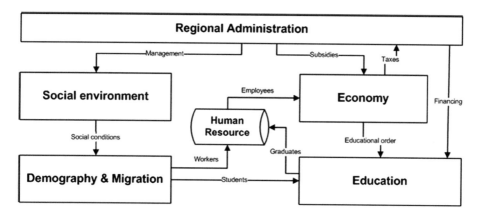

**Fig. 1.** Structure of the agent-based model of the regional human resource dynamics.

The module "Demography and Migration" generates agents representing population of the region, their grouping by households, resettlement by towns and districts. During modeling time agents get older, marry, divorce, die; they also migrate from the CIS countries (within the set quotes) and move to other regions.

Organizations of various economic sectors are created in the module "Economy". We set volume of production output and number of workplaces on the basis of statistical data [2]. Agents are assigned to workplaces in accordance with their qualifications; workable unemployed agents are assigned to the employment centers.

Education system in the model consists of educational institutions of various levels: school, secondary vocational and higher education. For each institution a list of educational places is formed, and agents of the corresponding age are assigned to them. Some details on this issue are presented in [12].

Social environment in the model includes housing, public transport, kindergartens and other parameters that affect quality of life in the region.

Regional administration implements events of the Program for regional human resource development.

The agent state diagram reflects its transition between states corresponding to different clusters. The state diagram of a city resident agent, shown in Fig. 2, corresponds to maturation, education, employment, retirement. Reverse transitions between clusters 4 and 10, 7 and 14 correspond to temporary loss or change of job.

Behavior of agents in the model is determined in education, employment and migration. In the educational sphere agents choose level and specialty of professional education. After graduating agent chooses to be an employee or a businessman and a sector of economy for his future job. Agents can change region of their residence; migration is directly related to education if agent is an entrant, or with employment, if he is a graduate or a qualified employee.

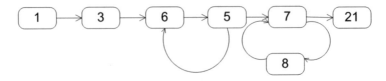

**Fig. 2.** State diagram of a city resident agent.

Dynamics of the model is determined by the impact of the Program's events on the regional human resource. To reflect influence of events on key indicators we use the following algorithm:

1. For a given event type, number of agent clusters connected with the event is calculated, the property of the agent that is changing after the event, and percentage of agents that are affected by an event of a given type are selected.
2. Depending on the administrative level of the event, the following steps are carried out for all, some or one locality in the region:
   2.1. Find the total number of agents in clusters, connected with the event, residing in this locality;
   2.2. Calculate the number of agents affected by the event;
   2.3. Change the specified property for the calculated number of agents.
3. Modeling of individual decision making by agents.
4. Recalculation of indicators of the Program.

The state and transition diagram reflects the process of assessing impact of the events on the target indicators for each agent group (Fig. 3).

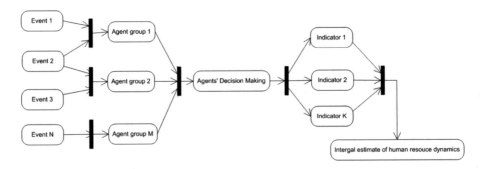

**Fig. 3.** Implementation of the Program events in the model.

Initially, the model consistently handles transitions to evaluate effect of the held events to change the values of target indicators in each agent group. Parameters of the transitions are probability characteristics of sensitivity of a particular agent group for various events, obtained on the basis of sociological surveys and statistical analysis. Further the agents make individual decisions with parameters changed under the influence of events. According to the results of the simulation, the values of the target indicators and the integral estimate of the regional human resource are formed.

The input of the model receives statistical information on the economic structures of the region, demographic and migration processes [2], plan of events of the Program of regional human resource development [3], and also assessment of the importance of factors influencing personal decisions, obtained through sociological surveys.

## 5   The Survey: Organization and Results

To obtain information about the parameters of social activity of the population and their simulation in the agent-based model of the regional human resource dynamics, a sociological survey was conducted in the Belgorod region. The purpose of the survey was to determine and assess factors influencing the decision to change the region of residence (Table 3).

**Table 3.**  Factors influencing the choice of region of residence.

| Group number | Type of factors in the group | Example of factors in the group |
|---|---|---|
| 1 | Economic factors | Personal income, average income per member of a household, living wage in the region |
| 2 | Personal factors | Presence of relatives in the region, personal assessment of the region |
| 3 | Infrastructure | Development of the public transport system, availability of medical services, outlets, kindergartens and schools |
| 4 | Ecology | Climate, proximity of industrial areas |

Two questionnaires were prepared for the survey: the first for employable adult residents, the second for students and schoolchildren. The questionnaires include the following sections:

Section 1. Affiliation of a resident to a social group (cluster): gender, age, place of residence (city/village), level of education, specialty, employment (employee/self-employed/unemployed).

Section 2. Significance (from 0 to 100) of each factor affecting the choice of region of residence (Table 3).

Section 3. Attitude towards changing region of residence (neutral/preference of native region/preference of another region or country).

Section 4. Evaluation of the impact of the Program for regional human resource development events on personal preferences and intentions in employment and migration.

At the first stage of the survey, 150 students of different ages living in towns and villages of the Belgorod region were interviewed. Results of the survey were processed by the following algorithm:

Step 1. Split the respondents into clusters in accordance with their answers to the Section 1 of the questionnaire. Representatives of three clusters were among the respondents: cluster 1 – schoolchildren living in the towns; cluster 2 – schoolchildren living in rural areas; cluster 3 – students. The number of representatives of each cluster among the respondents is presented in Table 4.

**Table 4.** Survey results in the Belgorod region.

| Cluster | | 1 | 2 | 3 |
|---|---|---|---|---|
| Number of representatives | | 45 | 40 | 65 |
| Position, % | leave | 69 | 50 | 61.5 |
| | neutral | 20.7 | 40 | 15.4 |
| | stay | 10.3 | 10 | 23.1 |
| Significance of the factor groups, % | 1 | 0.33 | 0.29 | 0.3 |
| | 2 | 0.2 | 0.26 | 0.23 |
| | 3 | 0.2 | 0.22 | 0.21 |
| | 4 | 0.27 | 0.23 | 0.27 |
| Influence of the events, % | E | 55 | 70 | 67 |
| | H | 14 | 5 | 65 |
| Intention, % | | 79 | 65 | 54 |

Step 2. Calculate significance of each group of factors (Table 3) within clusters, based on the data from the Section 2 of the questionnaire.

$$\alpha_{ij}^* = \frac{\alpha_{ij}}{\sum_{i=1}^{4} \alpha_{ij}}, \tag{1}$$

$$\alpha_i^* = \frac{\sum_{j=1}^{m} \alpha_{ij}^*}{m}, \tag{2}$$

where $\alpha_{ij}$ - significance of a factor group $i$, $i = \overline{1,4}$ for the resident $j$, $j = \overline{1,}$; $\alpha_{ij}^*$ - normalized individual significance of a factor group $i$; $\alpha_i^*$ - averaged significance of a factor group $i$ within the cluster.

Results of calculations for each cluster are presented in Table 4.

Step 3. Calculation of percentage distribution of residents' attitudes towards the choice of region of residence in each cluster.

Step 4. Evaluation of the influence of the Program events on decisions of representatives in various clusters. For students, events related to education (group E) were considered - creation of additional educational places, making relations with the regional enterprises; and improvement of living conditions (group H) - provision of housing for young professionals. Table 4 shows the percentage of respondents who were interested in implementing events of each group.

Step 5. Calculation of percentage of respondents with a firm intention to move to other regions in selected clusters. For clusters 1 and 2, percentage of respondents who were ready to move to another region for education was calculated. For cluster 3, percentage of respondents who did not want to return to their region after graduation was determined (Table 4).

According to the results of the survey, the following conclusions can be drawn:

1. In general, more than half of the respondents see their future outside the region and plan to leave for education and employment.
2. A large part of the respondents is interested in establishing communication with the regional enterprises and subsequent employment. Provision of housing in the region is not a serious incentive for the majority (more than 80%) of the respondents.
3. Respondents with a neutral position tend to stay in their native region the older they become: in clusters 1 and 2 (average age 15 years), this indicator is about 10%, in cluster 3 (average age 23 years) - already 23%.

For the Program of regional human resource development the target groups are residents with a neutral position. In favorable conditions, they would highly likely continue to live and work in the region.

## 6   Perspectives of the Study

Currently, a series of surveys are continuing among other categories of the population: employed, self-employed and unemployed. Analysis of data in these categories would make it possible to assess impact of economic and socially-oriented events of the Program of regional human resource development on the decisions of residents in employment and relocation to other regions. In total, it is planned to interview at least 500 respondents of different ages, gender and employment.

Obtained results would be loaded into the agent model of regional human resource dynamics, in particular, the calculated values would reflect attitudes and intentions in the agent groups corresponding to the clusters of respondents, and assess their liability to the Program's events. Within experimental studies on the agent-based model estimates of the impact of the Program events on decisions of residents would be obtained, resulting in the forecast of human resource dynamics in the Belgorod region.

Perspective direction of our research is conducting a series of surveys in various federal districts, which would allow to take into account influence of economic factors, social environment and climatic conditions on the attitudes of residents. The collected data would be integrated into the model of the Russian Federation spatial development and used to analyze effectiveness of measures of the Strategy of spatial development of the country.

**Acknowledgement.** The reported study was funded by RFBR according to the research project № 18-29-03049.

# References

1. Ministry of Economic Development of the Russian Federation official website. http://economy.gov.ru/minec/main. Accessed 22 Mar 2019
2. Russian Federation Federal State Statistics Service. http://www.gks.ru. Accessed 15 Mar 2019
3. Strategy of socio-economic development of the Belgorod region for the period up to 2025. http://docs.cntd.ru/document/428596289. Accessed 05 Dec 2018
4. Strategy of the spatial development of the Russian Federation for the period till 2025. http://static.government.ru/media/les/UVAlqUtT08o60RktoOXl22JjAe7irNxc.pdf. Accessed 27 Feb 2019
5. Barros, J.: Exploring urban dynamics in latin american cities using an agent-based simulation approach. In: Agent-Based Models of Geographical Systems, pp. 571–589. Springer, Dordrecht (2012)
6. Benenson, I., Omer, I., Hatna, E.: Entity-based modeling of urban residential dynamics: The case of Yaffo, Tel Aviv. Environ. Plan. B: Urban Anal. City Sci. **29**(4), 491–512 (2002)
7. Bonabeau, E.: Agent-based modeling: methods and techniques for simulating human systems. Proc. Natl. Acad. Sci. **99**(suppl 3), 7280–7287 (2002)
8. Epstein, J., Axtell, R.: Growing Artificial Societies: Social Science From the Bottom Up. MIT Press, Brookings Institution, Cambridge, Washington, D.C. (1996)
9. Epstein, J.: Modeling civil violence: an agent-based computational approach **99**(Suppl. 3), 7243–7250 (2002)
10. Feitosa, F.F., Le, Q.B., Vlek, P.L.G.: Multi-agent simulator for urban segregation (MASUS): a tool to explore alternatives for promoting inclusive cities. Comput. Environ. Urban Syst. **35**(2), 104–115 (2011). https://doi.org/10.1016/j.compenvurbsys.2010.06.001, http://www.sciencedirect.com/science/article/pii/S0198971510000608
11. Macy, M.W., Willer, R.: From factors to factors: computational sociology and agent-based modelling. Ann. Rev. Sociol. **28**, 143–166 (2002)
12. Mamatov, A.V., Konstantinov, I.S., Mashkova, A.L., Savina, O.A.: Information support system for regional human resource development. Amazonia Investiga **7**, 426–436 (2018)
13. Mashkova, A.L., Savina, O.A., Lazarev, S.A.: Agent model for evaluating efficiency of socially oriented federal programs. In: 11th IEEE International Conference on Application of Information and Communication Technologies (AICT), vol. 2, pp. 217–221. Institute of Control Sciences of Russian Academy of Sciences, Moscow (2017)
14. Semboloni, F., Assfalg, J., Armeni, S., Gianassi, R., Marsoni, F.: CityDev, an interactive multi-agents urban model on the web. Comput. Environ. Urban Syst. **28**(1), 45–64 (2004)
15. Tesfatsion, L.: Agent-based computational economics: growing economies from the bottom up. Artif. Life **8**(1), 55–82 (2002)

# IT Reliability-Based Model of Excellent IT Solutions for Controlling – Theoretical Approach

Agnieszka Bieńkowska$^{(\boxtimes)}$ ⓘ, Katarzyna Tworek ⓘ, and Anna Zabłocka-Kluczka ⓘ

Faculty of Computer Science and Management, Wroclaw University of Science and Technology, Wrocław, Poland
{agnieszka.bienkowska,katarzyna.tworek,
anna.zablocka-kluczka}@pwr.edu.pl

**Abstract.** The main of the article is to present a theoretical IT reliability-based model of excellent IT solutions for controlling. The problem of the evaluation of IT solutions for controlling is presented. IT reliability is proposed as one of the measures of those solutions. Its usefulness for differentiating groups of users is verified to establish the possibility for the development of agent model of excellent IT solutions for controlling based on IT reliability.

**Keywords:** Management · Controlling · IT reliability · IT solutions · Agent model

## 1 Introduction

Controlling is one of the methods of management support that brings benefits for the organization in the conditions of a changing or turbulent environment. However, the implementation of controlling in the organization requires proper development of its solutions. In theory, the invariably important research problem is the usefulness of controlling in business practice and shaping the solutions referring to it, described as the way in which substantive issues relating to controlling are settled in the organization, especially in the areas: functional, organizational and instrumental [2: 81]. After Ossadnik [31] three groups of controlling solutions can be mentioned: functional (referring to objectives, functions and actions undertaken by controlling), institutional (organizational) (referring to the organizational structure of controlling, i.e. controllers and responsibility centers) and instrumental (referring to tools enabling the controlling activities, including IT tools). Controlling solutions are diverse [2] and it was proven that it is reasonable to implement in the organization such controlling solutions (referring to each of the three mentioned groups), which would be characterized by a high level of excellence, related particularly to the functioning of the organization as a whole [2]. The low quality of controlling and the need for improvement measures in this area are discussed both in the literature and practice. It should be noted that the suggested low quality of controlling would refer to the "product" of controlling, which is the broadly understood as quality of services offered by the implementers of the idea of this

© Springer Nature Switzerland AG 2020
Z. Wilimowska et al. (Eds.): ISAT 2019, AISC 1052, pp. 221–231, 2020.
https://doi.org/10.1007/978-3-030-30443-0_20

method (quality of controlling outputs). The theorems about the necessity of improvement concern controlling solutions. It can be assumed that the high quality of controlling solutions affects the product quality of controlling itself, which means that there is a lasting connection between the quality of controlling solutions and the quality of its outputs [2]. The quality of controlling services, in turn, influences the broadly understood quality of management, understood as increasing its efficiency and adjusting to changes taking place inside and outside the organization as well as ensuring rationality of management [42]. As a consequence, the quality of controlling services determines the success of the organization, ensuring the continuity of the organization's operation and its development [1], as well as improving its competitiveness [24, 30]. In this context, it can be assumed that the quality of controlling solutions indirectly affects the organizational performance [2].

The search for excellent controlling solutions (perfect solutions, model solutions) or formulating recommendations concerning this notion has already been the subject of a scientific debate [2, 29, 41]. However, the model of excellent IT solutions for controlling have not been developed yet, especially taking into account the requirements of various categories of controlling users, although the subject of IT use as a tool is one of a key topics in modern controlling theory [4, 7, 22, 33, 43] and authors emphasize the importance of IT solutions of controlling, indicating them as a prerequisite not only for its proper implementation, but also for smooth functioning [11, 43].

The shape of excellent IT solutions of controlling should consider various contextual variables (as in the case of functional and organizational solutions). Attempts in this area in general terms were made in [2]. However, the presented approach seems to be insufficient, because in case of IT solutions of controlling design, the needs and expectations of various categories of their users should be considered, especially in the context of their tasks. Hence, the agent-based approach might be useful. The research [5] confirmed the mediating role of job performance for the relation between controlling use and organizational performance. Moreover, it was proved that IT reliability and UX are moderators of such a relation. The agent-based model will represent heterogeneity in the characteristics and behaviors of users [9], enabling the analysis concerning the influence of this heterogeneity on the shape of excellent IT solutions of controlling. Agents will be understood as various groups of IT solutions of controlling users (with their expectations, needs and tasks performed with IT solutions of controlling) separated by specific characteristics (e.g. role in organization, management level, type of interaction with IT solutions of controlling, and job characteristic) – called categories of users. Moreover, those categories of users will differ in case of the significance of their expectations for the final shape of IT solutions of controlling. Therefore, the main aim of the article is to propose an agent-based model of excellence of IT solutions of controlling in the organization.

## 2  IT Solutions for Controlling

Controlling, as one of the most often used methods of management in contemporary organization [6, 40], is understood as "management support (...) consisting of coordinating the process of solving specific management tasks, supervising and monitoring the course of their implementation, as well as participating in the performance of these tasks, mainly in the field of planning, controlling and providing information" [2: 289].

The implementation of controlling functions and tasks is not possible without the appropriate tools. However, controlling does not have its own set of tools. Most of those used within it were not developed for this method of management support but were created as a result of modification of existing management tools or were taken directly from other methods, especially from management accounting. Nowadays this purely substantive instruments of controlling "cannot be operated without the intensive support of information technology (IT)" [37: 4]. "Due to the huge amount of data processed and the need to make quick decisions, modern controlling is de facto doomed to computer aid" [15: 572], and "the implementation of controlling almost automatically connects with the implementation of the IT solution" [25], as it seems impossible to create, transform and transfer management information without appropriate IT solutions [22, 33]. "The implementation of appropriate IT modules is not only a support, but sometimes a necessary condition for the implementation of controlling solutions" [27: 180] and later their efficient functioning. Improperly selected (inappropriate) or poorly implemented IT solutions of controlling are perceived as an important barrier to the effective functioning of controlling in an organization [26].

IT solutions of controlling belong to the instrumental solutions of controlling. In a fact, IT is the factor, which completely changed the way contemporary organizations use management methods and techniques, including controlling, which seems to be especially susceptible to influences of IT solutions because of its information needs [3, 11, 43]. Ernst and Young [14] Report stated that IT solutions support is one of most important challenges ahead of controlling and its use in organization is not possible without that support. Moreover, there is an interesting two-way relation. On the one hand, the shape of instrumental controlling solutions (which include IT solutions of controlling) is undoubtedly a derivative of the assumed objectives and functions of controlling and its organizational solutions. On the other hand, the opportunities offered by IT can influence the scope and manner of implementing of controlling tasks in the organization. Essentially, IT solutions created for controlling needs should follow the following requirements:

- general, as well as all other IT solutions in the organization, concerning their reliability,
- specific (substantive), concerning the ability to support tasks of controlling [17, 22, 27].

The IT solutions of controlling differ in case of task complexity, which they are able to support – from simple information gathering tasks, through more complex but structured information and deviation analysis to complex and unstructured prediction analysis. Assuming that task complexity and structuralization is a basis for typology of IT solutions of controlling, they can be divided into two basic groups [17]: Accounting Information Systems (AIS) (supporting less complex and structured tasks) and Management Information Systems (MIS) (better for more complex tasks and definitely needed to unstructured tasks). Moreover, in case of AIS, three types of systems can be distinguished: Transaction Processing Systems (TPS), which allow for simple and structured tasks support (information management from daily business operations support), Financial Reporting Systems (FRS), which allow for more complex but still structured tasks support (gathering and analysis of financial statements information) and Management Reporting Systems (MRS), which allow for complex and structures tasks support (all available information processing and analysis for decision-making purposes and deviation analysis) [17, 41]. However, in order for IT solutions to be able to support unstructured tasks, MIS are needed, which are able to process financial and non-financial information in order to generate predictions of the future situation and provide more complex deviation analysis. Based on that typology and previous work of authors proposing a preliminary typologies of IT solutions for various management methods [41] and controlling itself [2], it is possible to develop the more detailed typology of IT solutions of controlling based on identified user requirements concerning tasks support needed from IT solutions. A detailed description of specific IT solutions used in management (as well as in controlling) is presented, among others, in the works of [10, 11, 15, 17, 21, 27, 35, 36] and can be used as a source of knowledge for such a typology development.

For the purpose of this development, it is important to remember that IT solutions of controlling have to meet the requirements of various categories of users (among them at least controllers and managers should be indicated) and meeting their needs is subordinated to the basic goal set for those solutions. The results of the research presented in [2] show that the IT solutions of controlling are most often used among controllers and managers, especially by top executives (over 90% of responses), as well as middle managers in the area of finance, production and sales & marketing (more than half of responses). On the one hand, IT solutions of controlling are supporting controllers in the implementation of controlling tasks (i.e. in generating cross-section information for management purposes), and on the other - supporting managerial staff in tasks concerning controlling organizational management (decision making). It is obvious that the needs in terms of information perception as well as the tasks and communication (typical criteria for agent differentiation) of the mentioned user categories are different [16]. It is confirmed by Weißenberger et al. [43], who propose Weißenberger and Angelkort model of IT solutions of controlling and underline that the analysis of those solutions should include at least those two points of view: managers and controllers. Moreover, they assume that it is impossible to analyze the integration level of IT solutions of controlling and especially the perceived quality of controlling outputs without including various categories of users and their perception of tasks complexity and structuralization, which should be supported by those solutions. In this project, identification and development of IT solutions of controlling will become the starting point for assessing their excellence with the use of IT reliability, quality of controlling outputs and organizational performance.

## 3   Quality of Controlling Solutions. Measuring the Quality of Controlling Solutions. Excellent Solutions of Controlling

The quality of controlling solutions is on the one hand crucial in the search for excellent controlling solutions and on the other in determination of the directions of its improvement. The quality of controlling solutions (functional, organizational and instrumental) is understood as the degree to which the set of inherent properties of controlling solutions meets the requirements of controllers (as they directly use these solutions, and thus are their direct users), and recipients of controlling services – other controlling stakeholders (in an indirect way) [2].

Excellent controlling solutions hence should be understood as the highest quality solutions ensuring high efficiency of this method, which indirectly also translates into high quality of management and high organizational performance [2]. From this it follows that there is no single excellent solution in the field of controlling. Uniform solutions can be distinguished for each of the groups of organizations with similar external conditions and internal characteristics as well as, as previously indicated, categories of user (which is particularly important in the process of determining excellent IT solutions of controlling). In this context, the model of excellence, as mentioned earlier, means a set of IT solutions of controlling characterized by the highest level of perfection assessed in particular parameters related to the functioning of the organization as a whole.

The key aspect of the evaluation of the quality of controlling solutions is the selection of the appropriate methods for measuring it. This measurement must enable the evaluation of the degree to which the set of properties related to controlling solutions meets the requirements. The literature has not yet developed a comprehensive method in this regard. Basically only Brokemper [8] directly mentions the measurement of the quality of controlling solutions. As the two main parameters of the evaluation, the author indicates costs and quality, while the evaluation of the cost parameter is carried out in accordance with the principles of benchmarking, and quality parameter from the point of view of auditors.

To assess IT solutions of controlling an indirect method should be chosen, based on the quality of controlling (as a parameter directly related to quality of controlling products), quality of management and organizational performance (as a comprehensive parameter, including the image of all parts of the surveyed entities). The quality of controlling affects the efficiency of the organization, with the quality of management being the mediator [2]. The whole is in turn the result of properly implemented controlling solutions (including IT solutions).

However, quality of controlling refers to the "product" of controlling, which is the broadly understood as quality of services offered by the implementers of the idea of this method (quality of controlling outputs), e.g. timeliness, credibility, unambiguity and substantive alignment to the formulated requirements of information provided by controlling, a budgeting system that meets both motivational and informational functions; as well as planning and control system aimed at a common goal for the organization as a whole. Quality of management is understood as "the degree to which the set of inherent properties of the management system meets the requirements (formulated in relation to the management – p. auhor's). It is understood as set of properties of the decision-making process aimed at achieving specific company goals and at the same time being related to its ability to satisfy identified and anticipated needs." [20: 95]. The organizational performance is a synthetic and multidimensional construct [34] related to the results of organization's functioning as a whole. Underlining the concept of Balanced Scorecard [19] it concerns following issues, like e.g. financial as well as customer outcomes, learning and growth or internal processes.

However, it was decided that the most important measurement method should be related to the substantive evaluation of IT solutions. It was decided to consider IT reliability, but only after verifying that as a measure, it gives the same (as to the direction) image, as the before-mentioned parameters (quality of controlling, quality of management and organizational performance). However, it provides more information, which can be used as a basis for analysis and improvement of IT solutions of controlling in organizations.

### 3.1  Reliability of IT in Organization as a Tool for Evaluating the Quality of IT Solutions of Controlling

Reliability of IT in organization is understood as measurable property of IT, useful for its control and management, identifying its quality level and pointing out potential problems [44] and it is directly linked to the efficiency of IT components, especially those critical to its proper operations. Therefore, it can be said that IT reliability in organization is a notion build by factors connected to 3 different IT theories. First one is Delone and McLean success model [13], second one is Lyytinen [23] 4 types of IT failure and third one is TAM model [12]. Therefore, in order to fully develop the notion of IS reliability it is crucial to identify factors that are constructs for each of 4 identified variables proposed in the IS reliability model. Based on current research [18, 28, 32], all factors potentially related to IS reliability in the context of above-mentioned 3 IT theories were identified and assigned to proposed 4 variables.

Model of IT reliability in organization has been developed, detailed description is published by Tworek [38–41]. The reliability of IT in organization consists of 4 factors: reliability of information included in IT in organization, reliability of support services offered for IT in the organization and reliability of system itself, which also includes the usability of this system. Each factor is built by series of items, listed in Table 1, together with the results of preliminary verification of the model. The model should be amended with user experience (UX) as a part of usage reliability, which will be useful for development of agent-based model concerning various categories of users (and their UX might be an important characteristic differentiating them).

**Table 1.** Items building four variables of IT reliability model. *(own work based on:* [43]*)*

| Variable: system reliability | Variable: usage reliability | Variable: information reliability | Variable: service reliability |
|---|---|---|---|
| • Usability | • Learnability | • Accessibility | • Assurance |
| • Security | • Memorability | • Searchability | • Empathy |
| • Hardware stability | • Efficiency | • Accuracy | • Responsiveness |
| • Availability | • Error proneness | • Relevance | • Failrate |
| • Replicability | • Acceptance | • Achievability | • Quality |
| • Saliency | • Responsiveness | • Portability | • Availability |
| • Compatibility | • UX | • Movability | |
| • Hardware performance | | | |
| • Adaptability | | | |

While considering the evaluation of controlling solutions, there is a need to indicate what exactly will be evaluated. In case of controlling, IT is an important part of those solutions and it is almost impossible nowadays to maintain any controlling functions in organization without it. As it was mentioned above, controlling seems to be especially susceptible to influences of IT solutions because of its information needs. Hence, nowadays – in the high dynamics of modern economic processes and the growing demand of management for a wide range of management information necessary to make key decisions – it is difficult to imagine the implementation and functioning of controlling without proper support in the form of IT. However, it is important to underline that the fact of using IT is not a factor differentiating organizations. Almost every organization, which has implemented controlling, also has IT solutions supporting that method. Hence, the new framework is needed to differentiate the IT solutions of controlling used by various organizations and analyze its actual potential to influence various areas of organization operations, including efficiency of controlling. In this project, IT reliability model developed by Tworek [38–41] is proposed as the framework used for the analysis of IT solutions of controlling.

Therefore, it seems (as previously indicated) that the key issue in the process of assessing IT solutions of controlling in organization might be IT reliability. Initial results of empirical study confirmed that the influence of IT reliability on the processes of improvement of the controlling itself, the time of its use in the organization, the quality of its outputs, as well as the results obtained due to its implementation is proven [3, 4]. As a result, the significant influence of IT reliability on controlling excellence can be assumed. Therefore, IT reliability has the potential to be useful in the process of evaluation of IT solutions of controlling in organizations.

## 4  Agent Model of Excellent IT Solutions for Controlling Based on IT Reliability

The agent-based excellence model should be understood as a set of excellent IT solutions of controlling dedicated to each category of users, characterized by the highest level of excellence assessed by particular parameters related to the functioning of the organization as a whole (IT reliability, quality of controlling outputs, quality of management and organizational performance).

Since the model will be created based on the excellent IT solutions of controlling in organization for each identified category of users, evaluated using IT reliability, it seems that preliminary verification of different assessment of IT reliability among various groups of users is needed.

The verification was performed as a part of empirical studies conducted among employees from organizations located in Poland and USA (in December 2018). It was the only condition limiting the sample (employees were surveyed regardless of their age, tenure, job position etc.) obtained from a respondents' panel from SurveyMonkey.

The research sample contained the employees from organizations operating in Poland and USA. 303 valid responses from Poland and 247 from USA were collected [5]. The sample cannot be considered as representative, since the population of employees in those two countries is finite but very large and the method of including employees in the sample do not supported its representativeness. However, it is sufficiently diversified (considering diversity of employees' characteristics and organizations characteristics as well) to be a basis for overall conclusions concerning the given topic. In order to verify the different perception of IT reliability among employees on managerial and non-managerial positions, descriptive statistics were calculated and the U Mann-Whitney analysis was performed. The results are presented in Table 2.

**Table 2.** U Mann-Whitney analysis of the perception of IT reliability among employees on managerial and non-managerial positions

| Variable | Employees | N | M | SD | t |
|---|---|---|---|---|---|
| IT reliability | Managers | 422 | 3,97 | 0,82 | t(548) = −3,426; p = 0,001 |
| | Non-managers | 128 | 3,66 | 0,93 | |

The preliminary verification shows that there is a statistically significant difference among employees on managerial and non-managerial positions in perception of IT reliability level. Therefore, the obtained results allow to assume that IT reliability will allow to differentiate the groups of users of IT solutions of controlling and be a basis for the development of agent model. Within individual categories of users, groups of solutions with growing excellence will be indicated and will build the agent-based model of excellent IT solutions of controlling in organizations, which in turn will be assessed based on various factors, indicated above and presented in Fig. 1.

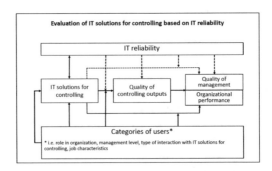

**Fig. 1.** Theoretical model of excellent IT solutions for controlling

## 5   Conclusions

The implementation of popular management methods is now common practice among Polish organizations. However, it often turns out that some of them are implemented only on paper, and some do not meet expectations formulated in relation to them. In recent years, allegations have also appeared in relation to controlling - one of the most frequently implemented methods of management support. This caused a heated discussion both in literature and practice regarding the need to assess the quality of controlling in the organization (i.e. the products meeting the needs and expectations of their internal recipients), as well as its effectiveness (measured by the ratio of benefits gained from the implementation of controlling and the cost of this implementation). For this reason, this paper addresses the issue of evaluating IT solutions for controlling that have a fundamental impact on the quality of controlling as well as on the quality of management and organizational efficiency. It has been proven - in the theoretical approach – that there is a need to measure the quality of IT solutions for controlling in case of reliability of these solutions, extended with issues related to UX. In this way, it was proposed to use the agent approach to designing and evaluating IT solutions for controlling, which would consider the varied needs and expectations of various stakeholder groups (primarily managers at various levels of management and non-managerial employees). That is mainly because their job performance is mediating the influence of controlling on the performance of the as a whole organization. It is obvious, however, that the proposed model should be empirically verified. The lack of comprehensive research in this area is a big limitation of this study. However, it is a starting point for further research in the field of controlling supported by IT solutions.

# References

1. Bea, F., Friedl, B., Schweitzer, M.: Allgemeine Betriebswirtschaftslehre-Bd. 2: Führung-Planung und Steuerung, Organisation, Controlling, Bilanzen, Kostenrechnung, Prognose, 9. Auflage, Stuttgart, UTB (2005)
2. Bieńkowska, A.: Analiza rozwiązań i wzorce controllingu w organizacji, Oficyna Wydawnicza Politechniki Wrocławskiej (PWr), Wrocław (2015)
3. Bieńkowska, A., Tworek, K., Zabłocka-Kluczka, A.: Information technology reliability influence on controlling excellence. Int. J. Digit. Account. Res. **19**, 1–28 (2019)
4. Bieńkowska, A., Tworek, K., Zabłocka-Kluczka, A.: IT reliability and its influence on the results of controlling - comparative analysis of organizations functioning in Poland and Switzerland. Inf. Syst. Manag. (2019, in press)
5. Bieńkowska, A., Tworek, K., Zabłocka-Kluczka, A.: Moderating role of UX and IT reliability in controlling influence on job performance and organizational performance, Raporty Instytutu Organizacji i Zarządzania Politechniki Wrocławskiej, Seria PRE, 9 (2019)
6. Bieńkowska, A., Zgrzywa-Ziemak, A.: Współczesne metody zarządzania w przedsiębiorstwach funkcjonujących w Polsce – identyfikacja stanu istniejącego. In: Hopej, M., Kral, Z. (eds.) Współczesne metody zarządzania w teorii i praktyce, Oficyna Wyd. PWr, Wrocław (2011)
7. Bogt, H.T., van Helden, J., van der Kolk, B.: New development: Public sector controllership —reinventing the financial specialist as a countervailing power. Public Money Manag. **36** (5), 379–384 (2016)
8. Brokemper, A.: Wir sind doch alle kundenorientiert? Sieben Schritte zum kunden- und marktorientierten Controlling, [w:] Strategische Steuer, Horvath P. (Hrg.), [online], Stuttgart (2000)
9. Brown, D., Robinson, D.: Effects of heterogeneity in residential preferences on an agent-based model of urban sprawl. Ecol. Soc. **11**(1), 46 (2006)
10. Czekaj, J.: Metody organizatorskie w doskonaleniu systemu zarządzania. WNT (2013)

11. Dani, A., Beuren, I.: Integration Level of Financial and Management Accounting Systems with the Accounting Convergence Process and the Effectiveness of Controllership, REPeC, Brasília, vol. 8, no. 3, art. 4, pp. 284–302 (2014)
12. Davis, F.D.: A technology acceptance model for empirically testing new end-user information systems: theory and results. Doctoral dissertation, MIT USA (1985)
13. Delone, W.H., McLean, E.R.: The DeLone and McLean model of information systems success: a ten-year update. J. Manag. Inf. Syst. **19**(4), 9–30 (2003)
14. Ernst & Young. The Changing Role of the Financial Controller Research Report. EY Publishing (2008)
15. Goliszewski, J.: Controlling.Koncepcje, zastosowanie, wdrożenie, Oficyna WKB, Warszawa (2015)
16. Grudzewski, W.M., Wilimowska, Z.: Od teorii do praktyki zarządzania. Czy zarządzanie jest nauką czy sztuką? Organizacja i kierowanie, nr 1/2017 (175), s. 11–50 (2015)
17. Hall, J.: Accounting Information Systems. South-Western Cengage Learning, Mason (2011)
18. Irani, Z.: Information systems evaluation: navigating through the problem domain. Inf. Manag. **40**(1), 11–24 (2002)
19. Kaplan, R.S., Norton, D.P.: The Balanced Scorecard: Translating Strategy into Action. Harvard Business School Press, Boston (1996)
20. Karaś, E.: Metoda oceny jakości zarządzania przedsiębiorstwem, PhD th, PWr, Wrocław (2006)
21. Kiełtyka, L.: Informatyczne przemiany zarządzania technologiami informacyjnymi w organizacjach. Przegląd Organizacji **3**, 26–29 (2011)
22. Lira, A.M.D., Parisi, C., Peleias, I.R., Peters, M.R.S.: Uses of ERP systems and their influence on controllership functions in Brazilian Companies. JISTEM-J. Inf. Syst. Technol. Manag. **9**(2), 323–352 (2012)
23. Lyytinen, K.: Different perspectives on information systems: problems and solutions. ACM Comput. Surv. (CSUR) **19**(1), 5–46 (1987)
24. Marciniak, S.: Controlling. Teoria, zastosowania. Difin, Warszawa (2008)
25. Młodkowski, P., Kałużny, J.: Próba rozwiązania problemu niedopasowania systemów informacyjnych do wymagań controllingu w przedsiębiorstwie, w: Sierpińska, M., Kustra, A. (eds.) Narzędzia controllingu w przedsiębiorstwie, pp. 22–31. Vizja Press&IT, Warszawa (2007)
26. Nesterak, J.: Systemy informatyczne w controllingu, Aktualne problemy funkcjonowania i rozwoju przedsiębiorstw. Kowalik, M., Sierpińska-Sawicz, A. (eds.) Wyższa Szkoła Finansów i Zarządzania w Warszawie, Warszawa, pp. 221–244 (2014)
27. Nesterak, J.: Controlling zarządczy. Projektowanie i wdrażanie, Oficyna WKB, Warszawa (2015)
28. Niu, N., Da Xu, L., Bi, Z.: Enterprise information systems architecture—analysis and evaluation. IEEE Trans. Ind. Inform. **9**(4), 2147–2154 (2013)
29. Nowosielski, K.: Sprawność procesów controllingowych. Istota. Przejawy. Determinanty, Wyd. UE we Wrocławiu, Wrocław (2018)
30. Nowosielski, S.: Centra kosztów i centra zysku w przedsiębiorstwie, Wyd. Akademii Ekonomicznej im. Oskara Langego we Wrocławiu, Wrocław (2001)
31. Ossadnik, W.: Controlling. Oldenbourg Verlag, München-Wien (1998)
32. Palmius, J.: Criteria for measuring and comparing information systems. In: Proceedings of the 30th Information Systems Research Seminar in Scandinavia IRIS 2007 (2007)
33. Peleias, I.R., Trevizoli, J.C., Côrtes, P.L., Galegale, N.V.: Pesquisa sobre a percepção dos usuários dos módulos contábil e fiscal de um sistema erp para o setor de transporte rodoviário de cargas e passageiros. Revista de Gestão da Tecnologia e Sistemas de Informação **6**, 247–270 (2009)

34. Richard, P.J., Devinney, T.M., Yip, G.S., Johnson, G.: Measuring organizational performance: towards methodological best practice. JOM **35**(3), 718–804 (2009)
35. Sierpińska, M., Niedbała, B.: Controlling operacyjny w przedsiębiorstwie. PWN, Warszawa (2003)
36. Szarska, : Jak controllerzy oceniają systemy informatyczne BI. Controlling **12**, 15–21 (2010)
37. Szukits, Á.: Management control system design–the effect of tools in use on the information provided. Vezetéstudomány/Budapest Manag. Rev. **48**(5), 2–13 (2017)
38. Tworek, K.: Model niezawodności systemów informacyjnych w organizacji. ZN Organizacja i Zarządzanie/Politechnika Śląska **88**, 335–342 (2016)
39. Tworek, K.: Information systems reliability in the context of higher education institutions. In: 10th Annual International Conference on Education and New Learning Technologies: Conference Proceedings, Palma de Mallorca (Spain), 2–4 July 2018. IATED Academy (2018)
40. Tworek, K.: The reliability of Information Systems in organization as a source of competitive advantage. Eur. J. Int. Manag. (2018, in press)
41. Tworek, K.: Aligning IT with Business. Springer, Cham (2019)
42. Weber: Wprowadzenie do controllingu, Oficyna Wydawnicza Profit, Katowice (2001)
43. Weißenberger, B.E., Angelkort, H., Holthoff, G.: MAS integration and controllership effectiveness: evidence of a preparer-user perception gap. Bus. Res. **5**(2), 134–153 (2012)
44. Zahedi, F.: Reliability of information systems based on the critical success factors-formulation. Mis Q. **11**, 187–203 (1987)

# Scrum Project Management Dynamics Simulation

Roman Pietroń[(⊠)] [iD]

Faculty of Computer Science and Management, Wrocław University of Science
and Technology, Wybrzeże Wyspiańskiego 27, 30-572 Wrocław, Poland
roman.pietron@pwr.edu.pl

**Abstract.** The paper presents dynamic aspects of software development project management. Particularly, a client influence or intervention risks, staff promotions, and their influences to software development project management with Agile/Scrum approaches and processes in terms of project performance dynamics and also economic results are considered. During the software development life cycle, the development team is expected to understand customer needs, but it is challenged by many constraints, e.g. narrow project schedule, customers' interventions and changes, short budget, or staff promotions. For this analysis a simulation model of software project management with application of continuous simulation by System Dynamics (SD) approach is developed. The paper shows dynamics of software project management caused by project risks and disturbances, particularly client interference or intervention, and also draws some conclusions regarding client intervention risk management during software contractor project performance, based on simulation experiments. The paper constitutes an original proposal of PM archetype simulation model and a research of client intervention risk influence to software project management due to application of Agile/Scrum elements within classical PMI/PRINCE2 methods management environments.

**Keywords:** Management · Project management · Simulation modelling

## 1 Introduction

Project management (PM) processes are having today increasing complexities with an increasing risk of projects' defects and failures caused by management approach and uncertainty. The statistics of a gap between project final results and expected (assumed) project results or expenditures during project realisations are estimated within a range of 40% up to 200%, and additionally many projects end up with defeats. Considerable efforts have been directed towards the improvement of the existing traditional methodologies and models, being proposed for project management. Inadequacies of network-based tools have been recognised, and improvement of estimating models, like the COCOMO, and teaching tools, like SCRUMIA [13], has also been the focus of attentions.

Generally, software development projects, similarly like other domain projects, are split into five managerial stages: initiation, planning, execution, control and closure.

© Springer Nature Switzerland AG 2020
Z. Wilimowska et al. (Eds.): ISAT 2019, AISC 1052, pp. 232–243, 2020.
https://doi.org/10.1007/978-3-030-30443-0_21

Each stage then has multiple checkpoints that must be met before the next stage begins. However, the shift from one stage to another is not a "smooth-sailing" but is fragmented, complicated and uncertain. The impact of uncertainties is a factor many software development companies have never thought about even though it is one of the most important, time and cost-consuming project result issues. Software development managers and engineers should be prepared for the occurrence of possible risks. An effective project management will not only eliminate risks but will provide a standard process and procedure to deal with them and help prevent issues such as: projects finishing late, exceeding budget or not meeting customer expectations, inconsistency between the processes and procedures used by projects managers, high stress levels with significant overtime, and unforeseen events impacting the project [15].

For the research purpose, a dynamic, continuous simulation policy design model of project management is developed with an application of System Dynamics (SD) method. This method, originated by Forrester [6], belongs to systems thinking and macroscopic continuous simulation modelling methodologies, and also risk analysis and management approach by scenario testing is to be applied. The main focus of this modelling research is to analyse an impact of customer (client) requirements change to project management, particularly in terms of PM approaches (classical vs. Scrum). It also includes a proposal of PM system archetype (or generic) structure and assumptions used for the better performance. Cause-effect and causal-loop relations (as influence and structure diagrams), quantitative SD models of these (as a set of mathematical differential equations translated into Vensim PLE simulation modelling package notation, and as a simulator interface), and some selected results of simulation one-, and multi-factor experiments are presented.

## 2 Research Method

### 2.1 System Dynamics Simulation of Project Management

Increasing role and importance of projects in modern management (management by projects), and still many failures in project management imply a search for new modelling approaches (methods and techniques) supporting project management (e.g. [4, 8, 10]). "Classical" methods based on operations research (OR) optimisations and analysis, even supported by historical organisational and managerial methods, present some disadvantages and limitations (Table 1). Project modelling by OR methods reduces project process complexity to a limited operational scope and range with a neglect of strategic and qualitative aspects of project processes. Additionally, "classical" project management methods and techniques are based on project elements repetitiveness – despite of project uniqueness and one-of-a-kind structure in occurrence. Therefore, a project is disaggregated to elementary structures and project logic fosters on linear and static improvement of elements.

These negative constraints perceived as disadvantages can be reduced (or even substantially weaken) by an application of simulation modelling (e.g. [5, 7]), particularly SD method (e.g. [9, 11, 16]). The SD method relies extensively on system's structure (particularly feedback loops and delays) in order to analyse and explain how

**Table 1.** An overview of classical techniques in project management.

| Method/technique/tool | Goal |
|---|---|
| Work break-down structure (WBS) | Basic defining the project work (activities). Project planning, scheduling, and cost estimation |
| Responsibility matrix | Organisational integration of project and responsibility allocation to each WBS |
| Gantt charting | Simple project plan mapping without an relation analysis between project activities and tasks |
| Network techniques: PERT, CPM, PDM, GERT, etc. | Network techniques for activity planning and scheduling deliver analytical data on relations between activities, critical paths, time slacks. It is a base to detailed cost estimating, resource allocations, risk evaluations, and determining the management method |
| Budget planning and cost analysis | Identification of capital requirements and needs for resources. The goal is to determine realistic budget estimates for standard project completion by measurement of milestones and overall project work structure completion |
| Project control: variance analysis, PERT/cost, profit, etc. | Project completion control by application of performance indicators. Additionally, it is to detect project risk and threats, and to initiate corrective actions |
| Agile/Scrum | Project completion control by application of sprints and dynamic activity completion monitoring |

system structure drives behaviour and leads to particular patterns of behaviour. Even some formal methods are being developed for an analysis of "structure-behaviour" relation (e.g. loop polarity dominance, behavioural analysis for loop dominance, pathway participation metrics, graph theory measurements), still practical analysis by simulation modelling and one- or multi-factor experimenting have largely been restricted to laboratory simple examples as guides to intuition. In SD modelling and analysis practice applied to PM modelling research, large-scale models with many loops are still analysed in a largely informal way, using trial-and-error simulation (Table 2). SD method of system modelling allows to analyse a managed process of project completion so as to: model the ways in which its information, action and consequences components interact to generate dynamic behaviour; diagnose the causes of faulty behaviour; tune its feedback loops to get better behaviour.

The basic viewpoint and associated methodologies of SD approach require a definition of a 'system'. PM system is a collection of parts organised for a purpose, and this system may also fail to achieve its purpose. Knowledge acquisition, system activities, decision making choices and learning consequences of choices need time – there are 3 delays in: data and information mining and acquisition to develop knowledge about project management (information about project stakeholders context, i.e. client and contractor relations), project work breakdown structure and resource requirements, project management decision making rules), and experiencing

**Table 2.** Differences between methods of project management issues modelling.

| Management requirements | Classical approach | Simulation approach |
|---|---|---|
| Task specification | + | − |
| Responsibility allocations | + | − |
| Planning | + | − |
| Resource management | + | + |
| Cost/budget estimation | + | + |
| Controlling/monitoring | + | + |
| Event influence evaluation | + | + |
| Decision results evaluation | − | + |
| *Post mortem* diagnosis | − | + |
| Identified factors | WBS logic<br>Cost of resources<br>Indirect cost<br>Availability of resources<br>Resource requirements | Project quality<br>Personnel qualifications<br>Experience<br>Productivity, motivations<br>Time pressure<br>Rework generation<br>Project reality evaluation<br>Cooperation with client |
| Management decisions | Cost-time relation<br>Task changes<br>Resource allocation to tasks<br>Changes in WBS logic | Human resources<br>New technologies<br>Quality assurance<br>Rework detection demand<br>Multi-project planning<br>Schedule changes |
| Unforeseen events | Completion time delays<br>Time limits<br>Resource limits<br>Uncertainty | WBS work scope changes<br>Quality changes<br>Information flow delays<br>Staff limits |
| Main estimates | Project completion time<br>Project cost<br>Allocation of resources | Project completion time<br>Project cost and profit<br>Needs, allocation of resources |

consequences of decisions in a project management process and system states (project productivity, project time, project finances, client satisfaction).

Although this is not a weakness, any formal tool that might help identify important structures in the model as they affect a particular mode of behaviour could be of enormous utility, particularly in large models trying to map complexity relations in management. SD method applied to PM research can be interpreted as a systems thinking approach and a branch of management science, which deals with the dynamics, and controllability of managed systems. SD method implementation in business and organisation systems' modelling and analysis usually focuses and addresses on the following basic research issues: circumstances in a system to use different policies in order to control its behaviour as time passes and circumstances change; system's policies design to become robust against change and ability to create and exploit opportunities and avoid, or defend itself against, setbacks; and information feedback structure design to ensure possible effective policies.

## 2.2    System Dynamics Simulation of Project Management

The first stage (Fig. 1) in SD application to PM modelling is to recognize the problem and to find out which people care about it, and why. It is rare for the right answers to be found at this stage, and one of the attractive features of SD as a management science methodology is that one is often led to re-examine the problem that one is attempting to solve. Secondly, at Stage 2, comes the description of the PM by means of an influence diagram, sometimes referred to as a "causal loop diagram" (CLD) or "cause-effect diagram". This is a diagram of the forces at work in the system, which appear to be connected to the phenomena underlying people's concerns about it. Influence diagrams are constructed following well-established techniques – basically "least-extension" technique. Having developed an initial diagram, attention moves to Stage 3, 'qualitative analysis'. The term simply means looking closely at the influence diagram in the hope of understanding the problem better. This is, in practical SD, a most important stage, which often leads to significant results (sometimes it is the end of modelling project). If qualitative analysis does not produce enough insight to solve the problem, work proceeds to Stage 4, the construction of a simulation model with "stock and flow diagram" (SFD). At this stage, we exploit the important property that the CLD diagram can be drawn at different levels of aggregation. It is not even necessary to show every single detail - if the CLD diagram has been properly drawn, the simulation model can be written from it without a separate stage of flow-charting. Stage 5 is where results based on quantitative analysis start to emerge. Initially, use is made of the insights from the bright ideas and pet theories from qualitative analysis. This stage represents exploratory modelling of the system's patterns of behaviour by laboratory experimenting with the aim of enhancing understanding and designing new decision rules for PM stakeholders.

Basically, SD simulation modelling method is an approach used to study a nonlinear system and feedback control in engineering, economic, social and human sciences. With the aid of computer continuous simulation, supported by *Iceberg Model* of systems thinking approach in collecting knowledge through the simulation runs, it is a powerful tool in understanding complex systems. SD is originally based on feedback control theory which includes both hard (quantitative) and soft (qualitative) approaches in analysing dynamic behaviours of the development and changes of a system. SD assists to improve decision making process and policy formation through its characteristics of incorporating all relevant cause-effect relationships as well as feedback loops in dynamic behaviour modes of systems. By developing a mathematical model as a set of differential equations solved by numerical integration (e.g. by Euler method) in an environment of computer simulation technologies, SD is capable to resolve a dynamic, inter-dependent, counter-intuitive and complex system such as problem of investigating the impact of PM method selection and project risk on project economics.

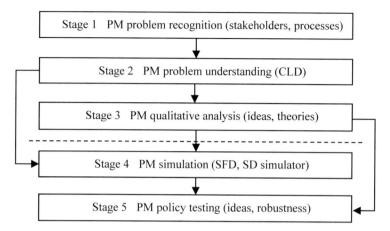

**Fig. 1.** The structure of SD approach to project management modelling. Source: own work based on: [3, p. 11] and [14, p. 86].

# 3   Project Management SD Model Development

## 3.1   Project Management Mental Model and Dynamic Hypothesis

The Scrum framework as one of the most adapted Agile principles (e.g. [1, 2]) and practices of PM, and its methodology facilitates the coordinated activity of programmers who break their work into small tasks that can be completed within fixed duration cycles or "sprints", tracking progress and re-planning in regular meetings in order to develop products incrementally. It is usually defined as a holistic approach to flexible, autonomous and dynamic teamwork with six main characteristics, namely built-in instability, self-organising project teams, overlapping development phases, multi-learning, subtle control, and organisational transfer of learning.

During the project completion by Agile/Scrum method three basic actors are to be identified: Scrum Master, Product Owner, and Project Team (Developers). This method of project management is based also on four elements of project control: Sprint Planning Meeting, Daily Scrum, Sprint Review, and Sprint Retrospective. Agile/Scrum project artefacts consists of: Project Backlog, Sprint Backlog, and Burndown Chart. Agile software development is commonly recognized as an enabler to accelerate software delivery, manage priorities changes and increase productivity. During the software development life cycle, the development team is expected to understand customer needs, but it is challenged by many constraints, e.g. narrow project schedule, customers' interventions and changes, or even budget limitation [12].

A causal loop is formed when a set of variables has been linked together in a connected path. There are two types of causal loop namely "reinforcing loop" (indicated by symbol R or plus sign) and "balancing loop" (indicated by symbol B or minus sign). Balancing loops generally (and always for $1^{st}$ and $2^{nd}$ order feedback loop) tend to stabilise the system while reinforcing loops always tend to destabilise the system. The loop is defined as positive (known as reinforcing loop) when the number of negative relationships is even (or multiplication of polarity signs within loop gives plus sign), otherwise the loop is negative one (known as balancing loop). Causal loop is also represented by an arrow headed line with sign "+" which means that a change in the influencing variable produces a change of the target variable in the same direction, while sign "−" means that a change in the influencing variable produces a change of the target variable in the opposite direction. The holistic analysis of causal loops relations is a helpful tool to predict the impact of factors in the system (sometimes this is quite difficult to find a loop dominance in structures with many different feedback loops).

Causal loop diagram of simple generic PM model in this research consists of the following parts (Fig. 2): causal loop diagram of main factors for work and rework activities, causal loop diagram of client influence factors (project scope change, negotiation attitude, trust for project performance), causal loop diagram of project results (productivity, finance, firm value).

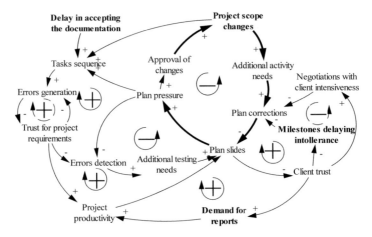

**Fig. 2.** Causal-loop diagram (CLD) of generic project management model.

## 3.2 Project Management SD Concept Model and Simulator

In this research, all factors presented in Fig. 2 were translated into stock and flow diagrams (Figs. 3 and 4) with an application of Vensim PLE software package to build the SD model and simulator of PM processes and economic consequences of projects completion. The development of SD model includes several types of variables such as stocks, flows, auxiliary variables, lookup functions, constants and connectors.

Stock, which is also known as level, acts as an accumulate (integration) reservoir of quantities (represented by rectangle) and describe the state variable of the system. The increasing flow (inflow) and decreasing flow (outflow) of a stock are also known as rates (represented by valve). The condition of the stock depends on the rates while the rates can be influenced by the other factors affecting inflow or outflow which are known as converters or auxiliaries (sometimes represented by circle). Finally, the connector that represents cause and effect links within the model structure is

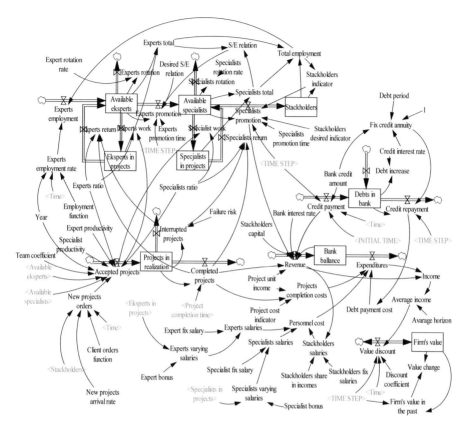

**Fig. 3.** Stock-and-flow diagram (SFD) of generic PM model – staff and performance.

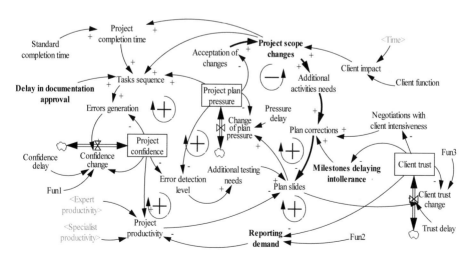

**Fig. 4.** Stock-and-flow diagram (SFD) of PM model – client influence risk.

represented by the single-line arrow. In the PM, model SD method basic paradigm is that a system's behaviour depends on underlying causal feedback structure, decision rules, amplifications and delays. CLD and SFD diagrams are used to represent cause-effect structure delays in information and physical flows. The dynamic equations of the model refer to nonlinear relations between identified basic variables and data estimated with an application of some empirical observations and pet theories regarding PM processes and systems.

The structure of a SD software PM model should capture the basic characteristics of the software development process and stakeholders' roles. Several techniques can be used to provide a rigorous definition of this process, which is also the basis for the development of traditional models, like the WBS and logical networks. The method proposed in this work to define the structure of the PM model is based on three basic principles: (1) dual lifecycle view of human resource management function in software engineering, (2) breakdown of the project into major sub-tasks, (3) dual life-cycle of work with defects, confidence, and trust within the engineering process, and (4) single high-level project and finance management. The proposed method is intended to support the development of new SD models, specialised in a specific project, and within a specific organisation.

## 4    Simulation of the PM Model

### 4.1    Scenario for First Experiments

Basic aim in first experiments of the developed SD model mapping PM is to analyse an influence of client on software development process dynamics, particularly to project completion time, software development productivity and financial results of software development company. The model is used as a "test laboratory" to assess the performance of the current plan and identify risks.

Calibrating the SD model to reproduce the *steady* behaviour requires the explicit definition of several metrics, which otherwise would remain "hidden" in the experiments. In fact, it is not to establish the desired results directly in the model; instead, the qualitative relationships of the "mechanics" within the model must be able to produce such behaviour. This uncovering of metrics is an important exercise as it helps to identify some unrealistic assumptions made in calibration stage, and suggests readjustments in the process. After this stage, the model is to be used for assessment of process sensitivity to risks through the analysis of "what-if" scenarios. In the next stage, disturbances must be introduced in the model and this provides explicit description of possible *unsteady behaviours*. Management policy alternatives can be tested, and further readjustments are carried out to reduce the process sensitivity. Another important output from the calibration is an estimation of the number of PM process inefficiencies that must be analysed in the future development stages, which provides awareness about the PM human resource management to avoid "over-optimism" in terms of project performance. After the model data has been eventually revised, the model can be used to investigate the causes for eventual deviations and to test whether alternative PM structure and control policies could have provided better results in terms of PM performance and financial measures.

## 4.2 Simulation Early Results

After model calibration stage, a set of experiments is to be run by introducing some changes in model parameters and client impact variable, as an exogenous function of time. In the basic experiment examples, we consider simple Step type of client intervention function (Figs. 5 and 6) and more realistic (empirical) function pattern (an increase of interventions with a magnitude till the half of project completion time, and then rather rapid decrease of client impact (Fig. 7). First experiments proved rather low sensitivity of project performance to client impact, which allow to formulate a preliminary hypothesis that internal PM system structure (particularly feedback loops, amplifications, and delays) determines PM system dynamic properties.

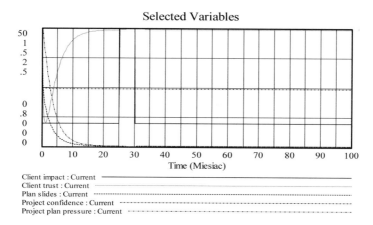

**Fig. 5.** Project slides, pressure and confidence in test experiment (Exp. 1) – Step function.

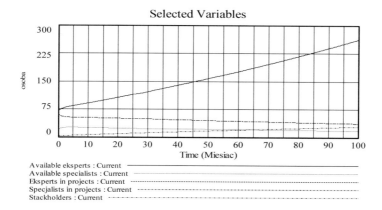

**Fig. 6.** Project human resources as levels in test experiment (Exp. 1) – Step function.

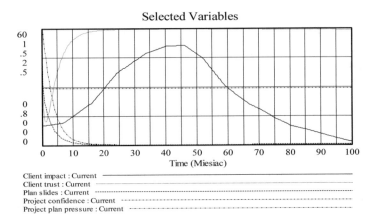

**Fig. 7.** Project slides, pressure and confidence in test experiment (Exp. 2) – empirical function.

While the software development firm (a contractor) reaches an agreement with the client regarding the milestone, management has then to decide how to pass this onto staff. Typically, management agrees a shorter deadline with development staff as both a contingency measure and to ensure a minimum level of schedule pressure and time tolerance. This involves deciding the policies to react to project slippage in order to avoid overrunning the schedule milestone, and to introduce human resources' capacities in order to minimise slippage with acceptable level of quality.

## 5  Conclusions and Final Remarks

The software development and implementation activities become complex, particularly because of the need of direct user participation in software design to generate a solution in an integrated way. The usage of Agile software development methodology in PM allows quick software release to users by reducing the time between the design and deployment, promoting both partial tests and deliveries with greater agility compared to classic PM methods. Successful experiences collected in software development projects' practices, which adopt the Agile/Scrum method of PM, allow to formulate some recommendations, e.g. face-to-face communication, iteration planning meeting and retrospectives facilitated by self-organizing cross-functional teams, and continuous integration with testing facilitates in short iterations and releases.

Traditional PM modelling methods, tools and techniques have proven inadequacy to deliver reliable information to PM staff. The systems thinking (and particularly holistic) perspective of SD approach provides a more appropriate and natural way of PM research and analysis. The pilot SD model and a "flight simulator" proposed in the paper allow to schedule adjustments to be managed and negotiated with the PM clients and other stakeholders, therefore downstream effects of introducing software new requirements can be minimized with increasing potentials of staff quality improvement.

# References

1. Beck, K., Beedle, M., van Bennekum, A., Cockburn, A., Cunningham, W., Fowler, M., Kern, J.: Manifesto for Agile Software Development (2001). http://agilemanifesto.org/
2. Ciric, D., Lalic, B., Gracanin, D., Palcic, I., Zivlak, N.: Agile project management in new product development and innovation processes: challenges and benefits beyond software domain. In: 2018 IEEE International Symposium on Innovation and Entrepreneurship (TEMS-ISIE), March 2018, pp. 1–9. IEEE (2018)
3. Coyle, R.L.: System Dynamics Modelling. A Practical Approach. Chapman & Hall, London (1996)
4. Duncan, W.R. (ed.): A Guide to the Project Management Body of Knowledge. PMI Publ. Newtown Square U.S.A. (1996)
5. Ferreira, S., Collofello, J., Shunk, D., Mackulak, G.: Understanding the effects of requirements volatility in software engineering by using analytical modeling and software process simulation. J. Syst. Softw. **82**(10), 1568–1577 (2009)
6. Forrester, J.W.: Industrial Dynamics. Productivity Press, Cambridge (1961). Available also from Pegasus Communications, Waltham
7. Ilaria Lunesu, M., Muench, J., Marchesi, M., Kuhrmann, M.: Using simulation for understanding and reproducing distributed software development processes in the cloud. Inf. Softw. Technol. **103**, 226–238 (2018)
8. Kunreuther, H., Slovic, P.: Challenges in Risk Assessment and Risk Management, vol. 545. SAGE Periodicals Press, Thousand Oaks (1996)
9. Lyneis, J.M., Ford, D.N.: System dynamics applied to project management: a survey, assessment, and directions for future research. Syst. Dyn. Rev. **23**(2–3), 157–190 (2007)
10. Magalhães Magdaleno, A., Lima Werner, C.M., Mendes de Araujo, R.: Reconciling software development models: a quasi-systematic review. J. Syst. Softw. **85**(2), 351–369 (2012)
11. Rodrigues, A., Bowers, J.: System dynamics in project management: a comparative analysis with traditional methods. Syst. Dyn. Rev. **12**(2), 121–140 (1996)
12. Scrum Alliance 2016. The State of Scrum Report 2017 Edn. (2017). https://www.scrumalliance.org/learn-about-scrum/state-of-scrum/2017-state-ofscrum
13. SCRUMIA – An educational game for teaching SCRUM in computer courses. J. Syst. Softw. **86**(10), 2675–2687 (2013)
14. Sterman, J.D.: Business Dynamics: Systems Thinking and Modelling for a Complex World. Irwin/McGraw-Hill, Boston (2000)
15. Vallon, R., da Silva Estácio, B.J., Prikladnicki, R., Grechenig, T.: Systematic literature review on agile practices in global software development. Inf. Softw. Technol. **96**, 161–180 (2018)
16. White, A.S.: A control system project development model derived from system dynamics. Int. J. Project Manag. **29**(6), 696–705 (2011)

# Big Data for Customer Knowledge Management

Celina Olszak and Arkadiusz Kisiołek$^{(\boxtimes)}$

University of Economics in Katowice, Katowice, Poland
{celina.olszak,arkadiusz.kisiolek}@ue.katowice.pl

**Abstract.** Contemporary organizations face numerous challenges, including effective customer knowledge management (CKM). A knowledge about customers, their behaviors, preferences and interests has become an important asset and a source of competitive advantage. It is said that effective CKM results mainly in increase of sales, customer satisfaction and loyalty. Customer knowledge management is a complex process based on appropriate acquiring, storing and analyzing of different, dispersed information resources as well as discovering a new knowledge about customers. Nowadays, more and more information about customers originate from Internet, social media, mobile devices and different data bases that are called big data (BD). An exploration of BD has become for many organizations an excellent opportunity to better know their customers and to gain a greater understanding of the way of CKM and the challenges that confront it. The main objective of this study is to investigate a role of BD for customer knowledge management.

**Keywords:** Big data · Knowledge management · Data mining · Text mining · Sentiment analysis

## 1 Customer Knowledge Management Issue

Contemporary organizations face numerous challenges, including effective Customer Knowledge Management (CKM). Knowledge about customers, their behaviours, preferences and interests has become an important source of competitive advantage in knowledge based economy [20]. In the literature there are different interpretations of CKM term. It is considered that CKM is a process [22], strategy [17] and management tool in contemporary organization. According to Kumer [22], CKM is a dynamic recycling process of acquiring and refining valuable customer data and sharing the generated customer knowledge across the organization. Huang [17] highlights that CKM is a strategy that focuses on the task of gathering information including finding effective ways to extract data from customers as well as to locate and retrieve information from other sources. In turn, [10] stresses that the aim of CKM is value creation for the customers. Sa'ad Al-Hyari (2016) claims that CKM is a process of capturing, sharing, transferring, and applying the data, information, and knowledge related with customers for organizational benefits. According to many authors [5, 10, 36] CKM is a process that consists of various steps, and its final result should be to make an appropriate decision. Table 1 presents various interpretations of "Customer Knowledge Management" term.

© Springer Nature Switzerland AG 2020
Z. Wilimowska et al. (Eds.): ISAT 2019, AISC 1052, pp. 244–253, 2020.
https://doi.org/10.1007/978-3-030-30443-0_22

**Table 1.** Selected definitions "Customer Knowledge Management" term.

| Author | Definition |
|---|---|
| [3] | CKM is a combination of knowledge management and customer relationship management (CRM) |
| [4] | CKM includes more than just knowledge from the customers and perceive it as a comprehensive approach for customer knowledge |
| [5] | CKM is a process by which organization generates wealth on the basis of its intellectual or knowledge-based organisational assets |
| [7] | CKM is about acquiring, sharing and enlarging the knowledge residing in customers contributing to both customer and corporate welfare |
| [9] | CKM is involved in the promotion of three types of knowledge: knowledge for the customer, knowledge from the customer and knowledge about the customer |
| [10] | CKM forms a continuous and stable strategic process, through which companies provide the opportunity for their customers to change from passive recipients of information to the state of becoming strong partners with sufficient knowledge |
| [15] | CKM can be defined as the systematic handling and management of knowledge collected at customer interaction points which are required for the efficient and effective support of business processes |
| [17] | CKM is a strategy that focuses on the task of gathering information including finding effective ways to extract data from customers as well as to locate and absorb information from other sources |
| [21] | CKM is creating the maximum value of knowledge and knowledge management at a strategic level |
| [22] | CKM is a dynamic recycling process of acquiring and refining valuable customer data by means of various paths and methods, and sharing the generated customer knowledge across the organization. Through this process the organization promotes and optimizes the customer relationships in the customer oriented organizational model, frame and environment |
| [30] | CKM is a concept utilising contemporary IT solutions in the field of relationships with customers management. Utilised IT solutions are perceived as basic tool to achieve interaction between company and its customers |
| [34] | CKM is an area of management where knowledge management instruments and procedures are applied to support the exchange of customer knowledge within an organization and between an organization and its customers, in order to improve CRM processes such as customer service, customer retention and relationship profitability |
| [35] | Basic principle of CKM is examining knowledge about client as a key category covering its forms such as: knowledge about customers, knowledge for customers, knowledge from customer and knowledge cocreated with customer. |
| [36] | CKM can be generally regarded as the processes of capturing, sharing transferring and applying data information as well as knowledge related with customers for organizational benefits |

Source: own-elaborated

For the purpose of this article, it was assumed that CKM is a complex process consisting of multiple stages. Among the most important ones can be mentioned: collecting, organizing, summarizing, analyzing, synthesizing of information and decision making. Collecting is the first stage of customer's knowledge management process. This stage is crucial for organisation because its effect should be accumulation of adequate, actual, appropriate and valid data, that will be utilised in subsequent stages of CKM [29]. Collected data should be arranged in structures and repositories used by organisation, e.g. data warehouses [37]. Data prepared in a such way creates basis for carrying out lots of analyses which aim is to i. a. detecting dependencies and patterns concerning customer' behaviours, theirs' preferences, habits or opinions. The results of data analysis are reports and dashboards that form the basis for acquired data interpretation, their synthesis and ultimately for transforming information into knowledge. The outcome effect of acquiring knowledge by customers is the management process of making an appropriate decision. Analysis of related literature indicates that CKM process is exceptionally complex and requiring commitment by many people: analytics, managers, statisticians, IT specialists. This involves a need for utilising different tools and information technologies as well as various distributed data sets and repositories, including Big Data [19].

## 2    Big Data - New Source of Knowledge

More and more information about customers comes from the Internet, social media, mobile devices and various databases called Big Data (BD) [31]. BD exploration became for many organizations the best opportunity to understand customers and results in deeper understanding of CKM and challenges facing them.

BD can be defined as comprehensive set of data enlisted from various fields like big volume, diversity, real time stream of data flow as well as significant variability and complexity. Such type of data requires an application of modern technologies, instruments and IT methods, so that in process of analysis, we can extract new, useful knowledge [24]. Data can be product from humans' activity or results of some operations like: measurements or computation, for example from various systems, devices or sensors. Among exemplary BD sources we can distinguish: online users transactions, electronic mail, video files, sound files, images, social networks posts, system's activity (so called logs), relationships between users coming from social networks, data incoming from sensors and electronics like Internet of Things devices [18].

The growing number of sources of data, such as circumstantial data, situational data, behavioural data, etc. makes BD presents endless opportunities to reveal patterns about the different types of customers and how they could be serviced in a more efficient manner [8]. The BD term can be applied to specific data sets characterized by traits, so called 7 V: Volume - The quantity of data is measured in peta- and zettabytes, Velocity- The meteoric speed of data emergence and the need to analyse it in a real time, Variety - The heterogenic nature of data, data can have different form and come from various devices and applications, Veracity - Data can be inconsistent, incomplete and inaccurate, Value - Significant value hidden in data, Variability - Variability mostly focuses on properly understanding and interpreting the correct meanings of raw data that depends on its context, Visualisation - Visualization refers to how the data is presented to the management for their decision making [11, 13, 23, 27, 28]. Basic value of BD manifests itself in possibility to discover underlying relationships between data sets, often coming from varying sources [1]. Discovering such dependencies requires utilising various techniques involving mainly text mining (text data analysis), machine learning, predictive analysis, data mining, statistics [38].

Summarizing this point of discussion it is worth highlighting that BD can be applied in many fields such as: retail, banking, education, trade, production, law, telecommunication or health care [14, 37]. It is also an excellent opportunity to improve the process of CKM. Exemplary area of application of BD and considering this process as an opportunity to set improvements in CKM have been presented in reports by Deloitte and they concern: pre-purchase & discovery, purchase & receipt, service & maintenance, repurchase, ownership & community [8].

## 3   Research Method

The aim of the study is the analysis of utilising BD tools in process of CKM. A research has been performed on samples of million records covering data about preferences of Spotify platform's clients, enriched by additional attributes coming from other data sets. These attributes cover e.g. information about most played songs for example danceability index or energy. Spotify is a Swedish service that streams music, podcasts and video content to its subscribers. The content it distributed is protected content that are directly obtained from the record labels of Media Companies that own them. The service is available in Western Europe, The Americas and Oceania. The music available on the service can be filtered by artist, album, genre, playlist or record label [2]. Basic functionalities of the platform are free, however there is a possibility to buy extended premium version covering additional functionalities. Service may be used on various devices – computers, mobile phones, tablets, TV sets and built-in car audio systems. After the end of the first quarter of 2018 number of service's users exceeded 170 million of which 75 million who paid for subscription. For the purpose of this article, a research has been conducted with use of R platform which was designed in 1993 as a tool that would show the possibilities of utilizing data analysis courses, but now is one of basic tools for data analysis [25]. R is a free, open source programming language and environment for statistical computing, data exploration, analysis and visualization [32]. To carry out research, multiple test methods have been utilised, especially: (1) time series analysis - statistical method of analyzing data from repeated observations on a single unit or individual at regular intervals over a large number of observations. Time series analysis can be viewed as an example of longitudinal designs. Classic time series analysis methods focus on decomposing series changes and building forecasting models [39]; (2) cluster analysis - seeks to identify natural subgroups in data with closer resemblance between items within a subgroup than between items in different subgroups. It represents a form of unsupervised learning, where the algorithm is not provided with known examples from a number of prespecified classes or other information on the nature of the classes [6, 16]; (3) text mining - interdisciplinary field of activity amongst data mining, linguistics, computational statistics and computer science. Standard techniques are text classification, text clustering, ontology and taxonomy creation, document summarization and latent corpus analysis [12]; (4) sentiment analysis – field of study that analyses people's opinions, sentiments, evaluations, appraisals, attitudes, and emotions towards entities such as products, services, organizations, individuals, issues, events, topics, and their attributes. It represents a large problem space [26].

## 4   Finding and Discussion

The research has been conducted accordingly to proposed CRM process approach. First stage of conducted research was analysis of titles of most frequently played songs in 2017 through text mining.

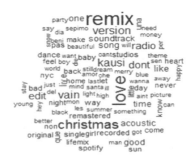

**Fig. 1.** Word cloud for the most-played songs. Source: own-elaborated

As shown in Fig. 1, among the most frequently repeated words one can distinguish "Christmas", "love" and "remix". Songs from the movies and TV series were among frequently played pieces as well, indicated by high position of word "soundtrack". Number of played songs increased every month whereas maximum value was reached in December, as shown at Fig. 2. It indicates a high platform user's activity in Christmas season and can prove conclusion drawn from tag·cloud (Fig. 1), where one of most frequently appearing words was "Christmas".

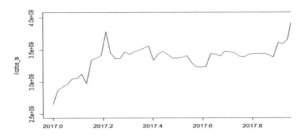

**Fig. 2.** The number of songs to listen to in 2017. Source: own-elaborated

Furthermore, 100 most frequently played songs in 2017 have been examined for indicators such danceability, energy, speechiness, acousticness, liveness and tempo. Definitions of these indicators have been shown in Table 2.

**Table 2.** Definitions of selected indicators describing songs included in study

| Measure | Definition |
| --- | --- |
| Danceability | Danceability describes how suitable a track is for dancing based on a combination of musical elements including tempo, rhythm stability, beat strength, and overall regularity |
| Energy | Energy is a measure from 0.0 to 1.0 and represents a perceptual measure of intensity and activity. Typically, energetic tracks feel fast, loud, and noisy |
| Speechiness | Speechiness detects the presence of spoken words in a track. The more exclusively speech-like the recording (e.g. talk show, audio book), the closer to 1.0 the attribute value. Values below 0.33 most likely represent music and other non-speech-like tracks |
| Acousticness | A confidence measure from 0.0 to 1.0 of whether the track is acoustic. 1.0 represents high confidence the track is acoustic |
| Liveness | Detects the presence of an audience in the recording. Higher liveness values represent an increased probability that the track was performed live |
| Tempo | The overall estimated tempo of a track in beats per minute (BPM). In musical terminology, tempo is the speed or pace of a given piece and derives directly from the average beat duration |

Source: [33]

Figure 3 shows levels of mentioned above measures for the most frequently played songs in 2017. These songs are characterized by similar levels of danceability and energy indicators. On the other hand, the biggest differences occur in speechiness and accousticness indicators. It might mean that the most important features of played songs are danceability and energy and the songs are similar in these terms.

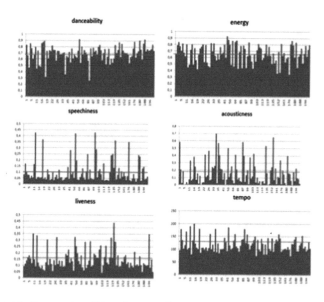

**Fig. 3.** Selected indicators describing the most frequently played songs in 2017. Source: own-elaborated

The next stage of analysis included data clustering of 100 most frequently played pieces with regard to their most important features. The result of this process was separation of two clusters which show songs similar to each other (Fig. 4).

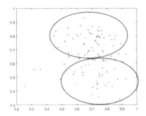

**Fig. 4.** Cluster analysis of the most-played songs based on energy and danceability indicator. Source: own-elaborated

Conducted analyses shows the holistic approach in regards to users' interests. Below are presented information about the most frequently played songs and their texts. For the purposes of this study, song by Tiesto – On My Way ft. Bright Sparks was selected (Fig. 5).

**Fig. 5.** Word cloud and sentiment analysis for text of song: Tiësto - On My Way ft. Bright Sparks. Source: own-elaborated

Sentimental level of song by Tiesto - On My Way ft. Bright Sparks is relatively high and the most frequently used words in this song are, "love", "say", "millionaire" and, "world". In the next step, YouTube users' comments were analysed covering studied song. On basis of this analysis it can be concluded that the song is generally liked by the majority of users and their opinion have sentimental character (Fig. 6).

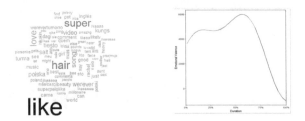

**Fig. 6.** Word cloud and sentiment analysis for YouTube users comments about Tiësto - On My Way ft. Bright Sparks. Source: own-elaborated

Conducted research let us to create a presentation of new approach towards CKM using BD tools in decision making process. Research discovered groups of similar songs most frequently played by Spotify users in 2017. These songs were characterized by similar levels of such measures as danceability and energy as well as including also analysis of users' opinions concerning played songs with use of text mining techniques including sentiment analysis. Opinions had moderate sentimental level and the majority of them were positive. Conducted analyses can be beneficial for managers, analytics dealing with studying customers' preferences and opinions in Web as well as for users themselves using various portals.

## 5 Conclusions

BD is increasingly and ever more willingly utilised by companies across the world. It can be used in many fields, including CKM. BD in CKM enables appropriate use of knowledge about customers in performed activities. One of the methods to discover such knowledge is text mining. It enables us to extract knowledge from text sources, so for example from comments left by the users, relations to customers opinions. Research points that text mining works for analyzing customers' preferences and evaluating their opinions. Results of analysis were visualised with use of tag cloud that enables accessible way of presenting study results. Under analysis emotional quotient of users' opinions were measured and performed the data clustering for the most frequently played songs as well. The results can be used in process of analysis and assessment of customer satisfaction regarding offered products or services, as well during decision making concerning performed marketing actions. Companies, including music sector such as Spotify can utilize presented analyses for appropriate matching of songs with users' preferences and getting to know their interests. An appropriate use of analyses results can benefit with greater interest concerning offered services. Analysis of customers' behaviours, opinions, preferences or habits seems to be very interesting, broad and challenging case to research. One example of research to further exploration can be a presentation of decision making process in CKM with use of random forest method and similar ones.

## References

1. ADMA: Best Practice Guideline: Big Data. A guide to maximising customer engagement opportunities through the development of responsible Big Data strategies. Association for Data-drivers Marketing & Advertising (2013). http://datascienceassn.org. Accessed 24 Feb 2019
2. Alexa: Spotify.com Traffic, Demographics and Competitors. https://www.alexa.com/. Accessed 05 Mar 2019
3. Boljanović, J., Stanković, J.: The role of knowledge management in building client relationships. Singidunum J. **9**(1), 16–28 (2012)
4. Bueren, A., Schierholz, R., Kolbe, L., Brenner, W.: Customer knowledge management - improving performance of customer relationship management with knowledge management. In: Proceedings of the 37th Annual Hawaii International Conference on System Sciences, pp. 1–10 (2004)

5. Bukovitz, W., Williams, R.: The Knowledge Management Fieldbook, p. 2. Financial Times – Prentice Hall, Pearsons Education Ltd., London (1999)
6. Chandler, R., Juhlin, K., Fransson, J., Caster, O., Edwards, I., Noren, G.: Current safety concerns with human papillomavirus vaccine: a cluster analysis of reports in VigiBase. Drug Saf. **40**, 81–90 (2017)
7. Cotobal, J.: E-intelligence tools in the customer consumption analysis introduction and background of customer intelligence's tool. Roczniki Ekonomii i Zarządzania 117–129 (2014)
8. Deloitte: Knowledge Management & Big Data. Making Smart Enterprise a Reality. Confederation of Indian Industry (2018). https://www2.deloitte.com/. Accessed 22 Feb 2019
9. Desouza, K., Awazu, Y.: What do they know? Bus. Strategy Rev. **16**(1), 42–45 (2005)
10. Du Plessis, M., Boon, J.: Knowledge management in eBusiness and customer relationship management: South African case study findings. Int. J. Inf. Manag. **24**(1), 73–86 (2004)
11. Erl, T., Khattak, W., Buhler, P.: Big Data Fundamentals: Concepts, Drivers & Techniques. Prentice Hall, Boston (2015)
12. Feinerer, I., Hornik, K., Meyer, D.: Text mining infrastructure in R. J. Stat. Softw. **25**(5), 1–54 (2008)
13. Fernández, A., del Río, S., López, V., Bawakid, A., del Jesus, M., Benítez, J., Herrera, F.: Big data with cloud computing: an insight on the computing environment, MapReduce, and programming frameworks. In: Pokornowski, M. (ed.) The Fourth V, as in Evolution: How Evolutionary Linguistics Can Contribute to Data Science, Theoria et Historia Scientiarum, vol. 11, pp. 48–49 (2015)
14. Frizzo-Barker, J., Chow-White, P., Mozafari, M.: An empirical study of the rise of big data in business scholarship. Int. J. Inf. Manag. **36**(3), 403–413 (2016)
15. Geib, M., Riempp, G.: Customer knowledge management. In: Abecker, A., Hinkelmann, K., Maus, H., Mu¨ller, H.J. (eds.) Geschaftsprozessorientiertes Wissensmanagement – Effektive Wissensnutzung bei der Planung und Umsetzung von Gescha¨ftsprozessen, pp. 393–417. Springer, Berlin (2002)
16. Hammou, F., Hammouche, K., Postaire, J.: Convexity dependent anisotropic diffusion for mode detection in cluster analysis. Neurocomputing **306**, 80–93 (2018)
17. Huang Li, E., Weng, X.: Applying customer knowledge management to alignment and integration of strategy maps. In: The Thirteenth International Conference on Electronic Business, 1–4 December, Singapore (2013)
18. IBM: Big Data Analytics. Employ the most effective big data technology. www.ibm.com/. Accessed 10 Feb 2019
19. Izhar, T., Shoid, M., Baharuddin, M., Mohamad, A., Ramli, A.: Understanding big data to improve knowledge management practices: gaps and limitations. Soc. Sci. **21**(2), 241–244 (2017)
20. Khan, Z., Vorley, T.: Big data text analytics: an enabler of knowledge management. J. Knowl. Manag. **21**(1), 18–34 (2014)
21. Kordvalouei, H., Vakili, Y., Moradi, M.: A survey on the relationship between knowledge sharing and organizational performance (case study: Bandarabas municipality). Arth Prabhand: J. Econ. Manag. **3**(6), 147–156 (2014)
22. Kumer Roy, T., Stavropoulos, C.: Customer knowledge management (CKM) in the e-business environment cases from Swedish bank. M.S. thesis in Business Administration, Lulea University of Technology, Sweden (2007)
23. Lahiru, F.: 7 V's of Big Data (2017). http://blogsofdatawarehousing.blogspot.com. Accessed 22 Feb 2019
24. Lau, R., Zhao, J., Chen, G., Guo, X.: Big data commerce. Inf. Manag. **53**(8), 929–933 (2016)

25. Lim, A., William, W.: R High Performance Programming. Packt Publishing, Birmingham-Mumbai (2015)
26. Liu, B.: Sentiment Analysis and Opinion Mining. Synthesis Lectures on Human Language Technologies (2012)
27. Manyika, J., Chui, M., Brown, B., Bughin, J., Dobbs, R., Roxburgh, C., Byers, A.: Big Data: The Next Frontier for Innovation, Competition and Productivity. McKinsey Global Institute (2011)
28. Olszak, C.M.: Toward better understanding and use of Business Intelligence in organizations. Inf. Syst. Manag. **33**(2), 105–123 (2016)
29. Patil, M., Yogi, A.: Importance of data collection and validation for systematics software development process. Int. J. Comput. Sci. Inf. Technol. (IJCSIT) **3**(2) (2011)
30. Pavičić, J., Alfirević, N., Žnidar, K.: Customer knowledge management: toward social CRM. Int. J. Manag. Cases **13**(3), 203–209 (2011)
31. Pedrycz, W., Chen, S.M.: Information Granularity, Big Data and Computational Intelligence. Studies in Big Data. Springer, Heidelberg (2015)
32. Ren, K.: Learning R Programming. Become an Efficient Data Scientist with R. Packt Publishing, Birmingham – Mumbai (2016)
33. Rico, N., Diaz, I.: Chord progressions selection based on song audio features. In: Cos Juez, F., Villar, J., la Cal, E., Saez, J., Corchado, E. (eds.) Hybrid Artificial Intelligent Systems. 13th International Conference, HAIS 2018, Oviedo, Spain, Proceedings (2018)
34. Rollins, M., Haliens, A.: Customer knowledge management competence: towards a theoretical framework. In: Proceedings of the 38th Hawaii International Conference on System Sciences (2005)
35. Rowley, J.: Eight questions for customer knowledge management in e-business. J. Knowl. Manag. **6**, 500–511 (2002)
36. Sa'ad Al-Hyari, H.: Customer knowledge management towards customer attraction from managers' perspective; a case study of Arab Bank in Amman City. Inf. Knowl. Mang. **6**(11), 47–57 (2016)
37. SAS2: Big Data. Czym są i dlaczego mają znaczenie? https://www.sas.com. Accessed 05 Mar 2019
38. Spotify: What is Spotify? https://support.spotify.com/. Accessed 05 Mar 2019
39. Velicer, W.F., Fava, J.L.: Time series analysis. In: Schinka, J.A., Velicer, W.F. (eds.) Handbook of Psychology: Research Methods in Psychology, Hoboken, NJ, vol. 2, pp. 581–606 (2003)

# Synergies Between Web-Based Open Innovation Platforms and Open Information Infrastructures

Ricardo Eito-Brun[✉]

Universidad Carlos III de Madrid, Getafe, Madrid, Spain
reito@bib.uc3m.es

**Abstract.** Open Innovation is defined as "the use of ideas and market knowledge – both internal and externals – to develop innovations" (CEN/TS 16655-5:2014) and it is based on this premise: "Valuable ideas can come from inside or outside the company and can go to market from inside or outside the company as well." Open Innovation has resulted in a set of collaborative tools to help involved agents develop networks where they can share ideas, exchange knowledge and propose new challenges and relies in what are called technology e-brokers or innovation markets. These tools consist of web-based platforms where companies can publish innovation challenges and other agents can analyze and propose potential solutions. They are used to capture external knowledge and to assemble it with the company's internal knowledge base, acting as intermediaries where innovation-involved agents can establish links and start collaboration activities. Although these collaborative innovation platforms or markets are extremely useful, their functionality could be improved with the integration of the content provided by open archives and institutional repositories. Using the integration opportunities offered by repositories, developers of Open Innovation platforms can reach additional content and information about innovators to support idea development and validation.

**Keywords:** Open innovation · Knowledge and Technology Transfer · Open Science · Open repositories

## 1 Introduction

Innovation management is a sub-discipline of management that studies the rules that govern the generation, diffusion and adoption of innovation, and the relationships between innovation inputs and outputs. Innovation management was traditionally understood as a linear model that comprised a sequence of activities – from basic research to serial production and market launch -, completed by a single entity. In the linear model, the achievement of innovations was determined by the planning, financing and execution of internal R&D activities or external technology acquisition.

Linear models were replaced by collaborative models based on feedback and interactions between different partners (Busse 2013; Harmelen 2012). Today, innovation management is seen as a non-linear, evolutionary, interactive process between the company and its environment that requires the close collaboration of different

© Springer Nature Switzerland AG 2020
Z. Wilimowska et al. (Eds.): ISAT 2019, AISC 1052, pp. 254–261, 2020.
https://doi.org/10.1007/978-3-030-30443-0_23

agents (Iordatii 2013). This evolution culminated in the Open Innovation model (Chesbrough 2003). Open Innovation states that valuable ideas may come from both inside and outside the company, and it is the result of knowledge specialization, availability of highly skilled workers, increasing capabilities of suppliers, and the difficulties of having a complete domain of all the aspects that need to be mastered in a successful innovation life cycle. Different entities or agents have a different level of participation in the generation of the knowledge streams that provide the inputs to create innovations (market, scientific and technical knowledge, and social knowledge) and the complex interfaces between them. The popularity of Open Innovation has led companies to consider the need of making a systematic planning of innovation and has extended the interest on innovation management to companies – like SMEs - whose characteristics made difficult setting up an innovation management plan based on dedicated R&D departments and costly investments on the development of products or services of uncertain success.

Between the different manifestations of the open innovation paradigm and its adoption by a wider number of companies we found several web-based platforms where companies can post challenges to an open audience. Some of the platforms successfully deployed to support collaborative innovation processes include, among others, InnoCentive, NineSigma, Brightidea, InnovationExchange, Atizo, YourEncore, Battle of Concepts or Yet2.com, just to name a few. These platforms can be defined as "technical infrastructures for knowledge sharing, discovering and social interaction" (Bygstad 2010; García-Barriocanal 2012).

Innovation platforms like those cited before, are mainly based on the Web 2.0 paradigm, offering companies the possibility of openly distributing and forecasting "business challenges" and collect ideas from outsiders. In general, companies collect and analyze these ideas to identify potential improvements in existing products or requirements that may lead to the development of new products or services. The platforms offer functionalities to collect ideas and complete their assessment, with a clear focus on open, distributed collaboration. But the requirements of open innovation management processes may go beyond these Web 2.0 capabilities. In particular, there are two areas that are subject of improvement:

- Giving support to the staff in charge of analysing and assessing the solutions proposed in response to challenges.
- Helping innovation managers making informed decisions on the feasibility of the solutions proposed by the potential partners, and their actual capability.

In other words, once the answers to the posted challenges are collected and screened, innovation managers must take informed decisions on the proposed solutions' feasibility. This is likely to depend on the internal capabilities of the people, group or entity that proposed the solution. When moving forward on the innovation management process, information about the capabilities and experience of potential partners also becomes quite relevant, as decision makers must decide on the most appropriate ways to move from ideation to conceptualization. This step in the innovation management process will probably need additional assessments of the potential partners' capabilities.

This paper presents a summary of a solution developed with the aim of exploring and analysing the role of open APIs (Application Programming Interfaces) to complement web-based Open Innovation platforms with additional support data. The research is aimed to provide an answer to the information needs of innovation analysts responsible of making decisions on innovations' and potential partners' feasibility.

The proposed approach is considered particularly useful for Universities and Knowledge and Technology Transfers Agencies (KT&T), as the proposed solution can help them leverage their competitiveness and capability to participate in open innovating processes. This is due to two reasons: first, these institutions can easily promote and give visibility to their activities and achievements, in order to become partners of major companies in open innovation processes; secondly, the tool can be useful to find qualified partners to develop projects where complementary competences and skills are needed.

## 2   The Open Information Ecosystem

Although not related to the Open Innovation concept, in the last years our society has also assisted to the emergence and progressive consolidation of a new approach to share knowledge, information and data: the Open Information model. This model was initially tied to the academic world. Academics started sharing reports and documents through web-based platforms known as open archives or digital repositories. The open information model was characterized by the following:

- An open publication model where authors can easily publish research results.
- An open access approach, as information and research results are available to the community for free, with no cost.
- Removal of the traditional barriers to access information: cost, difficult findability of the information, etc.
- The development and homogeneous use of a set of technical protocols that automate data collection, aggregation and sharing (e.g., Open Archives Initiative – Protocol for Metadata Harvesting or RDF)

The evolution of this new paradigm to publish information has followed different steps:

- Institutions launched their corporate open repositories, where their staff and employees could publish their results of their research activities; this situation led to the existence of "information silos", with a high number of distributed, local document-based repositories that were hard to exploit;
- The development of the technical protocols cited above led to the creation of ppen databases or repositories that aggregated documents and document metadata harvested from the local, institutional repositories. Both subject-specific and general-purpose databases became popular.
- The initial focus of these repositories – centered on documents – moved to a wider vision and repositories started to incorporate additional information about researchers, research groups and projects. The incorporation of data-sets to the

repositories, and not only documents, is also one of the results of this change in their initial scope.

This evolution has led to a situation where, instead of the initial term – "open access" -, it is feasible to talk about "open information". Today, academic repositories are moving from the document-centered "institutional repository" to CRIS (Current Research Information Systems) where data about research teams and their achievements is openly shared; Major initiatives like the OpenAire portal are becoming more and more important to give access to results of research programs funded by the European Commission, like FP7 or Horizon2020. In fact, some of these projects request the participant companies the publication of their project conclusions in open access, through these repositories (special clause 39).

These efforts are contributing to the creation of an open information infrastructure that – combined with the technical capabilities and protocols currently available for data exchange and aggregation – can be further exploded beyond data or document retrieval needs. In fact, the availability of these platforms is called to play a significant role in the development and improvement of our current R&D and innovation capabilities, due to these reasons:

- Finding potential partners in both the academic or entrepreneurial world may become easier.
- Innovation analysts can use the data available in these open information portals to assess the capabilities of potential partners for the ideation or conceptualization activities.
- Innovation analysts working with web-based open innovation platforms can use these portals to collect additional information about the researchers or research groups posting solutions to challenges; this background information may be extremely useful when assessing the quality and feasibility of the proposed ideas.

## 3   Innovation's Information Needs

To validate our hypothesis about the potential synergies between web-based open innovation platforms and the open information infrastructures, the research team conducted interviews with R&D managers and analysts from Philips, Bombardier Transportation, AstraZeneca and Lilly. The interviews aimed to do an initial identification of the information that analysts consider useful when assessing the novelty and feasibility of ideas and the reliability of their partners. Interview findings led to conclude on the relevance of finding expertise on the problem area or on the technical solution proposed to solve the challenge at hand. Searching for expertise was considered one of the key activities when setting up partnerships and teams to work on the challenge that will lead to an innovation.

The conclusions of research gives the possibility of identifying the following information items that contribute to the assessment of ideas and partners:

- Company details, including lines of business and activities.
- Areas of knowledge, competences and skills.

- Description of the company facilities and resources.
- Description of projects (including product development) in which the company has participated.
- Entities and partners (universities, research centres, companies) the company has worked with in collaborative projects.
- References and clients, whenever possible with the context of the projects they have collaborated.
- Publications and data sets generated by the entity staff.
- Patents granted to the entity.

These data items shall contribute to a resulting metadata infrastructure that can be used in two different ways in the context of a web-based innovation supported process: (a) to help companies identify partnership opportunities in a global context with sound criteria, and (b) to help companies assess the potential relevance of incoming ideas sent in response to "innovation challengers", based on the level of confidence attached to the entity providing the idea.

The final information infrastructure incorporates metadata for recording and transmitting information about "business challenges", although this point is satisfactorily covered by open innovation collaborative platforms. Regarding this point, current functionalities of these platforms could be extended to incorporate "contracting opportunities" characterized by a set of objectives, target price, etc. In other words, the solution proposed in response to a challenge could be presented at a separate location were other entities could: (a) start a process to find potential projects or challengers to create a team, and (b) be identified as potential partners by other companies using the innovation platform.

The identification and analysis of these needs led to the development of a prototype platform that combines the capabilities of three different tools: (a) web-based open innovation platforms, focused on the publication of challenges and solution; (b) business and experts directories, focused on providing contact details and basic administrative information; and (c) content and data repositories (CRIS, open archives) where staff and research groups experience and competencies can be known and assessed.

The proposed solution leverages the capabilities of standard experts and business directories:

- Business directories usually exclude SMEs, as they focus on larger companies or in companies working in a specific geographical area.
- Information in these directories focuses on financial data, and provides just generic activity codes that do not have the level of detail needed to establish partnerships based on the knowledge, competences, skills and expertise of the companies.
- Collaboration opportunities in the open innovation context should not be restricted to companies, but include other agents like universities, academics, public research officers or knowledge and technology transfer agencies.

## 4 Envisioned Solution and Technologies

To validate the research hypothesis and analyse the feasibility of the envisioned approach, the research team completed the development of a prototype platform aimed to support the activities of innovation and R&D analysts. Specific requirements of the tool included:

- Maintaining a core dataset with information about experts and entities for a specific area of knowledge.
- Offering dynamic interfaces with portals in the open information infrastructure and patents databases to collect data about researchers, companies or research groups on demand.
- Offering interfaces to web-based open innovation platforms, so analysts working with them could easily interact with our repository.

The last point becomes relevant, as the proposed tool does not aim to replace existing open innovation platforms, but complement them with additional capabilities to their users. The proposed solution can be seen as an intermediate tool, in the middle between the "open innovation platforms" and the "open information infrastructures".

The need of interfacing with different tools led to the adoption and implementation of Application Programming Interfaces (APIs) published by different open repository tools. If the current open innovation platforms are mainly based on Social Web, or Web 2.0 technologies, the automatic capture and forecasting of data and the automation of data exchange between platforms makes necessary to move on the data and application integration direction. Activities subject of data exchange automation include, among others:

- identification of partners and collaborators in particular knowledge areas,
- identification of opportunities or challenges suited to the skills and competences of the SME,
- assessing the potential value of companies based on their previous experience and
- Assessing the potential quality of the ideas submitted by a company.

In the case of (c) and (d), the availability of the company data in an easy to process format would allow conducting a preliminary filtering of entities and ideas.

A decision was taken to focus the content of the prototype on the medical engineering area. This decision led to the identification, data collection and loading about experts, research groups, companies and conferences, including their publications, projects and patents. A sample triple repository – supported by the VIVO tool – was built with data about 320 persons and entities. This repository can be searched using APIs and the Semantic Web SPARQL standard language, what makes it accessible from third party tools. Available access points include both persons and entity names, geographic location, areas of expertise, projects, and free text. The tool flexibility allows more complex queries, like getting the list of persons related to a company involved in projects making use of specific technologies. To improve the relevance of the results, controlled vocabularies (MESH, Medical Subject Headings) have been incorporated to describe the person and entities competencies. Controlled vocabularies are in fact one of the main components of the proposed metadata infrastructure, as the description of business, technological challenges and the skills and competences need to be described using a common vocabulary to allow the full automation of the data searching tasks.

The second pillar of the technical solution is the capability of searching for content (documents, projects, data sets of patents) related to a specific expert or entity at external sources. The feasibility of this functionality has been validated with interfaces to search for patents in the European Patent Office semantic data set, and projects, documents and data sets in OpenAire. Both repositories offer access to their data through the SPARQL query language, what makes possible to build a data pipeline through web-based connectors and to generate a consolidated view of the distributed information about a specific expert, company or research entity.

## 5  Validation of the Tool

The validation of the proposed approach was conducted through walkthroughs with staff from the Knowledge and Technology Transfer Agency of the Universidad Carlos III de Madrid. The main benefits they identified are summarized as follows:

- Open Innovation platforms are mainly based on the Web 2.0 paradigm, offering companies the possibility of openly distributing and forecasting "business challenges" and collect ideas from outsiders. They can be further improved by opening capabilities to search experts in open repositories like those hold by Universities and public research institutions. This aspect was positively valued.
- Searching for expertise by calling the APIs offered by typical tools used to manage open repositories was also considered useful and a way to give visibility to the researchers, who can be not only identified, but also notify whenever an open innovation opportunity is raised in the open innovation platform.

## 6  Conclusions

Open innovation relays on the capability of finding and building partnerships with external experts and entities. In a knowledge-based, global economy, finding partners and having the data to make an informed assessment of their capabilities becomes a complex task. Existing web-based open innovation platforms focus on putting challengers and solvers in contact. The proposed solution is aimed to help innovation analysts in these complex tasks, by facilitating access to information that may help them make informed decisions on the feasibility of the proposed solutions and the capabilities of potential partners. The solution acts as an intermediary layer that can be easily integrated on top of existing open innovation platforms, allowing a seamless integration with the ecosystem of tools that made up the "open information infrastructure": open archives, repositories and patent databases. The use of APIs and Semantic Web technologies like SPARQL end points makes possible to streamline the connections between all these tools and build a virtual place to explore the current data ecosystem to leverage the companies' innovation capabilities.

# References

Bygstad, B., Aanby, H.: ICT infrastructure for innovation: a case study of the enterprise service bus approach. Inf. Syst. Front. **12**, 257–265 (2010)

Busse, M., et al.: Innovation mechanisms in German precision farming. Precis. Agric. **15**, 1–24 (2013)

Chesbrough, H.: Open innovation: how companies actually do it. Hardware Bus. Rev. **81**(7), 12–14 (2003)

García-Barriocanal, E., Sicilia, M.A., Sánchez-Alonso, S.: Social network-aware interfaces as facilitators of innovation. J. Comput. Sci. Technol. **27**(6), 1211–1221 (2012)

van Harmelen, F., et al.: Theoretical and technological building blocks for an innovation accelerator. Eur. Phys. J. **214**, 183–214 (2012)

Iordatii, M., Venot, A., Duclos, C.: Designing concept maps for a a precise and objective description of pharmaceutical innovations. BMC Med. Inform. Decis. Making **13**(10), 1–8 (2013)

# Security of Enterprise Information Resources in Cyberspace

Aldona Dereń[1], Danuta Seretna-Sałamaj[2(✉)], Jan Skonieczny[1],
and Zofia Kondracka[1]

[1] Wydział Informatyki i Zarządzania, Politechnika Wrocławska,
Wrocław, Poland
{aldona.deren, jan.skonieczny,
zofia.kondracka}@pwr.edu.pl
[2] Państwowa Wyższa Szkoła Zawodowa w Nysie, Instytut Finansów,
Nysa, Poland
danuta.seretna-salamaj@pwsz.nysa.pl

**Abstract.** The aim of the article is to present the issues of security of enterprise information resources in cyberspace as key intangible assets conditioning its development and competitiveness. The implementation of this aim is based on a literature analysis and conducted surveys among companies operating in both the private and public sectors in Poland. In this survey have participated thirty one enterprises classified according to the type of the industry in which they operate and the size of enterprises measured by the number of employees employed in them. The research covered enterprises belonging to the IT sector and companies that do not provide typical IT services, but are active in many industries like investment, banking, logistics, car, bicycle, clothing, cosmetics, aviation and catering as well as in real estate.

**Keywords:** Cyberspace · Information · Security · Enterprise

## 1 Introduction

Technological progress of the last few decades has given rise to the development of a new type of social space that is cyberspace. It is identified with the Internet as a virtual space in which connected computers and other digital media (telephones, tablets, radio, television) communicate with each other in different places of the real world [14, pp. 23–38].

Modern computer techniques allow the processing of countless information and data every day, both by individuals and by organizations. The Internet has become the fastest and most easily accessible source of information, but also a place of economic activity, especially the provision of many services.

The aim of the article is to present the security of enterprise information resources (data, information, knowledge) in cyberspace as key intangible assets conditioning its development and competitiveness. Data, information and knowledge collected as a result of the digitization of business events implemented in the enterprise within the entire logistics chain, often functioning in the global space, create information

© Springer Nature Switzerland AG 2020
Z. Wilimowska et al. (Eds.): ISAT 2019, AISC 1052, pp. 262–275, 2020.
https://doi.org/10.1007/978-3-030-30443-0_24

resources that constitute a valuable asset for various stakeholder groups. In this situation, it is necessary to undertake activities aimed at developing and implementing a security system for these resources in the new dimension of human activities such as cyberspace. This included security has an impact on the overall activity of the company, especially on the adopted business model, in which information resources have an important strategic potential. These resources require safe management, understood as the process of identifying, organizing and protecting them in cyberspace.

The objective of this work was based on an analysis of literature and conducted surveys among companies operating in both the private and public sectors in Poland. Thirty-one enterprises took part in the study. They have been classified in the industry in which they operate and the size of the organization calculated on the basis of the size of employees. The surveyed enterprises included enterprises belonging to the IT sector and companies that do not provide typical IT services, but are active in the investment, banking, logistic, car, bicycle, clothing, cosmetics, aviation, and catering industries as well as those involved in real estate trading.

## 2   Information Resources of the Company

The value of each organization is determined by the resources it owns. In addition to classic resources, such as: land, capital and work, in the information civilization, the information resources and their security are of special value for the enterprise.

The notion of information resources is widely understood by the authors as an available amount of information and other resources, such as: knowledge, reputation, abilities, intellectual property, which are used to make strategic decisions.

In the literature on the subject, information is defined as the content of a message sent from the sender to the recipient, expressed in the appropriate language or code. Information can be remembered (transferred in time), transferred, forwarded, communicated (moved in space), it should organize the system to which it refers. Information received by the recipient should allow him to better adapt to the outside world by focusing his behavior [12, pp. 9–10].

In terms of the theoretical conception of a communication system, information is an element of this system and can be defined as a source of information sent from the sender to the recipient, expressed in the appropriate language or code. The information can be remembered (transferred in time), transferred, forwarded, communicated (moved in space). It should organize the system it refers to. Information received by the recipient should allow him to better adapt to the outside world by targeting his behavior [11, pp. 225–229].

On the other hand, management in the concept of information should be understood as knowledge (accurate, current, complete and appropriate) describing the possibilities of an individual or organization. In addition to knowledge to the information resources of the company, you can include reputation, abilities and intellectual property.

Reputation is the total information about the company, formulated by internal and external observers, based on the assessment of its activity in the financial, social and environmental aspects [1, p. 26]. This information is created on the basis of the observers' own experience with the given organization, information from other entities, in various ways related to the organization, and also communicated by itself to the environment. Reputation is an aggregated information containing the assessment of the company, based on the perception of various stakeholder groups.

On the other hand, the concept of ability should be understood as cognitive skills of synthesizing new information in order to use it in the process of enterprise functioning [5, 23].

Information resources also include intellectual property understood as the total of products of creative human work as a result of synthesis and creative reflection of collected and processed information, which due to its uniqueness and high economic, economic and development importance deserve protection.

Information resources understood in this way are of interest to various groups of stakeholders communicating in cyberspace. The condition of proper functioning of this communication is safe distribution and exchange of information resources.

## 3  The Concept of Cyberspace

The concept of cyberspace in terms of semantics is a conceptual hybrid of the English words cybernetics and space, which in free translation means cyber space. Originally this term was created for the needs of science fiction literature, where in 1982 it was used by W. Gibson in the novel called Burning Chrome, defining it as "*A consensual hallucination experienced daily by billions of legitimate operators, in every nation, by children being taught mathematical concepts... A graphical representation of data abstracted from the banks of every computer in the human system. Unthinkable complexity. Lines of light ranged in the nonspace of the mind, clusters and constellations of data*" [6, p. 59].

The vision of cyberspace was also created using computer graphics, using three-dimensional and interactive elements moving in a common environment, preserving the laws of gravity or collision of objects, thanks to which it gained features specific to the real world, which could not only be seen but also to touch and shape using appropriate prepared computer interface.

These artistic approaches to cyberspace have given rise to discussions on the definition of a new social space. However, it has not been possible to clearly determine what cyberspace is. One of the most-cited definition of cyberspace in the literature is the one formulated by the US Department of Defense: "A global domain within the information environment consisting of the interdependent network of information technology infrastructures and resident data, including the Internet, telecommunications networks, computer systems, and embedded processors and controller" [4].

So, cyberspace is a global domain of the information environment consisting of interdependent networks created by information technology (IT) infrastructure and data contained therein, including the Internet, telecommunications networks, systems computers, as well as embedded processors and controllers" [10, p. 10].

Thus, the essence of cyberspace is to create a parallel environment that is a new dimension of human decisions and actions, where the data record will reflect the information possessed by the user, organization or state.

However, from the social point of view, cyberspace refers not only to the users themselves, but to the intangible results of their interaction, as well as to the relationship between the software and available services. That is why cyberspace is defined as an environment of open communication between connected ICT devices, which identifies with people from around the world, various organizations [10, p. 10].

## 4 Threat to the Security of Enterprise Information Resources in Cyberspace

Security is a key element in the functioning of cyberspace. information collected and processed in this space. However, as yet, there is no clear and precise definition of security in cyberspace, i.e. cybersecurity. Perhaps it makes a blanket of the very concept of security which, depending on the context presented, takes on different meanings.

Security related to cyberspace is variously defined. The National Initiative for Cybersecurity Careers and Studies (NICCS) defines them as a guarantee that "(…) information or communication systems and information contained in them are protected or protected against damage, unauthorized use, modification or use" [3, p. 108]. On the other hand, the Cybersecurity Strategy of the Republic of Poland for 2017–2022 presents this term as: "resilience of IT systems, at a given level of trust, to all activities violating the availability, authenticity, integrity or confidentiality of stored, transferred, processed data or related services offered or available through these networks and IT systems" [13].

According to M. Cavelty, cyber security should be defined in the context of preventing damage, protection and in the perspective of restoring the ability to properly operate computers, electronic communication systems or communication services taking place in cyberspace. The second element is the protection of information contained in the electronic communication space, in order to ensure confidentiality and at the same time authenticate persons authorized to do so [2, pp. 19–23].

From the entrepreneur's point of view, cyber security is the state (achieved sense of security) and the process (ensuring a sense of security). The second approach is more practical, reflecting the natural, dynamic nature of security. In this sense, the security of a given entity is the field of its activity, the content of which is to provide opportunities for survival, that is, existence and freedom to implement its own interests in a dangerous environment, in particular by taking advantage of opportunities, facing challenges, reducing risk and counteracting all kinds of threats to the subject and his interests [9, p. 20].

Threats to information resources of an enterprise in cyberspace mean the potential cause of an undesirable incident, which may result in disruptions and damage caused to the information resources collected by the organization in its IT systems.

According to the definition adopted in the PN-ISO/IEC 27001 standard, an information security incident should be understood as a single event or series of undesired or unexpected events related to information security that pose a significant probability of disrupting business activities and jeopardize the security of information [15].

Threats to information security in cyberspace evoke so-called cyber attacks, cybercrimes and cyberterrorism. The first of these - cyber attacks are aimed both at material resources and intangible networks, and indirectly at people, groups, institutions and even countries. Table 1 presents the basic types of cyber attacks targeted at information resources owned by the enterprise. These cyber attacks have a diverse nature, but their destructive power is directed primarily to information resources, which are a key resource of the enterprise [8].

Cybercrime is a conscious unlawful act committed in cyberspace aimed at violating the stability and integrity of information resources collected in information systems. These actions may take the form of:

- violation of access rights to information resources (hacking);
- data capture;
- modification of data using malware (e.g. Trojan);
- falsification of passwords and electronic documents, payment cards, etc.;
- computer espionage.

Cyber terrorism is a kind of diversion taking place in the domain of cyberspace, which consists in intercepting and using enterprise information in order to disrupt business processes, which may even result in the elimination of a company from a market game. An example of cyberterrorism can be:

- falsification of documents;
- opening fictitious accounts and accounts;
- valuable redirecting to another address;
- counterfeiting and counterfeiting of trade marks;
- manipulation on financial exchanges;

Ensuring the security of information resources in an enterprise against the above-mentioned threats is a process focused on reducing the risk of a cyber attack or in the event of its occurrence, limiting the negative effects. It's important to take care of security throughout the organization to know where the potential weakest fire can be. On the other hand, security requires constant compromises, because the higher its level, the more financial, technical and human resources should be involved in its functioning.

An effective enterprise information security program should be tailored to the subject of its business. Standards and legal norms applicable in particular industries should specify basic security measures that set a minimum level of protection. Creating such a program is a complex undertaking, consisting of many decisions and requiring the involvement of all employees. This program must take into account four key aspects:

- risk analysis;
- analysis of financial consequences;
- planning actions in crisis situations
- business continuity planning planning.

The risk analysis includes the probability of damage and the extent of potential damage caused to the information resources held, while the analysis of financial

**Table 1.** Types of cyber attacks (Source: [7].)

| Type of cyber attack | Characteristics |
|---|---|
| Malware | Malicious software that performs on behalf of a third party without the user's consent and knowledge |
| Man in the Middle | This is a type of attack involving the participation of a third party, e.g. in a transaction between an online store and a client. The purpose of such attacks is to intercept information or cash (e.g. to obtain the data necessary to log in to the electronic banking system) |
| Cross site scripting | This is a type of attack consisting in placing a special code on the website, whose clicking by the user causes redirection to another website (e.g. to the competition's website) |
| Phishing | It is an attack consisting in making attempts to take over passwords serving the user to log on to, for example, social networks, to which access allows attackers to obtain personal data of the user |
| DDoS | The purpose of this attack is to block the user's ability to log into the website by simultaneously logging in to the same page as many users. The artificial motion called in this way strengthens the interest of users, for example, a product available in the online store |
| SQL Injection | It is an attack consisting in exploiting the gaps in the security of, for example, the application by criminals, and allowing non-authorized persons to obtain personal data |
| Ransomware | This is a type of attack aimed at taking over and encrypting user's data in order to make the same data available to the user in the next step, provided he brings a "ransom" |
| Malvertising | This type of attack allows criminals to reach users who view trusted websites through media that are shared on advertising websites, and then install malware without the user's knowledge and consent on user's devices |
| Ataki typu APT (Advanced Persistent Threats) | It is a complex, long-term and multistage action directed against specific individuals, organizations or companies. They consist in gathering for months by the attackers information about the employees of the organization, so that only after some time they can proceed to the planned attack. The programs and tools they use are created and used in a way that minimizes the chance of detecting unwanted activity by the targeted organization. Criminals create false profiles of entire teams from the company they are going to attack and impersonate its employees. They scrupulously build virtual identities, providing credibility to the scenarios that they then use to gain the trust of the staff or collaborators of the targeted organization. When such a gap is created, there are often repeated data thefts" |

consequences determines the scope of the security plan of those resources, which are crucial from the point of activity. Planning actions in crisis situations amounts to determining the best method to minimize the effects of damages, and maintaining business continuity concerns the methods of its undisturbed operation regardless of the situation. These four aspects must be an integrated whole that will ensure adequate protection.

## 5    Security of Information Resources in the Light of Own Research

The type of threats related to the security of information resources in IT networks induces a deeper research reflection on how this situation is shaped in enterprises operating in Poland. The surveys covered enterprises operating both in the private and public sectors in Poland. Thirty-one enterprises took part in the study. They were classified according to the industry in which they operate and the size of the organization calculated on the basis of the size of employees. The surveyed enterprises included enterprises belonging to the IT sector and companies that do not provide typical IT services, but are active in the investment, banking, logistic, car, bicycle, clothing, cosmetics, aviation, and catering industries as well as those involved in real estate trading.

The complexity of information security issues in the company has led the authors to choose the key issues for the development of basic recommendations in the field of security management of these resources. Therefore, the authors focused on their research on two cognitive aspects: threats to the security of information resources and the condition and condition of the adopted security system in the surveyed enterprises.

Regarding the first aspect, it was important to gain knowledge about the number and frequency of sources of threats and incidents and cyber attacks registered during the year. On the other hand, the second aspect focuses on the search for knowledge about the security of information resources in the surveyed organizations by indicating the condition, level and maturity of security areas in the adopted cyber security system in the enterprise. In addition, the subject of the analysis were the methods of reaction of the surveyed enterprises to cyber attacks and the identification of those areas in the enterprise that are most vulnerable to cyber attacks (The questionnaire surveys discussed were carried out and included in the magisterial work carried out by Mr. Z. Kondracka, "Information security in IT systems of enterprises". The work was carried out in the field of study: Management for engineers at the Faculty of Computer Science and Management at the Wroński University of Technology. The work was prepared under the supervision of Phd. Aldona Małgorzata Dereń from the Department of Management Infrastructure, Faculty of Computer Science and Management, Wroclaw University of Technology).

Regarding the issue of the number and frequency of annual risks and incidents, the study showed that three-fourths of the businesses were of interest to cybercriminals

(74%). On the other hand, 26% of respondents claim that they were not subject to incidents and cyber attacks within the scope of information resources.

The respondents were not able to clearly indicate which source was the most dangerous. cyber threats. Especially single hackers were distinguished (it was indicated by more than 45% of respondents), which may be the result of the myth of the "super hacker" myth, which consists in presenting a given cyber attack as one person's actions, although in reality it is much more necessary to implement it people. Attacks on mass scale for large enterprises that are noticed, require a lot of effort and commitment of many people, so a single hacker would not be able to do this. In practice, a well-organized criminal group usually hides behind the actions of a single hacker (it indicated a criminal group of over 35% of the respondents).

However, as regards the types of cyber threats caused by cyber attacks, according to the respondents the two dominant threats, i.e. data leakage (54.8% of respondents) and data extortion (51.6% of respondents). Other threats include: data theft by employees (38.7%); advanced targeted attacks (38.7%); attacks using application errors (29%); data leaks due to theft or loss of devices (16.1%); burglaries to mobile devices (9.7%).

As part of the survey, respondents were asked to assess the maturity of individual areas of the adopted cybersecurity system. The results were presented in the form of a spider diagram, which illustrates how strong its individual elements are. The further the ratings are from scratch, the more intrusive is the cyber security system. Every collapse of the chart is a potential weak point that can be used by criminals to enter the organization. The cybersecurity system is as strong as its weakest link is strong, from which it follows that if there is at least one weak spot, then other safeguards, even the strongest ones, will not protect the organization against cybercriminal attack. The responses of individual enterprises were interpreted according to the sector of activity and the size of the company. It is clearly visible difference between the results for enterprises in the IT sector and enterprises from other sectors. Corporations that provide broadly understood IT services are more aware of the threats resulting from cyberspace, which may result from the fact that information resources are of primary importance to them. This does not mean, however, that the cyber security system in all categories is at the same level of maturity.

The difference between enterprises in the IT sector and others for large organizations is relatively small. There is also a visible tendency that a group of enterprises that do not provide IT services is so developed that their security has reached a very advanced level. These are probably companies with a high market position, which they achieved thanks to their products or services and they realize that their information resources should be protected and found in their budget.

In the case of enterprises whose size is defined as average, there is a clear difference. The security maturity of the organizations from the enterprise sector that do not provide IT services is more often closer to the middle of the chart than from the IT sector. This may result from the allocation of funds for other purposes, which at the moment may provide a greater profit and lack of awareness of the importance of securing the available information resources. This is probably a situation when the company is dynamically growing and the management does not see the need to invest in collateral yet, because it believes that it is not necessary at this stage.

The situation is very similar for small enterprises, because the IT sector in no way intersects the zero point. However, in the case of enterprises from outside the IT sector, this point is repeated, probably due to lack of knowledge and awareness of the existence of threats and a sufficient budget for building an efficient cyber security system.

When it comes to micro-enterprises, the IT sector shows a very good result in every area of cyber security, which can be equal to those in large enterprises. On the other hand, micro enterprises from the non-IT sector have the worst assessed system maturity, which may reflect the low value of information resources they have, and at the same time insufficient budget for building a cyber security system. The described relationships are presented in Charts. 1, 2, 3, 4, 5, 6, 7 and 8.

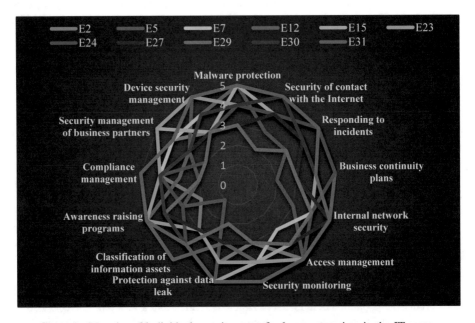

**Chart 1.** Maturity of individual security areas for large enterprises in the IT sector.

However, when it comes to the choice of the company's activities and reactions to cyber attacks, according to respondents, responding (51.6%) reacts based on the rules and plans of crisis management adopted in the enterprise, which are implemented at the time of the cyber attack. Other ways of reacting are presented in the chart below.

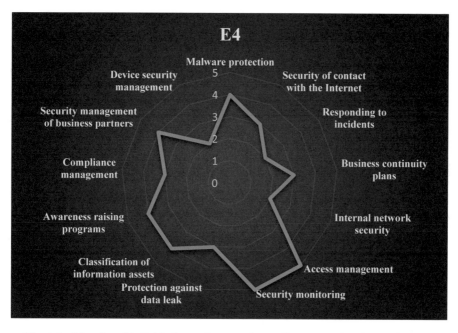

**Chart 2.** Maturity of individual security areas for medium enterprises in the IT sector

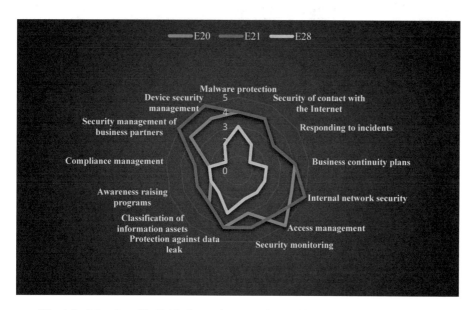

**Chart 3.** Maturity of individual security areas for small enterprises in the IT sector

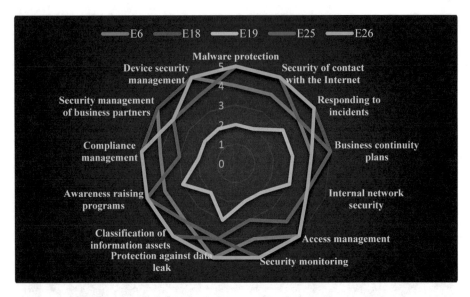

**Chart 4.** Maturity of individual security areas for micro enterprises in the IT sector

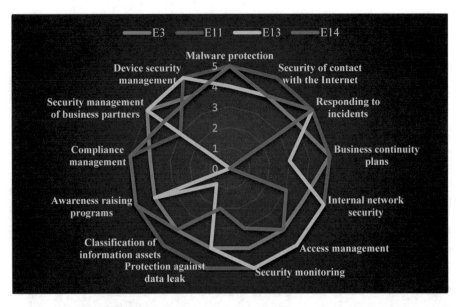

**Chart 5.** Maturity of individual security areas for large enterprises from sectors other than IT

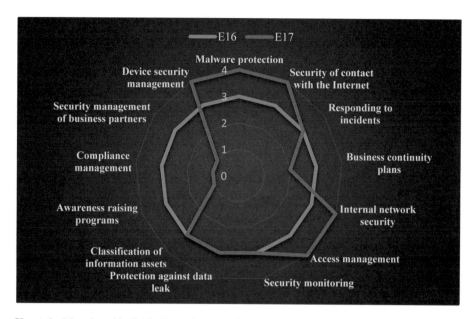

**Chart 6.** Maturity of individual security areas for medium enterprises from sectors other than IT

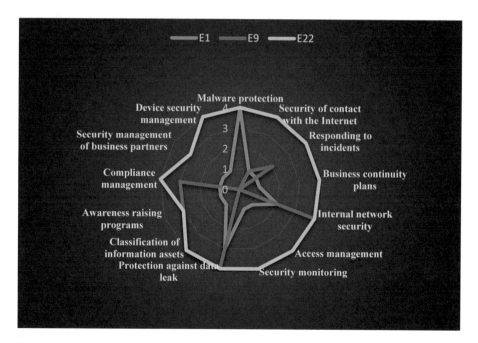

**Chart 7.** Maturity of individual security areas for small enterprises from sectors other than IT

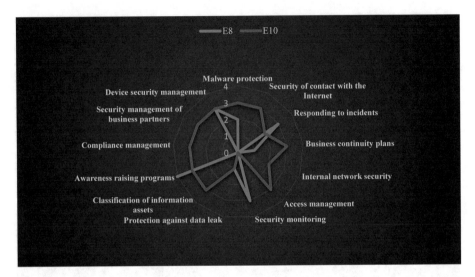

**Chart 8.** Maturity of individual security areas for micro enterprises from sectors other than IT

## 6 Conclusion

Enterprise security is a state that gives it a sense of certainty and guarantees business continuity with acceptable risk. In addition, it offers opportunities for its further development, including based on information resources. Nowadays, when cyberspace is the basic sphere of company's activity, both in the public and private sectors, all activities undertaken in the real and virtual world are interrelated. This new approach influences the perception of cyberspace as a new area of functioning, which due to access to information resources of the organization must be subject to safeguards. It is connected with the necessity to develop and implement a security system that will enable undisturbed operation and management of business processes on the market.

The presented results of the survey indicate that although entrepreneurs in Poland perceive the problem of cyber threats, they insufficiently undertake activities aimed at creating and implementing solutions in the field of information resources security.

## References

1. Barnett, M.L., Jermier, J.M., Lafferty, B.A.: Corporate reputation: the definitional landscape. Corp. Reput. Rev. **9**(1), 26–38 (2006)
2. Cavelty-Dunn, M.: Cyber-Security and Threat Politics. US Efforts to Secure the Information Age, London (2008)
3. Chmielewski, Z.: Polityka publiczna w zakresie ochrony cyberprzestrzeni w UE i państwach członkowskich. Studia z Polityki Publicznej **10**(2), 103–128 (2016)
4. Department of Defense Dictionary of Military and Associated Terms. https://fas.org/irp/doddir/dod/jp1_02.pdf. Accessed 1 Apr 2019

5. Dyer, J., Gregersen, H., Christensen, C.M.: DNA innowatora. ICAN Institute Warszawa (2012)
6. Gibson, W.: Neuromancer, Katowice, Wyd. Książnica (2009). (tłum. P. Cholewa)
7. Gwoździewicz, S., Tomaszycki, K.: Prawne i społeczne aspekty cyberbezpieczeństwa, Warszawa (2017). (Międzynarodowy Instytut Innowacji « Nauka – Edukacja – Rozwój » w Warszawie 2017)
8. Kowalewski, J., Kowalewski, M.: Zagrożenia informacji w cyberprzestrzeni, Cyberterroryzm, Oficyna Wydawnicza PW, Warszawa (2017)
9. Koziej, S.: Bezpieczeństwo: istota, podstawowe kategorie i historyczna ewolucja, Biuro Bezpieczeństwa Narodowego, Warszawa (2011)
10. Madej, M., Terlikowski, M.: Bezpieczeństwo teleinformatyczne państwa, Warszawa (2009)
11. Meyer B.: Informacja w procesie obsługi ruchu turystycznego. Ekonomiczne Problemy Turystyki (7) (2006)
12. Mynarski, S.: Elementy teorii systemów i cybernetyki. PWE, Warszawa (1979)
13. Strategia Cyberbezpieczeństwa RP na lata 2017–2022 (Cybersecurity Strategy of the Republic of Poland for 2017–2022). https://www.gsmservice.pl/19333,Strategia_Cyberbezpieczenstwa_RP_na_lata_2017__2022news.html. Accessed 11 May 2019
14. Tarkowski, A.: Internet jako technologia i wyobrażenie. Co robimy z technologią, co technologia robi z nami? In: Marody, M., Nowak, A. (eds.) Społeczna przestrzeń Internetu. Wydawnictwo Szkoły Wyższej Psychologii Społecznej "Academica", Warszawa (2006)
15. Wójcik, A.: System Zarządzania Bezpieczeństwem Informacji zgodny z ISO/IEC 27001 (cz. 1). https://www.zabezpieczenia.com.pl/ochrona-informacji/system-zarz%C4%85dzaniabezp iecze%C5%84stwem-informacji-zgodny-z-isoiec-27001-cz-1. Accessed 14 May 2019

# Analytical Forms of Productions Functions with Given Total Elasticity of Production

Guennadi Khatskevich[1]([envelope]), Andrei Pranevich[2]([envelope])[ORCID],
and Yury Karaleu[1]([envelope])

[1] School of Business of Belarusian State University, Obojnaya st. 7,
220004 Minsk, Belarus
g.a.khatskevich@gmail.com, yykorolev@sbmt.by
[2] Yanka Kupala State University of Grodno, Ozheshko str. 22,
230023 Grodno, Belarus
pranevich@grsu.by

**Abstract.** In this paper we completely classify production functions with given total elasticity of production (or elasticity of scale). The analytical form of the two-factor production function with given total elasticity of production is indicated. Classes of two-factor production functions that correspond to given (constant, linear, linear-fractional, exponential, etc.) total elasticity of production are obtained. Also, we give some generalization for multi-factor production functions with given total elasticity of production. The new production functions, which we have introduced in this article, integrates various well-known production functions such as Cobb-Douglas, CES or ACMS, Lu-Fletcher, Liu-Hildebrand, VES, and Kadiyala. The results may be useful in economic modelling of production at the regional and country levels.

**Keywords:** Production function · Total elasticity of production

## 1 Introduction

The idea of a production function is fundamental for economic analysis. It and its allied concept, the utility function, form the twin pillars of neoclassical economics [1]. Roughly speaking, the production functions are the mathematical formalization of the relationship between the output of a firm (industry, economy) and the inputs that have been used in obtaining it. In fact, a two-factor production function is defined as

$$Y = F(K, L) \tag{1}$$

where $K$ is the quantity of capital employed, $L$ is the quantity of labour used, $Y$ is the quantity of output, and the nonnegative function $F$ is a continuously differentiable function on a domain $G$ from the first quadrant $\mathbf{R}_+^2 = \{(K, L) : K > 0, \ L > 0\}$.

The two-factor production function (1) expresses a technological relationship. It describes the maximum output obtainable, at the existing state of technological knowledge, from given amounts of factor inputs.

Z. Wilimowska et al. (Eds.): ISAT 2019, AISC 1052, pp. 276–285, 2020.
https://doi.org/10.1007/978-3-030-30443-0_25

At the present time, production functions apply at the level of the individual firm and the macroeconomy at large. At the micro level, economists use production functions to generate cost functions and input demand schedules for the firm. The famous profit-maximizing conditions of optimal factor hire derive from such microeconomics functions. At the level of the macroeconomy, analysts use aggregate production functions to explain the determination of factor income shares and to specify the relative contributions of technological progress and expansion of factor supplies to economic growth [2–5].

Among the family of production functions (1), the most famous is the Cobb-Douglas production function. It was introduced in 1928 by the mathematician Cobb and the economist Douglas in the paper "A theory of production" [6]

$$Y = A K^\alpha L^\beta, \tag{2}$$

where $A$ is a positive constant which signifies the total productivity factor. We note that in the original definition of Cobb and Douglas it is required that $\alpha + \beta = 1$, but this condition has been later relaxed [7].

The Cobb-Douglas production model was generalized in 1961 by the economists Arrow, Chenery, Minhas, and Solow [8]. They introduced the so-called CES function (Constant Elasticity of Substitution production function)

$$Y = A \left(\alpha K^\gamma + \beta L^\gamma\right)^{\rho/\gamma}, \quad A > 0, \quad \alpha, \beta, \rho > 0, \quad \gamma \neq 0, \ 1. \tag{3}$$

Nowadays the Cobb-Douglas production function (2) and the CES production function (3) are widely used in economics to represent the relationship of output to inputs. Note also that CES production function includes as a special case the Cobb-Douglas production function and many other famous production models, like a linear production model, a multinomial production function or Leontief function.

Concerning the history of the development of the theory of production functions see the papers [1, 9]. For further results concerning new production models in economic see recent articles [10–17].

For production function (1), we recall some economic-mathematical indicators:

1. The output elasticity of capital (labour) is defined as

$$E_K(K,L) = \frac{K}{F(K,L)} \frac{\partial F(K,L)}{\partial K} \quad \left(E_L(K,L) = \frac{L}{F(K,L)} \frac{\partial F(K,L)}{\partial L}\right);$$

2. The total elasticity of production (or elasticity of scale)

$$E(K,L) = \lim_{t \to 1} \frac{t}{F(tK,tL)} \frac{\partial F(tK,tL)}{\partial t} = E_K(K,L) + E_L(K,L)$$

3. The Hicks elasticity of substitution [18, 19] is

$$\sigma^H(K,L) = \frac{\frac{\partial F}{\partial K} \cdot \frac{\partial F}{\partial L} \left( K \frac{\partial F}{\partial K} + L \frac{\partial F}{\partial L} \right)}{KL \left( 2 \frac{\partial F}{\partial K} \cdot \frac{\partial F}{\partial L} \cdot \frac{\partial^2 F}{\partial K \partial L} - \left( \frac{\partial F}{\partial K} \right)^2 \frac{\partial^2 F}{\partial L^2} - \left( \frac{\partial F}{\partial L} \right)^2 \frac{\partial^2 F}{\partial K^2} \right)} .$$

For instance, the functions (2) and (3) are productions functions with constant Hicks elasticity of factors substitution. The Cobb-Douglas production function (2) has unit elasticity of substitution i.e. $\sigma^H(K,L) = 1$. The CES production function (3) has the Hicks elasticity of factors substitution $\sigma^H(K,L) = 1/(1-\gamma)$. L. Losonczi proved [10] that a twice differentiable two-factor homogeneous production function with constant Hicks elasticity of substitution is either the Cobb-Douglas production function or the CES production function. This result complements the main propositions of the classical works [8, 20] and is consistent with known results on the classification of production functions [3]. The analogue for multi-factor production functions was proved by Chen in [12]. These results were recently generalized [14, 15] for quasi-homogeneous production functions with constant Hicks elasticity of substitution.

Vilcu and Vilcu classified homogeneous production functions with constant elasticity of labour and capital [13]. Their classification generalized some results by Ioan and Ioan concerning to the sum production function [11].

The aim of this paper is to identify all production functions with given total elasticity of production. The paper is organized as follows. In Sect. 2, we formulate our main result (Theorem 1) for two-factor production functions and obtain partial cases of production functions with given (constant, linear, linear-fractional, exponential, etc.) total elasticity of production. In Sect. 3, we prove Theorem 1. And in the last section of the paper, we generalize Theorem 1 for arbitrary number of inputs $n \geq 3$.

## 2   Two-Factor Production Function

### 2.1   Main Result

**Theorem 1.** *Suppose $E(K,L)$ is the total elasticity of production for some production technology. Then this production technology can be described by one of the production functions of the form*

$$F_\phi(K,L) = \phi\left(\frac{K}{L}\right) \exp\left(\int \frac{E(C_1 L, L)}{L} \, dL\right)_{|C_1 = \frac{K}{L}} \tag{4}$$

*where $\phi$ is arbitrary nonnegative continuously differentiable function on $(0; +\infty)$.*
Using Theorem 1, from some given total elasticities of production, we obtain the corresponding classes of production functions (see Table 1).

Suppose for some production technology we know the total elasticity of production $E(K,L) = \delta = const$. Then the production function has the form

**Table 1.** The form of the production function with given total elasticity of production

| No. | Total elasticity of production $(\alpha,\ \beta,\ \delta \in \mathbf{R},\quad f, g \in C(G))$ | The analytical form of the production function |
|---|---|---|
| 1. | $E(K,L) = \delta$ | $F_\phi(K,L) = L^\delta \varphi\!\left(\frac{K}{L}\right)$ |
| 2. | $E(K,L) = \alpha K + \beta L + \delta$ | $F_\phi(K,\ L) = L^\delta \varphi\!\left(\frac{K}{L}\right) \exp(\alpha K + \beta L)$ |
| 3. | $E(K,L) = f(\alpha K + \beta L)$ | $F_\phi(K,L) = \varphi\!\left(\frac{K}{L}\right) \exp\!\left(\int \frac{f(\xi)}{\xi} d\xi\right)_{\mid \xi = \alpha K + \beta L}$ |
| 4. | $E(K,L) = f(K) + g(K)$ | $F_\phi(K,L) = \varphi\!\left(\frac{K}{L}\right) \exp\!\left(\int \frac{f(K)}{K} dK + \int \frac{g(L)}{L} dL\right)$ |
| 5. | $E(K,L) = f\!\left(\frac{K}{L}\right)$ | $F_\phi(K,L) = \varphi\!\left(\frac{K}{L}\right) \exp\!\left(\ln L \cdot f\!\left(\frac{K}{L}\right)\right)$ |
| 6. | $E(K,L) = K^\alpha f\!\left(\frac{K}{L}\right),\ \alpha \neq 0$ | $F_\phi(K,L) = \varphi\!\left(\frac{K}{L}\right) \exp\!\left(\frac{1}{\alpha} K^\alpha f\!\left(\frac{K}{L}\right)\right)$ |
| 7. | $E(K,L) = L^\beta f\!\left(\frac{K}{L}\right),\ \beta \neq 0$ | $F_\phi(K,L) = \varphi\!\left(\frac{K}{L}\right) \exp\!\left(\frac{1}{\beta} L^\beta f\!\left(\frac{K}{L}\right)\right)$ |
| 8. | $E(K,L) = f(K^\alpha L^\beta),\ \alpha \neq -\beta$ | $F_\phi(K,L) = \varphi\!\left(\frac{K}{L}\right) \exp\!\left(\frac{1}{\alpha+\beta} \int \frac{f(\xi)}{\xi} d\xi\right)_{\mid \xi = K^\alpha L^\beta}$ |

$$F_\phi(K,L) = L^\delta \phi\!\left(\frac{K}{L}\right)$$

and we have the following statements:

- if $\phi(\xi) = A\xi^\alpha$, then we obtain the Cobb-Douglas function (2) with $\beta = \delta - \alpha$;
- if $\phi(\xi) = A(\alpha \xi^\gamma + \beta)^{\delta/\gamma}$, then we get the CES production function (3) with $\rho = \delta$;
- if $\phi(\xi) = A(\alpha \xi^\gamma + (1-\alpha)\beta \, \xi^{-\sigma(1-\gamma)})^{\delta/\gamma}$, then we have the Lu-Fletcher function [21]

$$F(K,L) = A\left(\alpha K^\gamma + (1-\alpha)\beta \left(\frac{K}{L}\right)^{-\sigma(1-\gamma)} L^\gamma\right)^{\delta/\gamma},$$

which for $\sigma = 0$, $\beta = 1$ becomes the CES production function;

- if $\phi(\xi) = A\left((1-\alpha)\, \xi^\gamma + \alpha \, \xi^{\sigma\gamma}\right)^{\delta/\gamma}$, then we obtain the Liu-Hildebrand function [22]

$$F(K,L) = A\left((1-\alpha) K^\gamma + \alpha K^{\sigma\gamma} L^{(1-\sigma)\gamma}\right)^{\delta/\gamma},$$

which for $\sigma = 0$ becomes the CES production function;

- if $\phi(\xi) = A\left(a\, \xi^{\alpha+\beta} + 2b\, \xi^\alpha + c\right)^{\delta/(\alpha+\beta)}$, then we have the Kadiyala function [23]

$$F(K,L) = A\left(a K^{\alpha+\beta} + 2b K^\alpha L^\beta + c L^{\alpha+\beta}\right)^{\delta/(\alpha+\beta)},$$

where $a + 2b + c = 1$, $a, b, c \geq 0$, $\alpha(\alpha+\beta) > 0$, $\beta(\alpha+\beta) > 0$. Since $b = 0$, we see that the Kadiyala production function is the CES production function. If $c = 0$, then the Kadiyala production function generates directly the Lu-Fletcher function. For $a = c = 0$, we obtain the Cobb-Douglas function. Finally, for $\alpha = 1/(\mu\gamma) - 1$, $\beta = 1$, $\mu = 1/(1-a)$, and $c = 0$, we get the VES production function [24, 25]

$$F(K,L) = A K^{(1-\mu\gamma)\delta}(L + (\mu-1) K)^{\mu\gamma\delta}.$$

## 2.2   Proof of Theorem 1

Let $E(K,L)$ be the total elasticity of production for some production technology. Then the production function (1), which is corresponding to this production process, is a solution to the first-order partial differential equation

$$K \frac{\partial F(K,L)}{\partial K} + L \frac{\partial F(K,L)}{\partial L} = F(K,L) \cdot E(K,L). \tag{5}$$

The ordinary differential system in the symmetric form

$$\frac{dK}{K} = \frac{dL}{L} = \frac{dF}{E(K,L)F} \tag{6}$$

corresponds to the partial differential equation of the first order (5).

Integrating the differential equation $\frac{dK}{K} = \frac{dL}{L}$, we obtain

$$d \ln K = d \ln L, \quad d \ln \frac{K}{L} = 0, \quad \ln \frac{K}{L} = \tilde{C}_1, \quad \frac{K}{L} = C_1,$$

where $C_1 = \exp \tilde{C}_1$ and $\tilde{C}_1$ are arbitrary real constants. Therefore, the rational function $\xi_1(K,L) = \frac{K}{L}$ is a first integral of the differential system (6).

Since $\frac{K}{L} = C_1$ or $K = C_1 L$, it follows that

$$\frac{dL}{L} = \frac{dF}{E(K,L)F}, \quad \frac{dF}{F} = \frac{E(C_1 L, L)}{L} dL, \quad \ln F = \int \frac{E(C_1 L, L)}{L} dL + \tilde{C}_2,$$

$$F = C_2 \exp\left(\int \frac{E(C_1 L, L)}{L} dL\right), \quad F \exp\left(-\int \frac{E(C_1 L, L)}{L} dL\right) = C_2,$$

where $C_2 = \exp \tilde{C}_2$ and $\tilde{C}_2$ are arbitrary real constants. Therefore, the function

$$\xi_2(K,L) = F \exp\left(-\int \frac{E(C_1 L, L)}{L} dL\right)\Big|_{C_1 = \frac{K}{L}}$$

is a first integral of the differential system (6).

Using the functionally independent first integrals $\xi_1$ and $\xi_2$ of system (6), we get the general solution to the differential Eq. (5) in the form

$$\Phi\left(\frac{K}{L}, \ F \exp\left(-\int \frac{E(C_1 L, \ L)}{L} dL\right)_{|C_1=\frac{K}{L}}\right) = 0,$$

where $\Phi$ is arbitrary continuously differentiable function. If we take functions $\Phi$ such that this equation is solvable with respect to the second argument (see the implicit function theorem, for example, in [26]), then from this equation, we get the solution to the Eq. (5) in the explicit form (4).                                                                □

## 3   Multi-factor Production Function

Let $F$ be a production function with n inputs $x_1, \ldots, x_n$, $n > 2$. Then, the output elasticity with respect to a certain factor of production $x_i$ is defined as

$$E_i(x) = \frac{x_i}{F(x)} \frac{\partial F(x)}{\partial x_i} \equiv \frac{\partial(\ln F(x))}{\partial(\ln x_i)}, \quad i = 1, \ldots, n,$$

and the total elasticity of production is given by

$$E(x) = \lim_{t \to 1} \frac{t}{F(tx)} \frac{\partial F(tx)}{\partial t} \equiv \sum_{i=1}^{n} E_i(x).$$

Now, we are able to prove the following result, which generalized Theorem 1 for arbitrary number of inputs.

**Theorem 2.** *Let $E(x)$ be the total elasticity of production for some production technology. Then this production technology can be described on the domain $G \subset \mathbf{R}_+^n$ by one of the production functions*

$$F_\phi(x) = \phi\left(\frac{x_1}{x_n}, \ldots, \frac{x_{n-1}}{x_n}\right) \exp \int \frac{E(C_1 x_n, \ldots, C_{n-1} x_n, x_n)}{x_n} dx_n \bigg|_{C_i = \frac{x_i}{x_n}, \ i=1,\ldots,n-1}, \quad (7)$$

*where $\phi$ is arbitrary nonnegative continuously differentiable function on $\mathbf{R}_+^{n-1}$.*

*Proof.* Suppose for some production technology we know the total elasticity of production $E(x)$. Then the production function, which is corresponding to this production process, is a solution to the first-order partial differential equation

$$\sum_{i=1}^{n} x_i \frac{\partial F(x)}{\partial x_i} = F(x) E(x). \quad (8)$$

The ordinary differential system in the symmetric form

$$\frac{dx_1}{x_1} = \frac{dx_2}{x_2} = \ldots = \frac{dx_n}{x_n} = \frac{dF}{E(x)\,F} \tag{9}$$

corresponds to the quasilinear partial differential equation of the first order (7).

Integrating the differential equations $\frac{dx_i}{x_i} = \frac{dx_n}{x_n}$, $i = 1,\ldots,n-1$, we obtain $\frac{x_i}{x_n} = C_i$, where $C_i$ is arbitrary real constant, $i = 1,\ldots,n-1$. Therefore, the rational functions $\xi_i(x) = \frac{x_i}{x_n}$, $i = 1,\ldots,n-1$, are first integrals of the differential system (9).

Since $x_i = C_i x_n$, $i = 1,\ldots,n-1$, we see that from $\frac{dx_n}{x_n} = \frac{dF}{E(x)\,F}$ it follows that

$$\frac{dF}{F} = \frac{E(C_1 x_n,\ldots,C_{n-1}x_n,x_n)\,dx_n}{x_n}, \quad \ln F = \int \frac{E(C_1 x_n,\ldots,C_{n-1}x_n,x_n)\,dx_n}{x_n} + \tilde{C}_n,$$

$$F = C_n \exp \int \frac{E(C_1 x_n,\ldots,C_{n-1}x_n,x_n)\,dx_n}{x_n}, \quad F \exp\left(-\int \frac{E(C_1 x_n,\ldots,C_{n-1}x_n,x_n)\,dx_n}{x_n}\right) = C_n,$$

where $\tilde{C}_n$ and $C_n = \exp \tilde{C}_n$ are arbitrary real constants. Therefore, the function

$$\xi_n(x) = F \exp\left(-\int \frac{E(C_1 x_n,\ldots,C_{n-1}x_n,x_n)\,dx_n}{x_n}\right)$$

is a first integral of the differential system in the symmetric form (9).

Using the functionally independent first integrals of the system (9), we can build the general solution to the first-order partial differential Eq. (8) in the form

$$\Phi\left(\frac{x_1}{x_n},\ldots,\frac{x_{n-1}}{x_n},\ F \exp\left(-\int \frac{E(C_1 x_n,\ldots,C_{n-1}x_n,x_n)}{x_n}dx_n\right)\bigg|_{C_i = \frac{x_i}{x_n},\ i=1,\ldots,n-1}\right) = 0,$$

where $\Phi$ is arbitrary continuously differentiable function. If we take $\Phi$ such that this equation is solvable with respect to the last argument (see the implicit function theorem in [26]), then we get the solution to the Eq. (8) in the explicit form (7).  □

The following assertions are immediate consequences of Theorem 2.

**Corollary 1.** *If the total elasticity of production $E(x) = \alpha_0$, $\alpha_0 \in \mathbf{R}$, then this production technology can be described by one of the production functions*

$$F_\phi(x) = x_n^{\alpha_0}\,\phi\left(\frac{x_1}{x_n},\ldots,\frac{x_{n-1}}{x_n}\right).$$

From Corollary 1, we obtain the following statements:

- if $\alpha_0 = 1$ and $\phi(\xi) = \sum_{i=1}^{n-1} \alpha_i \xi_i + \alpha_n$ for all $\xi = (\xi_1, \ldots, \xi_{n-1}) \in \mathbf{R}_+^{n-1}$, then we have

  the linear production function $F(x) = \sum_{i=1}^{n} \alpha_i x_i$ for all $x \in \mathbf{R}_+^n$, $\alpha_i \in \mathbf{R}_+, i = 1, \ldots, n$;

- if $\phi(x) = A\xi_1^{\alpha_1} \cdot \ldots \cdot \xi_{n-1}^{\alpha_{n-1}}$ for all $\xi \in \mathbf{R}_+^{n-1}$, then we get the generalized Cobb-

  Douglas production function $F(x) = A \prod_{i=1}^{n} x_i^{\alpha_i}$ with $\alpha_n = \alpha_0 - \sum_{i=1}^{n-1} \alpha_i$;

- if $\phi(x) = A \left( \sum_{i=1}^{n-1} \alpha_i \xi^{\gamma} + \alpha_n \right)^{\alpha_0/\gamma}$ for all $\xi \in \mathbf{R}_+^{n-1}$, then we have the generalized CES

  production function [20, 27] $F(x) = A \left( \sum_{i=1}^{n} \alpha_i x_i^{\gamma} \right)^{\alpha_0/\gamma}$ for all $x \in \mathbf{R}_+^n$.

**Corollary 2.** *If the total elasticity of production* $E(x) = \sum_{i=1}^{n} \alpha_i x_i + \alpha_0$, $\alpha_i \in \mathbf{R}$, $i = 0, \ldots, n$, *is a linear function, then this production technology can be described by one of the production functions* $F_\phi(x) = x_n^{\alpha_0} \, \phi\left(\frac{x_1}{x_n}, \ldots, \frac{x_{n-1}}{x_n}\right) \, \exp\left( \sum_{i=1}^{n} \alpha_i x_i \right)$.

**Corollary 3.** *Suppose for some production we have the total elasticity of production* $E(x) = u\left( \sum_{i=1}^{n} \alpha_i x_i \right)$, $\alpha_i \in \mathbf{R}$, $i = 1, \ldots, n$, *where u is a continuous function. Then this production technology can be described by one of the production functions*

$$F_\phi(x) = \phi\left(\frac{x_1}{x_n}, \ldots, \frac{x_{n-1}}{x_n}\right) \, \exp \int \frac{u(\xi)}{\xi} d\xi \, \Big|_{\xi = \sum_{i=1}^{n} \alpha_i x_i}.$$

**Corollary 4.** *Suppose for some production we have the total elasticity of production* $E(x) = \sum_{i=1}^{n} u_i(x_i)$, *where $u_i$ are continuous functions. Then this production technology can be described by one of the production functions of the form*

$$F_\phi(x) = \phi\left(\frac{x_1}{x_n}, \ldots, \frac{x_{n-1}}{x_n}\right) \, \exp\left( \sum_{i=1}^{n} \int \frac{u_i(x_i)}{x_i} dx_i \right).$$

**Corollary 5.** *Suppose for some production we have the total elasticity of production* $E(x) = u(x_1^{\alpha_1} \cdot \ldots \cdot x_n^{\alpha_n})$, $\alpha_i \in \mathbf{R}$, $i = 1, \ldots, n$, $\beta = \sum_{i=1}^{n} \alpha_i \neq 0$, *where u is a continuous function. Then this production technology can be described by one of the production functions* $F_\phi(x) = \phi\left(\frac{x_1}{x_n}, \ldots, \frac{x_{n-1}}{x_n}\right) \, \exp\left( \frac{1}{\beta} \int \frac{u(\xi)}{\xi} d\xi \right) \Big|_{\xi = x_1^{\alpha_1} \cdot \ldots \cdot x_n^{\alpha_n}}.$

In particular, if the elasticity of production $E(x) = \alpha_0 x_1^{\alpha_1} \cdot \ldots \cdot x_n^{\alpha_n}$, then we get the production function $F_\phi(x) = \phi\left(\frac{x_1}{x_n}, \ldots, \frac{x_{n-1}}{x_n}\right) \exp\left(\frac{\alpha_0}{\beta} x_1^{\alpha_1} \cdot \ldots \cdot x_n^{\alpha_n}\right)$.

For further results on the quasi-sum, quasi-product and homothetic production functions see [28]. Many other results concerning production models in economics from the viewpoint of geometry can be found in the survey [29].

**Acknowledgements.** This research was supported by the *Belarusian State Program of Scientific Research "Economy and humanitarian development of the Belarusian society"*.

# References

1. Humphrey, T.M.: Algebraic production functions and their uses before Cobb-Douglas. Econ. Q. **1**(83), 51–83 (1997)
2. Shephard, R.W.: Theory of Cost and Production Functions. Princeton University Press, Princeton (1970)
3. Kleiner, G.B.: Production Functions: Theory, Methods, Application. Finansy i statistika Publ., Moscow (1986)
4. Barro, R.J., Sala-i-Martin, X.: Economic Grows. McGraw-Hill, New-York (2004)
5. Gorbunov, V.K.: Production Functions: Theory and Construction. Ulyanovsk State University Publ., Ulyanovsk (2013)
6. Cobb, C.W., Douglas, P.H.: A theory of production. Am. Econ. Rev. **18**, 139–165 (1928)
7. Douglas, P.H.: The Cobb-Douglas production function once again: its history, its testing, and some new empirical values. J. Polit. Econ. **5**(84), 903–916 (1976)
8. Arrow, K.J., Chenery, H.B., Minhas, B.S., Solow, R.M.: Capital-labor substitution and economic efficiency. Rev. Econ. Stat. **3**(43), 225–250 (1961)
9. Mishra, S.K.: A brief history of production functions. IUP J. Manag. Econ. **4**(8), 6–34 (2010)
10. Losonczi, L.: Production functions having the CES property. Acta Mathematica Academiae Paedagogicae Nyiregyhaziensis **1**(26), 113–125 (2010)
11. Ioan, C.A., Ioan, G.: A generalization of a class of production functions. Appl. Econ. Lett. **18**, 1777–1784 (2011)
12. Chen, B.-Y.: Classification of $h$-homogeneous production functions with constant elasticity of substitution. Tamkang J. Math. **2**(43), 321–328 (2012)
13. Vilcu, A.D., Vilcu, G.E.: On homogeneous production functions with proportional marginal rate of substitution. Math. Probl. Eng. **2013**, 1–5 (2013)
14. Khatskevich, G.A., Pranevich, A.F.: On quasi-homogeneous production functions with constant elasticity of factors substitution. J. Belarus. State Univ. Econ. **1**, 46–50 (2017)
15. Khatskevich, G.A., Pranevich, A.F.: Quasi-homogeneous production functions with unit elasticity of factors substitution by Hicks. Econ. Simul. Forecast. **11**, 135–140 (2017)
16. Vilcu, G.E.: On a generalization of a class of production functions. Appl. Econ. Lett. **25**, 106–110 (2018)
17. Khatskevich, G.A., Pranevich, A.F.: Production functions with given elasticities of output and production. J. Belarus. State Univ. Econ. **2**, 13–21 (2018)
18. Hicks, J.R.: The Theory of wages. Macmillan Publ., London (1932)
19. Allen, R.G.: Mathematical Analysis for Economists. Macmillan Publ., London (1938)
20. Uzawa, H.: Production functions with constant elasticities of substitution. Rev. Econ. Stud. **4**(29), 291–299 (1962)

21. Lu, Y.C., Fletcher, L.B.: A Generalization of the CES production function. Rev. Econ. Stat. **50**, 449–452 (1968)
22. Liu, T.C., Hildebrand, G.H.: Manufacturing Production Functions in the United States. Cornell University Press, Ithaca (1965)
23. Kadiyala, K.R.: Production functions and elasticity of substitution. South. Econ. J. **38**, 281–284 (1972)
24. Revankar, N.S.: A class of variable elasticity of substitution production functions. Econometrica **1**(39), 61–71 (1971)
25. Sato, R., Hoffman, R.F.: Production function with variable elasticity of factor substitution: some analysis and testing. Rev. Econ. Stat. **50**, 453–460 (1968)
26. Il'in, V.A., Pozniak, E.G.: Fundamentals of mathematical analysis. Nauka, Moscow (2000)
27. McFadden, D.: Constant elasticities of substitution production functions. Rev. Econ. Stud. **30**, 73–83 (1963)
28. Chen, B.-Y.: On some geometric properties of quasi-sum production models. J. Math. Anal. Appl. **2**(392), 192–199 (2012)
29. Vilcu, A.D., Vilcu, G.E.: A survey on the geometry of production models in economics. Arab J. Math. Sci. **1**(23), 18–31 (2017)

# Outstanding People and Works in the Area of Computer Science Enhancing Quality of Life. The Evidence from Prestigious American, European and Asian Computer Science Awards

Łukasz Wiechetek[(⊠)] [iD]

Faculty of Economics, Maria Curie-Sklodowska University, Lublin, Poland
lukasz.wiechetek@umcs.pl

**Abstract.** Information technology is one of the fastest developing business sector and research area. People use IT solutions based on computer science both for entertainment, education and work. Modern organizations cannot operate effectively without application of IT tools for data acquisition, storage, communication, visualization, but also for linking digital world with the real life or the real business processes. Works in the field of computer science are the foundation of fast and balanced development of IT sector. They are the accelerators of transformation from traditional, real economy into digital business and digital society.

The main aim of the article is to explore what nations, people and works played the most significant role in development of information technology during the previous decades. Also what works are crucial today to ensure stability of present digital natives. The author analyzes the five prestigious awards in computer science to find out what works were honored for the past hundred years. In the first part of the article the characteristics of the explored awards is presented. The second part of the paper contains quantitative analysis of award recipients (n = 224) and prizes comparison.

The performed literature review and the analysis of the awarded works allow to conclude that during the last decades the main drivers for IT development, noticed by awards committees were related both to the theory, hardware, software development and application areas. The vast majority of laureates came from (was born in) United States (62%), United Kingdom (8,5%) and Japan (4,9%). At the beginning of twenty first century huge amounts of data are generated not only by people but also autonomous objects, so the most crucial in computer science was to prepare systems and networks for secure processing (transferring) huge amounts of data in real time and find methods to assess the accuracy and quality of the information. Therefore, new awards like IEEE Internet Award were established to honor the works closely related to the development of the Internet and computer networks.

**Keywords:** Computer science · IT Award · Turing Award · Millennium Prize · Kyoto Prize · IEEE Medal of Honor · IEEE Internet Award · IT evolution · Computer science evolution · Important works in IT

© Springer Nature Switzerland AG 2020
Z. Wilimowska et al. (Eds.): ISAT 2019, AISC 1052, pp. 286–298, 2020.
https://doi.org/10.1007/978-3-030-30443-0_26

# 1  Introduction

Computer science can be defined as a study on processes interacting with data, so using the algorithms for digital data processing. The main fields of computer science are theoretical basis (algorithms, data structures, computation theory, coding theory, or programming language theory), computer systems (architecture, performance analysis, networks, parallel an distributed systems, cryptography and databases) application of the computers (visualizations, graphics, user interaction or artificial intelligence). Finally very important field is also software engineering including analyzing, designing, modifying and maintenance of high quality and effective applications. Nowadays computer science is one of the main accelerator of modern economies and societies.

According to Rapaport [1] the greatest pioneers of computer science that defined the smallest possible language that can be used in any procedure for any computer were Francis Bacon (philosophical logic) Gottfried Wilhelm Leibniz (differential and integral calculus), George Boole (Boolean algebra), Alan Turing (Turing machine), Claude Elwood Shannon (information theory) and Samuel Morse (Morse code). Today's science, business and everyday live would be inconceivable without the computers, networks and the Internet [2]. Therefore, in the next part of the article the author presents prestigious computer science awards from America, Asia and Europe: IEEE Medal of Honor, Turing Award, Kyoto Prize in Advanced Technology, IEEE Internet Award, IEEE Millennium Technology Prize to show the most important achievements in this field that have sustainable impact on quality of life and people well-being.

Analyzed awards were established in different moments of time and concern all fields of computer science: theory, computer systems, computer applications and software engineering. Basing on the characteristics of people and works that were honored with those awards the author prepared the analysis showing what nations, what people, and what works had the outstanding contributions to the advancement of computer science and information technology. The analysis contains also the comparison of explored awards and shows the evolution of computer science interest over the time.

# 2  Prestigious Computer Science Awards

Computer Science is nowadays important field of human activity, therefore, to honor the most outstanding works in this area many awards were established. The awards have different range: global, regional or local. Some of them are given by IT companies other by non-profit organizations [3]. They address both theoretical, practical, software and hardware area of IT. There are also different works/areas for the prizes, some like Turing Award or IEEE Medal of Honor are more general, appreciate more technical importance of given work, others like IEEE Internet Award value an exceptional contribution or an extraordinary career in a very precise area.

In this article author explores five prestigious awards in computer science: Turing Award, Millennium Technology Prize, Kyoto Prize in Advanced Technology, IEEE Medal of Honor and IEEE Internet Award in order to find out what specialists received the prize, what subjects were the most important and how the importance of achievements changed over the time. Analyzed awards were chosen because they are recognized all over the world, were established in different moments at the beginning of the XX century, in the middle, but also at the end of 90's and the beginning the XXI century. Chosen awards are also granted by the committees, organizations located both in America, Asia and Europe.

## 2.1   IEEE Internet Award

The IEEE Internet Award was established in 1999. The main sponsor of the award is Nokia Corporation. It is presented to individual or up to three recipients for the huge contributions to development of internet technology, mobility and user applications [4]. The awarded person(s) receives certificate, medal and honorarium. The award is presented at main IEEE events. Up to 2019 the award was granted nineteenth times (2007 not awarded) to twenty six researchers, inventors or engineers with huge contributions to internet technology development. The first honored in 2000 were Paul Baran, Donald W. Davies, Leonard Kleinrock and Lawrence Roberts for preeminent contributions in packet switching networks. The last (up to the present) IEEE Internet Award was granted in 2019 to Jennifer Rexford for fundamental contributions to large computer networks [5].

## 2.2   IEEE Medal of Honor

The IEEE Medal of Honor was established in 1917. The main sponsor of the award is the IEEE Foundation. It is presented to individual for an exceptional contribution(s) or an extraordinary career in the IEEE fields of interest [6]. The awarded person receives gold medal, a bronze replica, a certificate and honorarium. The evaluation process considers: achievement significance, its originality, impact on society and the profession, but also related publications or patents. The award is presented at the annual IEEE Honors Ceremony. The first Medal of Honor was granted in 1917 to Edwin Howard Armstrong in "In recognition of his work and publications dealing with the action of the oscillating and non-oscillating audio". The last award (up to the present) was granted in 2019 to Kurt E. Petersen for development and commercialization of innovative micromachining and micro-electro-mechanical systems [7].

## 2.3   Kyoto Prize in Advanced Technology

The Kyoto Prize in Advanced Technology is granted annually for global achievement in advanced technology and presented by Inamori Foundation. It is granted to individuals but in some cases a single Prize may be shared among more than one person. The laureate is presented with a diploma, a gold Kyoto Prize medal, and 100 million yen. The Price was first awarded in 1985 and it is granted on a rotating basis to the researchers in the fields of electronics. biotechnology and medical technology, materials science and engineering and information science [8]. In this article the author analyses only two categories electronics and information science. Up to the end of 2018 the award (in electronics and information science) was granted eighteen times to twenty three researchers The first Kyoto Prize was in electronics. It was granted in 1985 to Rudolf Emil Kalman for the modern control theory establishment. The last award (up to the present) was granted in 2017 to Takashi Mimura for invention of the high electron mobility transistor [9].

## 2.4    Millennium Technology Prize

Millennium Technology Prize has been awarded every second year since 2004 by Technology Academy Finland for groundbreaking technological innovations that increase the quality of life, promote sustainable development, create new socioeconomic value or stimulate future R&D in science and technology. The Prize is presented by the president of Finland and the prize sum is 1 mln EUR [10].

Up to the end of 2018 the award was granted nine times. The first awarded in 2004 was Tim Berners-Lee for open World Wide Web system. The last award in 2018 was granted to T. Suntolafor for technology that enables manufacture of nanoscale thin material layers for microprocessors and digital memory devices [11].

## 2.5    Turing Award

Turing Award is granted annual by the Association for Computing Machinery. It is given since 1966 and recognized as the highest distinction in a field of computer science [12]. The award is named after British mathematician Alan Turing who was a pioneer in a computer architecture, formalization of computing [14]. Since 2014 the prize value is $1mln and the award is financially supported by Google. The first Turing Award was granted in 1966 to Alan J. Perlis for the outstanding contribution in the area of computer programming techniques and construction of the compiler. The last (up to now) award was presented in 2017 to John L. Hennessy and David A. Patterson for systematic, quantitative approach to the design and evaluation of computer architectures and enduring impact on the microprocessor industry [13]. The general characteristics of five explored computer science awards was shown in Table 1.

**Table 1.**  Comparative characteristics of computer science awards (alphabetical order).

| No. | Name | Description | Reward ($) | Awarded by | since | Web page |
|-----|------|-------------|------------|------------|-------|----------|
| 1. | IEEE Internet Award | Annually technical field award for up to three recipients, for the outstanding contributions to the advancement of Internet technology | – | Institute of Electrical and Electronics Engineers | 1999 | https://www.ieee.org/about/awards/technical-field-awards/technical-field-awards-internet.html |
| 2. | IEEE Medal of Honor | Annual prize given for a particular contribution that forms a clearly exceptional addition to the science and technology of concern to IEEE | – | Institute of Electrical and Electronics Engineers | 1917 | http://www.ieee.org/about/awards/medals/medal-of-honor.html |

*(continued)*

**Table 1.** (*continued*)

| No. | Name | Description | Reward ($) | Awarded by | since | Web page |
|-----|------|-------------|------------|------------|-------|----------|
| 3. | Kyoto Prize in Advanced Technology | Awarded once a year in the four fields: electronics, biotechnology, materials science and information science | About 0,5 mln | Inamori Foundation | 1985 | http://www.kyotoprize.org |
| 4. | Millennium Technology Prize | Awarded every two years. Given for innovations that have sustainable impact on quality of life and people well-being | 1 mln | Technology Academy Finland | 2004 | http://taf.fi/millennium-technology-prize/ |
| 5. | Turing Award | Annual prize given for lasting and major technical importance to the computer field | 1 mln | Association for Computing Machinery | 1966 | http://amturing.acm.org |

# 3   What Was and Is Important in the Computer Science

## 3.1   Methodology

**Research Questions.** The author performed the analysis of chosen five computer science awards in order to find the answers to the following research questions:

- What was and is important in the area of computer science?
- What countries and people had the largest contribution to IT development?
- What are the differences between works, people awarded by various prizes?

**Research Procedure.** The main reasons for choosing specific awards were: prestige and worldwide popularity, covering all the fields of computer science, representing main geographical regions of computer science development (America, Asia, Europe), but also a long history that allows for comparison over the time. The award web page was source of data not only about honored works but also about awarded people and their career. The personal details of the award recipients were downloaded from the Wikipedia and the recipients private websites (Fig. 1).

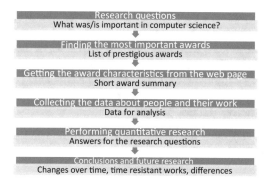

**Fig. 1.** The research procedure.

Data was collected in March 2019 form the awards' webpages and stored in tabular from. The author collected data about 224 rewarded engineers, inventors and researchers. The final analysis was carried out using Microsoft Excel and the visualization was performed in Tableau visualization software. In this paper only basic descriptive statistics are presented, formal statistical analysis of the collected data, showing significances of differences would be prepared during the next research.

## 3.2 Data Analysis

The characteristics of awards recipients was shown in Table 2.

**Table 2.** The characteristics of awards recipients.

| Characteristic | | n | % |
|---|---|---|---|
| Gender | Female | 10 | 4,5% |
| | Male | 214 | 95,5% |
| Age | <20;30) | 1 | 0,46% |
| | <30;40) | 3 | 1,39% |
| | <40;50) | 23 | 10,65% |
| | <50;60) | 74 | 34,26% |
| | <60;70) | 54 | 25,00% |
| | <70;80) | 53 | 24,54% |
| | <80;90) | 8 | 3,70% |
| Award | IEEE Internet Award | 26 | 11,61% |
| | IEEE Medal of Honor | 99 | 44,20% |
| | Kyoto Prize in Advanced Technology | 23 | 10,27% |
| | Millennium Technology Prize | 9 | 4,02% |
| | Turing Award | 67 | 29,91% |

The most of the awarded people were men 95,5%. The average age of honored person was about 61 years, the most numerous group was between 50 and 60 years, so very experienced researchers, inventors but still active. There was only one engineer that obtained the prize before the age of 30 and eight winners over the age of 80. The youngest laureate was 27, the oldest 85. There can be concluded that the high level IT is not only the domain of young people and that the committees appreciated the works usually when they survive the test of time.

**Country of Origin.** To show the country of origin of the laureates the collected data was visualized with Tabloeau Desktop software (Fig. 2).

**Fig. 2.** Country of birth of the laureate.

The author took into account the place of birth of the laureates. However many of them have moved to the United States and they had American citizenship. Some of the recipients were born before World Wars that changed the borders especially in Europe in that case also the country of birth is different than the place of residence. The vast majority of the recipients were born in the USA - 62%, significant role played also the inventors, engineers and researchers from United Kingdom, and Japan, respectively 8,5% and 4,9%. Among countries that had four or more winners were also: China, Italy, Netherlands, Canada, Hungary, Israel and India.

**Comparison Between the Prizes.** The biggest group are laureates from America. That group was almost twice bigger than the groups of recipients from Europe and Asia put together (Fig. 3).

America due to the Unites States have the most of laureates in four out of five awards. Only the Millennium Technology Prize which is presented by Technology Academy Finland (Europe) awarded mostly people from Europe. There can be noticed that the awards comities reward a little more often outstanding works from their region, so IEEE and Association for Computing Machinery awards have more recipients from America, Inamori Foundation (Kyoto Prize) more often appreciated the work of Asians and Technology Academy Finland usually rewarded Europeans (Figs. 3, 4).

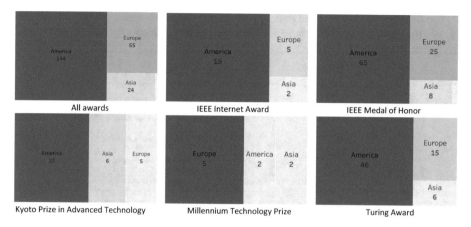

**Fig. 3.** Number of award recipients by continent and the award.

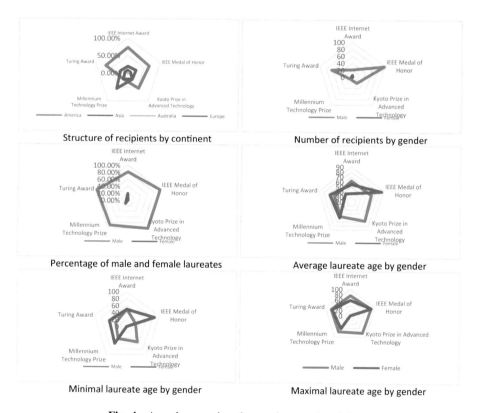

**Fig. 4.** Award comparison by continent and recipient.

The author couldn't find the date of birth of 8 analyzed laureates. They all received the IEEE Internet Award. Therefore the analysis of the laureates age was based on 216 recipients (Fig. 5).

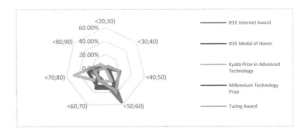

**Fig. 5.** Age of award recipients.

The oldest awards with the biggest numbers of recipients are IEEE Medal of Honor (99) and Turing Award (67). Much younger with only 23 awarded people is Kyoto Prize in Advanced Technology. Kyoto prize was in 100% granted to male recipients, also almost 99% of Medals of Honor were granted to men. Awards with the highest percentage of female laureates are Millennium Technology Prize (11%) and IEEE Internet Award (19%). The average age of whole prizes recipients was about 61. The oldest were recipients of Kyoto Prize (M = 68). The average age of female laureates was higher than male (respectively 63 and 61 years). The youngest rewarded was man Edwin H. Armstrong he received the IEEE Medal of Honor at the age of 27. The oldest laureate was Mildred Dresselhaus, she received IEEE Medal of Honor at the age of 85.

**The Most Important Areas.** The most important areas for the development of computer sciences were derived from the rationales of the analyzed awards. The author put all the rationales into Microsoft Excel table and assigned them to the categories. The list of categories was used to create the word cloud with world clouds generator [15] (Fig. 6).

**Fig. 6.** The most frequently awarded areas of computer science research.

The most important works building the basements of IT were radio, programming languages (environments, and theory), mechanisms and tools for communications that gave rise to the development of modern computer networks and mobile technologies.

They occurred more than 17 times. The important works were related also to computer and network architecture, artificial intelligence, cryptography and optoelectronics. Those works allowed for building cheap, fast, reliable and secure channels for communication. Often awarded were also the engineers, researchers that created the hardware foundations (transistors, semiconductors, microprocessors and electronic circuits). Many of laureates worked also on data storage developing more effective database systems or faster random access memory, but also improving the effectiveness of the algorithms. We can conclude that analyzed awards cover whole spectrum of both theoretical end empirical important areas of computer science. They appreciate all fields of information technology from theoretical foundations, through hardware components and finally software solutions. Committees rewarded wide spectrum of works that are needed for fast and sustainable development of computer science and information technology. During the decades the importance of various works has evolved. The researchers had also more and more theoretical knowledge, but also technological possibilities. This evolution can be observed also in the areas for which prizes were awarded (Fig. 7).

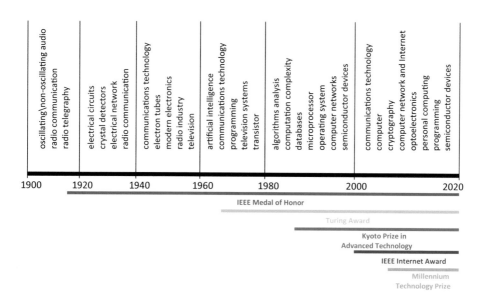

**Fig. 7.** Evolution of award-winning areas between 1917 and 2019.

The oldest of the analyzed awards is the IEEE Medal of Honor (granted since 1917). From the 1917 till 1966 only the laureates of this award were taken into account. Form 1900 till 1920 the most awarded works were related to the theoretical end practical aspects of radio communication. In the next two decades from 20's to 40's the award was still granted for works on radio communication, measurement and technology, but also for development of electrical networks, crystal detectors and electrical circuits.

In the period from 1940 to 1960, the recipients were awarded mostly for development of communication technology and the radio industry. The committees also appreciated the work on electron tubes, modern electronics also the contribution to the development of television.

From 1960 till 1980 the author took into account two awards IEEE Medal of Honor and Turing Award. During this period we can observe a lot of laureates still working on the communications technology, but the most noticed area seems to be programming (techniques, languages, compilers) and numerical analysis. Many honored works concerned also artificial intelligence and development of foundations of modern electronics – transistor. Awarded were also works in the area of television systems.

In 1985 Kyoto Prize in Advanced Technology was established. Next at the end of XX century IEEE due to the fast development and the huge impact of the Internet established IEEE Internet Award. In the period from 1980 till the end of the century the analysis is based on four awards: IEEE Medal of Honor and Internet Award, Turing Award, Kyoto Prize. During this period honored were contributions to algorithm analysis, works on computation complexity. Committees appreciated also effort towards building more effective operating systems and databases. Honored were also achievements in the area of computer network. In the hardware segment awards were granted for works on semiconductor devices.

At the beginning of the XXI century (2004) Technology Academy Finland established Millennium Technology Prize. In the first decades of this century we can observe fast development of global communication. Therefore, the awards committees honored researchers working on communications technology, optoelectronics and cryptography. In addition to appreciating the contribution to the development of computer networks and the Internet still very highly valued works were these addressing programming and development of semiconductor devices (industry).

Over a hundred years of granting the awards in the area of computer science shows that the most important areas for fast but stable computer science and information technology are related to both theory and practice, hardware and software, tests and applications.

# 4 Conclusion

Data about awarded works granted for the outstanding contributions in the field of computer science, available publicly on the prize web pages can be successfully used to know what was the most important for the development of the computer science and information technology for the last century.

Analysis of IT development based on data about prestigious awards recipients is not perfect. Some important works could be omitted, however, due to a very wide range of research, in various fields and rapidly changing knowledge and technologies in the area of computer science prestigious awards can be a source of reliable, synthetic information on particularly important works and scientific achievements.

The analysis performed by the author concerned only five chosen awards, this may limit the generalizability of the conclusions. Nowadays, due to rapid computer science and IT sector development many IT related awards were established [3]. Committees, awards promotors try to objectively assess the achievement level of candidates, however the quantitative analysis indicate that the awards comities reward a little more often outstanding works from their region.

The vast majority of computer science awards winners were men 95% (n = 214) of the honored researchers. They were born in mainly the United States (62%), United Kingdom (8,5%) and Japan (4,5%), This three countries can be seen as main centers of computer science development.

The average age of prize recipients was about 61. The oldest were recipients of Kyoto Prize (M = 68). The average age of female laureates was slightly higher than male (respectively 63 and 61 years). The youngest rewarded was Edwin H. Armstrong (27 years). The oldest laureate was Mildred Dresselhaus (85 years). We can state that working in the field of computers science must be hard and long-lasting to be honored. Achievements are appreciated only when they survive the test of time.

The text analysis of justifications for the prize indicates that over the 100 years of the computer science development main honored works were related to radio communication, programming (languages, environments), computation theory, but also mechanisms and tools for communications. The important works were related also to computer and network architecture, artificial intelligence, cryptography and optoelectronics.

During the decades the importance of various works has evolved. At the beginning of XX century the most honored were works in the area of ratio communication. In the middle of the century works on transistors, semiconductors, programming languages, television and artificial intelligence. Finally, at the turn of the 20th and 21st century the awards were granted for works on microprocessors, algorithm analysis, databases, personal computing, and the development of computer networks.

# References

1. Rapaport, W.J.: What Is Computation? (1992). https://cse.buffalo.edu/~rapaport/computation.html. Accessed 17 Mar 2019
2. Henderson, H.: Encyclopedia of Computer Science and Technology. Infobase Publishing, New York (2009)
3. IEEE Awards: List of external awards are those awards not exclusive to IEEE but of interest to IEEE. https://www.ieee.org/about/awards/external-awards.html. Accessed 11 Mar 2019
4. IEEE Internet: IEEE Internet Award webpage. https://www.ieee.org/about/awards/technical-field-awards/technical-field-awards-internet.html. Accessed 11 Mar 2019
5. IEEE Internet list: IEEE Internet Award recipient list. https://www.ieee.org/content/dam/ieee-org/ieee/web/org/about/awards/recipients/internet_rl.pdf. Accessed 11 Mar 2019
6. IEEE Medal: IEEE Medal of Honor web page. https://www.ieee.org/about/awards/medals/medal-of-honor.html#medal-recipients. Accessed 19 Mar 2019
7. IEEE Medal list: IEEE - IEEE Medal of Honor Recipients. https://www.ieee.org/content/dam/ieee-org/ieee/web/org/about/moh_rl.pdf. Accessed 21 Mar 2019
8. Kyoto Prize: Kyoto Prize in Advanced Technology web page. https://www.kyotoprize.org/en/about/fields/. Accessed 19 Mar 2019
9. Kyoto Prize list. https://www.kyotoprize.org/en/laureates/by_categories/advanced_technology/. Accessed 19 Mar 2019

10. Millennium: Millennium Technology Prize web page. https://taf.fi/millennium-technology-prize/. Accessed 18 Mar 2019
11. Millennium winners: Millennium Technology Prize winners. https://taf.fi/millennium-technology-prize/winners/. Accessed 18 Mar 2019
12. Turing Award: Turing Award web page. https://amturing.acm.org/. Accessed 17 Mar 2019
13. Turing Award list. https://amturing.acm.org/byyear.cfm. Accessed 17 Mar 2019
14. Turing bio: Alan Turing biography. https://www.biography.com/people/alan-turing-9512017. Accessed 19 Mar 2019
15. Wordcloud - free online word cloud generator. https://www.wordclouds.com. Accessed 17 Mar 2019

# Development and Validation of a New Tool to Measure the Profile of Strategic Competencies in the Organizational Practice

Beata Bajcar[(⊠)] [ID]

Faculty of Computer Science and Management, Wrocław University of Science and Technology, Wybrzeże Wyspiańskiego 27, 50-370 Wrocław, Poland
beata.bajcar@pwr.edu.pl

**Abstract.** In this study there was presented the validity process of a new tool for measure of strategic competencies. The Strategic Thinking and Behaviour Questionnaire (STBQ) includes 11 competencies: Activity, Creativity, Flexibility, Persistence, Risk preference, Self-efficacy, Analysis, Consequences prediction, Globality, Long-term planning, and Strategic evaluation. The research in a sample of 1503 workers indicated initial support for the validity and reliability of the STBQ. Internal consistency of the STBQ subscales was between .70 to .86. and high level of the validity in the context of cognitive abilities. The STBQ can be used to diagnose a profile of strategic competencies as a predictors set of the individual effectiveness in the task and goal-oriented activity in the work and management situations in the organizational practice.

**Keywords:** Strategic competencies · Strategic thinking · Strategic behaviour · Measurement

## 1 Introduction

The concept of strategic thinking is relatively well-established in management studies, which focus on its corporate aspects (e.g. strategic planning, strategic management, organization strategy, etc.), while neglecting its individual character. However, strategic thinking is undoubtedly an individual characteristic [1, 2]. Moreover, the current literature increasingly emphasizes the need for strategic thinking in organizations – as a set of managerial competencies [3–5], but also as an ability required in work at their lower levels [6]. Some human decisions and problem solving require strategic thinking, and applying different mental processes [7]. This universalization of the strategic thinking notion make its key attributes and mechanisms for the work activity.

In the definition of strategic thinking, management studies refer to the core cognitive processes involved in the creation, implementation and evaluation of behavioural strategies [3, 8]. Even from the organizational perspective, strategic thinking has a hidden individual dimension, which can be defined as a narrow or broad category. The narrow definition emphasizes the processes of generation, and creative, innovative, synthetic and intuitive thinking [8]. The broad definition of strategic thinking

© Springer Nature Switzerland AG 2020
Z. Wilimowska et al. (Eds.): ISAT 2019, AISC 1052, pp. 299–309, 2020.
https://doi.org/10.1007/978-3-030-30443-0_27

functionally combines the processes of synthesis, creativity, intuition and divergent thinking with the rational processes of problem solving (such as reflective, analytical, critical and convergent thinking) [2, 3, 9, 10]. Irrespective of the adopted level of analysis, one has to admit that a strategist should be characterized by self-efficacy, readiness to take risks, endurance, and persistence in pursuing strategic aims, innovativeness, pragmatic behaviour, as well as motivation to learn and develop. Strategic thinking allows managers to look at problems horizontally, from a broad perspective, and over a long time span [8, 11]. However, faced with the problems and challenges of professional life, not only managers, but also many workers may have to think strategically. Undoubtedly, strategic thinking can be compared to a core cognitive mechanism regulating human purposive behaviour and problem solving processes in insecure and dynamic environments [12, 13]. It can therefore be claimed that strategic competencies set is a universal construct and plays a key role in the self-management of work and task situations.

In leadership theories concerned with effectiveness, cognitive competencies and, in particular, strategic thinking, have been strongly emphasized, mainly owing to the fact that in real leadership situations, leaders must focus on processes associated with creating, evaluating, altering and implementing strategies. Therefore, strategic thinking plays a vital role in diverse management fields, including marketing strategy, strategic management, and human resources management organizations at the individual, group, and organizational levels [1, 6]. From the individual perspective, thinking and behavioural strategies are treated as dispositions (strategic competencies) or as processes (various self-regulation mechanisms). Also, strategic thinking can be considered in the context of the competence approach [5, 15] as the set of competencies include the ability to judge the situation well, to set aims and develop behaviour strategies, to predict the consequences of certain situations and behaviours, to display innovativeness, to respond to changes quickly, to cognitively reorganize the situation, to monitor the effectiveness of the strategy implemented, and, if need be, to change the behavioural strategy [16, 17]. Clearly, the concept of strategic thinking refers to that of cognitive styles [7, 18, 19], and accentuates a rational and conscious use of core competencies and core intellectual, motivational, and behavioural potentials of an individual [20]. Strong analogies may be observed between strategic competencies and metacognitive capabilities of an individual, and self-regulation mechanisms in behaviour, such as the ability to distance oneself from the situation, to think in different time perspectives, to match the suitable strategy to the problem, and, above all, to use one's intellectual resources (to solve a problem) at the right time and appropriately to the situation [16, 18].

In complex and insecure environments, any person may have to deal with shortages of information and available behaviour schemes [18] by mentally exercising behaviour strategies and methods of their implementation. The strategist's mind activates and integrates various cognitive operations and actions [2, 10, 18] which safeguard an individual's efficacy in various organizational activity. In the strategic thinking model, an individual integrates mechanisms of analysis, evaluation, and development of action plans with mechanisms for their implementation and the end result of a behaviour [21, 22]. The model relates to the deliberative and implementation nature of mechanisms involved in realizing purposive behaviours in work [23]. This integration of functionally different mechanisms is possible owing to the volitional and self-regulation processes involved in developing and implementing plans for achieving particular aims. Such a mechanism subscribes to the rational behaviour model [13] and constitutes a precondition of efficacious behaviour in all situations that involve task solving. In this paper the broad definition of strategic thinking was adopted, encompassing both strictly cognitive abilities with the motivational and behavioural factors. It is worth noting that some indices of strategic thinking are dispositional, while others are acquired and may be learned [24]. Therefore, strategic thinking is treated as a multidimensional set of competencies that determine efficient action and can be trained [15].

In the specialist literature, various conceptualizations can be found of strategic competencies as an element in the corporate management system and as a precondition of an organization effectiveness [2, 4, 19, 25, 26]. There are also some tools available to measure this construct, such as Strategy Profiler [27], Strategy Preference Indicator [28], Strategic Thinking Questionnaire [29] or Strategic Leader Development Inventory [25]. These methods are limited to investigating the strategic thinking in managers' work exclusively and include complex constructs as the dimensions of strategic thinking. The proposed tool for strategic thinking measurement involves core abilities and characteristics. An in-depth analysis of the issue of strategic thinking and competencies in management theories and in psychology has led to creating a tool to diagnose the profile of indices of strategic thinking and behaviour in different situations in the life of an individual. The author of the questionnaire adopted the broad definition of strategic thinking [2, 10], as well as general psychological concepts (specified above) accounting for the dispositional nature of strategic thinking indices. Because of the multitude of its aspects and dimensions, strategic thinking plays a key role in an individual's goal-oriented activity and in efficient problem solving in numerous domains, such as education, work, leadership, human resources development, personal and organizational counselling, and management practice.

# 2  Method

## 2.1  Participants

A total of 1503 workers completed the Strategic Thinking and Behaviour Questionnaire (STBQ), employed in various organization. First, in the preliminary study involved 243 persons (148 females, 95 males; $M = 28.4$, $SD = 10.6$), and in the validation study the 1260 workers participated (681 females, 579 males; $M = 41.2$, $SD = 10.2$). The relationships between the STBQ scales and cognitive abilities were conducted on a sample of 263 participants (155 females, 108 males; $M = 33.97$, $SD = 14.63$, age range 20–78 years).

## 2.2  Development of the STBQ

The preliminary version of the STBQ contained 160 close-ended statements, rated on a 5-point Likert scale (1 = absolutely disagree, 5 = absolutely agree). Initially, the complexity and linguistic transparency of the STBQ items were investigated ($N$ = 149). Unclear items for the subjects were simplified or removed. Subsequently, the content validity of the items was assessed by 5 competent judges (persons with psychometric expertise). In the final version of the STBQ, there are 110 items with highest CVR (content validity ratio) values [30]. Finally, the STBQ includes 11 sub-scales (per 10 items), diagnosing various strategic competencies: (1) *Activity* diagnoses activity orientation and the tendency to take initiative, e.g. 'I always undertake new challenges'; (2) *Flexibility* measures the tendency and ease in introducing changes in activities, as well as openness to and acceptance of novelty, e.g. 'Applying sudden change in action comes easily to me'; (3) *Creativity* measures the ability to search for, generate, and implement new ideas and non-stereotypical, innovative solutions, e.g. 'At work, I like to create new projects, and solutions'; (4) *Persistence* measures persistence and perseverance in progressing with activities undertaken and achievement of goals, e.g. 'I pursue set targets even if I encounter difficulties'; (5) *Risk preference* measures the level of risk preference, as well as the ease of coping with difficulties and uncertainty, e.g. 'I often take risks in life in order to achieve something in the future'; (6) *Self-efficacy* diagnoses an individual's attitude to efficacy and achievement of goals, e.g. 'I am an efficacious person'; (7) *Analysis* measures the ability to analyse strengths and weaknesses of a problem and the skills to optimize the problem-solving process, e.g. 'Before I make a choice I try to analyze all possibilities'; (8) *Globality* diagnoses the ability to think in broad terms, and the ease of assuming many different perspectives in the evaluation of goals, tasks, and activities, e.g. 'I consider every problem holistically'; (9) *Consequences prediction* measures the ability to predict own and other people's activities and consequences, from the personal and global perspective, e.g. 'Before I make a decision, I reflect upon its consequences'; (10) *Long-term planning* measures the tendency to plan and structure activity over a long period of time, e.g. 'At work, I like to plan long-term'; and (11) *Strategic evaluation* diagnoses skills to relevantly assess one's own actions with respect to personal and economic costs and available resources (as time, energy capacity, or finances), e.g. 'Before I embark on a task, I deliberate if I can cope with it'.

Due to numerous correlations between the STBQ scales the exploratory factor analysis was conducted (Principal axis method, Kaiser and Cattell criteria, Promax rotation) to extract second-order factors of the STBQ dimensions. The 1-, 2-, 3-, and 8–factorial models in a confirmatory factor analysis were tested [31]. The best fit to data was obtained the 2-factorial model (explaining 71% of variation), assessed with the use of the generalized least squares method. The revealed model was a good fit to the empirical data ($\chi^2/df = 1.56$, *RMSEA* = .03) [32]. In this model, all the standardized regression coefficients are statistically significant (see Fig. 1).

The first factor explained 53% of variation and is defined by Analysis, Globality, Consequences prediction, Long-term planning, and Strategic evaluation. The meaning content of this factor represents the very essence of strategic thinking [21]. The second factor, called strategic thinking, explained 17.9% of variation and consisted of the following dimensions: Activity, Creativity, Flexibility, Persistence, and Risk preference. These dimensions represent clearly more behavioural and motivational aspects of activity and for this reason the factor was called strategic behavioural competencies.

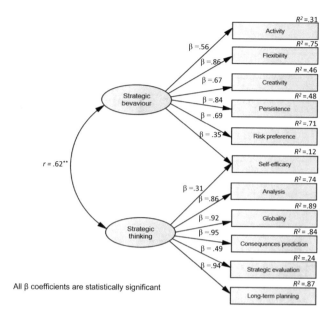

**Fig. 1.** Two factorial structure of STBQ competencies – the CFA results.

The Self-efficacy scale loaded in both factors with comparable loading values. Both factors are mutually correlated ($r = .62, p < .001$), explaining 71% of the variance. To conclude, the 2-factorial solution of the STBQ constitutes an empirical implication for a separate treatment of cognitive and behavioural indices of strategic competencies.

## 3 Results

### 3.1 The Reliability of the STBQ

The descriptive statistics, psychometric parameters of the STBQ scales, and correlation with age and cognitive dimensions resented in Table 1.

All the STBQ data had symmetrical distribution and oscillating around mean values (with kurtosis and skewness between −1 and 1). The STBQ scales displayed a high internal consistency [33] both in preliminary and validation studies. The test-retest procedure in the 2-week-period showed a high correlation of the subscales between 1st and 2nd measures ($N = 87$), indicating high stability of the STBQ [34]. Moreover, the STBQ items discrimination index was identified. The Pearson correlation analysis between items and the total scores in the STBQ subscales ranging between .20 and .72 (few items failed to achieve the .30 threshold). This means that the degree of items discrimination was acceptable [35]. To sum up, the STBQ achieved a high level of reliability

**Table 1.** Descriptive statistics, reliability coefficients, and correlations of STBQ with age and cognitive abilities.

| STBQ subscales | $M$ | $SD$ | $\alpha^1$ | Test-retest[2] | Age[2] | E-I[2] | S-N[2] | T-F[2] | J-P[2] |
|---|---|---|---|---|---|---|---|---|---|
| Activity | 37.40 | 6.29 | .81 | .80** | −.11** | .35** | −.26** | −.03 | −.16** |
| Flexibility | 36.99 | 6.03 | .84 | .90** | −.06* | .40** | −.32** | −.08 | −.26** |
| Creativity | 37.10 | 6.21 | .83 | .93** | −.06* | .33** | −.32** | −.07 | −.14** |
| Persistence | 36.21 | 5.40 | .77 | .88** | −.06* | .27** | −.06 | .01 | −.14** |
| Risk preference | 32.54 | 7.18 | .84 | .81** | −.10** | .32** | −.28** | .05 | −.20** |
| Self-efficacy | 40.46 | 4.67 | .80 | .86** | −.01 | .19** | −.02 | .07 | .13 |
| Analysis | 39.83 | 4.74 | .78 | .84** | .07* | −.05 | .25** | .26** | .35** |
| Consequences prediction | 37.62 | 5.18 | .76 | .89** | .07* | −.12* | .24* | .23** | .28** |
| Globality | 39.73 | 5.64 | .86 | .80** | .05 | −.02 | .11* | .18** | .27** |
| Long-term planning | 33.20 | 6.62 | .84 | .86** | −.06* | .05 | .07 | .07 | .26** |
| Strategic evaluation | 37.40 | 6.29 | .76 | .90** | .05 | −.17** | .12* | .17** | .23** |

*Note.* $N = 1260$, [1] Cronbach's $\alpha$; [2] Pearson correlation coefficient. E-I = Extraversion–Introversion; S-N = Sensing-Intuition; T-F = Thinking-Feeling; J-P = Judging-Perceiving.

## 3.2    The Validity of the STBQ

To verify the convergent validity of the STBQ, the correlation analysis between the STBQ dimensions and cognitive abilities were conducted (see Table 1). To measure the cognitive abilities a Polish version of Myers-Briggs Type Indicator (MBTI) [36], which contains 4 dichotomous dimensions: Extraversion-Introversion (E-I) describes where a person focus your attention; Sensing-Intuition (S-N) describes how a person takes in information; Thinking-Feeling (T-F) describes how a person make decisions; and Judging-Perceiving (J-P) expresses personal orientation to the outside world. The reliability coefficients of the MBTI subscales in this study ranged from .75 to 85.

From the analysis resulted that all dimensions of strategic behavioural competencies (*Activity, Flexibility, Creativity, Risk preference, Persistence*) correlated with Extraversion and Intuition in respect of information reception and processing. Activity-oriented strategists act quickly and spontaneously, dislike complex and routine procedures, and prefer novelty and diversity. Such individuals are prepared for changes, easily adjust to the situation, undertake many issues at the same time, and leave matters open to be able to change something in the future. Strategic thinking dimensions (*Analysis, Globality, Long-term planning, Consequences prediction, Self-efficacy, and Strategic evaluation*) positively correlated with Thinking-Feeling and Judging-Perceiving in the information processing style. This proves that strategists are analytical and sequential, rely on facts, and prefer logical order and exactness, and also well-thought and planned activity. No relationships were found between the STBQ scales with the Extraversion-Introversion and Sensing-Intuition dimensions. Moreover, age significantly but weakly correlated with STBQ dimensions.

To sum up, the STBQ dimensions achieved a high degree of the validity in the context of cognitive abilities. This demonstrates that strategic competencies are to a large extent internally conditioned and dispositional. This finding corroborates the results of research to date claiming that professional efficacy and task efficacy depend on emotional stability, reflectiveness and the ability to deal with difficult situations, as well as meticulousness and good organization and enduring motivation to achieve aims and complete tasks). The results of the analysis prove that the specificity of strategic thinking and behaviour dimensions relates to different cognitive, motivational and behavioural processes. Strategic thinking dimensions represent convergent, logical, and analytical mechanisms, while strategic behaviour dimensions are cognitive-motivational and behavioural in character and relate to divergent processes [2, 10].

### 3.3    Individual Profiles of Strategic Competencies

To verify the validity of the STBQ, the differences analysis of their subscales was conducted in the dependence of gender. The ANOVA results showed that men displayed a significantly higher level of Activity [$F(1,1258) = 6.02, p = .014$], Flexibility [$F(1,1258) = 7.45, p = .006$], Persistence [$F(1,1258) = 16.77, p < .001$], and Risk preference [$F(1,1258) = 34.88, p < .001$]; as compared with women. This means that men are more open to and tolerant of change and more easily adjusted to dynamic environments. In addition, men were more persistent in undertaking activities and in achieving goals. As compared with women, men have a significantly higher level of risk preference, better cope with insecurity and unpredictable situations. The results demonstrated that gender is a differentiating criterion for the profile of strategic behavioural competencies. These differences portray an evolutionally determined image of men as individuals prepared for challenges and capable of high endurance in dealing with adversities. Probably, behavioural strategic competencies of men and women are conditioned by different social expectations concerning fulfilment of gender roles in the social and professional life. No gender differences were observed in cognitive strategic competencies, which would imply other sources of their determination than gender differences. These findings indicate that the strategic behavioural competencies are conditioned by differences in the structure and functions of brain components [16, 18].

To verify the discriminant validity of the SBTQ, the $k$-means cluster analysis was conducted [37]. As a result, four subgroups differing in the patterns of strategic competencies have been extracted: (1) Activists, (2) Thinkers, (3) Passivists, (4) Strategists. Each group is characterized by a different level of strategic competencies (see Fig. 2).

As can be seen in Fig. 2, the first cluster are Thinkers – characterized by high scores in strategic thinking and low scores with respect to the strategic behaviour dimensions, expressing deliberative thinking style as a set of strategic competencies. This means that Thinkers limit their activity to the sphere of thinking and conceptual interest, without any behavioural consequences and idea implementation. The second cluster – Activists – had high scores in strategic behaviour and low scores in strategic thinking dimensions. They are strongly activity-oriented, more creative and flexible, and also manifest a higher threshold of Persistence and Risk preference. This means that Activists focus on the implementation and maintenance of an activity, at the same time decreasing the intensity of conceptual work. The above configuration of the STBQ

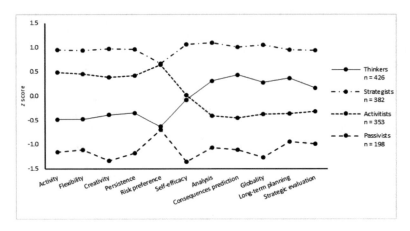

**Fig. 2.** The STBQ profiles – clustering results.

dimensions expresses the implementation style in the strategic competencies profile. The next cluster included persons with a high level in all the dimensions of strategic thinking and behaviour, namely Strategists. Their set of strategic competencies expresses a pragmatic quest for efficiency and for adapting the activity to the situation requirements. Furthermore, they have a great ability to integrate cognitive activity with actuated behaviour. Strategists easily integrate the effects of analytical processes with practical implementation. Therefore, they can easier adjust to a particular situation and they deal with obstacles fast and effectively. Strategic competencies indicate a high intellectual potential, which may lead to high efficacy of the activity. The last cluster – Passivists – was characterized by low scores in all dimensions of strategic thinking and behavioural competencies. The most passive style expresses the subject's tendency to preserve the *status quo*, to go with the flow, or to await the course that events will take.

The revealed patterns of strategic competencies vary in the intense and configuration of the skills. The differences may derive from both the individual characteristics and various profiles of work activity and taking tasks. This is very important for recruitment, evaluating human potential in organizational efficiency, and career planning.

### 3.4    Strategic Competencies Profiles of Managers and Employees

To obtain a fuller picture of the discrimination power of the STBQ, the measure in managers' sample was compared with the measure in employees sample. The hypothesis was that managers displayed a significantly higher level of strategic competencies. The study was conducted among 297 top- and mid-level managers from large and small domestic and international companies, aged 25–65 ($M = 42.57$, $SD = 11.68$). An analysis of variance demonstrated that managers scored significantly ($p < .001$) higher on all STBQ scales compared with the general population (see Fig. 3).

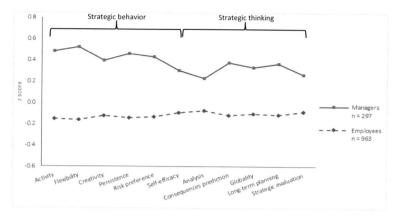

**Fig. 3.** The STBQ dimensions in managers and employees samples.

The results proved that managers are highly competent with regard to conceptual activity, as well as analytical and strategic thinking, as required in corporate management. In all likelihood, the professional activity of managers entails dealing with complex problems, which fosters frequent application of various cognitive mechanisms and engagement in creative, innovative and long-term tasks. Thus, the STBQ may be regarded as a good diagnostic tool for measuring strategic competencies, allowing to estimate a particular profile of a person's activity and the level of efficacy in solving individual or corporate problems. Moreover, the majority of managers are strategists (46% of respondents), although there are other strategic competence patterns, varying in the degree of adaptability [17]. The differences may derive from both the individual characteristics and from various profiles of human activity and taking tasks. It is very important for recruitment, evaluation of managerial potential in organizational efficiency and path career planning.

## 4   Conclusions

The presented validation results demonstrated that the STBQ is a tool for measuring strategic competencies with a high degree of reliability, internal and external validity. The diversity of the STBQ makes it possible to diagnose a multidimensional set of competencies useful in solving complex cognitive, emotional, and interpersonal problems, as well as in dealing with information deficit, with lack of available behaviour schemes, with interpersonal conflicts, with insecurity, and with complex and new tasks in work and management situations. A numerous relationships between strategic competencies and cognitive abilities suggest that model of the strategic competencies may constitute a set of predictors of efficiency in complex and new tasks solving, in work and management settings. However, strategic competencies are treated mainly as cognitive resources of an organization. This bears important implications for the validity of evaluating managerial and work efficiency, and for the process of recruitment, selection, and support for individual employees career development in an organization and in the changing environment. Strategic thinking competencies, as a high-order cognitive skills configuration, constitute an intellectual potential of an organization and play an important role in human resources management, and in building the strategy and competitive advantage of the organization [6, 38, 39]. Moreover, human resources professionals should help develop the strategic thinking competencies of the workers by facilitating the processes of professional development and career path achievement [14, 40]. Strategic competencies as an individual potential help in coping on the labour market and in the disclosure entrepreneurial behaviour for all workers. Empirical results indicated that the organizations that created and implemented strategic thinking culture were characterized by high level of organizational and economic development [4, 9].

# References

1. Bonn, I.: Improving strategic thinking: a multilevel approach. Leadersh. Organ. Dev. J. **26** (5), 336–354 (2005)
2. Liedtka, J.M.: Strategic thinking: can it be taught? Long Range Plan. **31**, 120–128 (1998)
3. Heracleous, L.: Strategic thinking or strategic planning. Long Range Plan. **31**(3), 481–487 (1998)
4. Moon, J.B.: Antecedents and outcomes of strategic thinking. J. Bus. Res. **66**, 1698–1708 (2013)
5. Goldman, E., Scott, A.R., Follman, J.M.: Competency models for assessing strategic thinking. J. Strategy Manag. **9**(3), 258–280 (2016)
6. Goldman, E.: The power of work experiences: characteristics critical to developing expertise in strategic thinking. Hum. Resour. Dev. Q. **19**(3), 217–239 (2008)
7. Gallén, T.: The cognitive style and strategic thinking. In: Proceedings of the Leadership and Myers-Briggs Type Indicator, Washington DC, USA, pp. 25–30 (1999)
8. Mintzberg, H.: The rise and fall of strategic planning. Harvard Bus. Rev. **72**(1), 107–114 (1994)
9. Bonn, I.: Developing strategic thinking as a core competency. Manag. Decis. **39**(1), 63–76 (2001)
10. Wilson, I.: From scenario thinking to strategic action. Technol. Forecast. Soc. Chang. **65**(1), 23–29 (2000)
11. Raymond, G.A.: Elements of strategic thinking. Int. Stud. Rev. **10**, 315–317 (2008)
12. Carver, C.S., Scheier, M.F.: Attention and Self-regulation: A Control Theory Approach to Human Behavior. Springer, New York (1981)
13. Fishbein, M., Ajzen, I.: Belief, Attitude, Intention, and Behavior: An Introduction to Theory and Research. Addison-Wesley Publishing Company, Reading (1975)
14. Goldman, E.F., Scott, A.R., Follman, J.M.: Organizational practices to develop strategic thinking. J. Strategy Manag. **8**(2), 155–175 (2015)
15. Raven, J.: Competence in Modern Society: Its Identification, Development and Release. Oxford Psychologists Press, Oxford (1984)
16. Sternberg, R.J.: WICS: a model of leadership. Psychol.-Manag. J. **8**(1), 29–43 (2005)
17. Bajcar, B.: Are all managers strategists? Thinking and behavioural styles of Polish managers. In: Marek, T., Karwowski, W., Frankowicz, M., Kantola, J., Zgaga, P. (eds.) Human Factors of a Global Society: A System of Systems Perspective, pp. 525–538. CRC Press, London (2014)

18. Nosal, C.S.: Psychologia myślenia i działania menedżera (Psychology of manager's thinking and action). AKADE, Wrocław (2001)

19. Dragoni, L., Oh, I.S., Vankatwyk, P., Tesluk, P.E.: Developing executive leaders: the relative contribution of cognitive ability, personality, and the accumulation of work experience in predicting strategic thinking competency. Pers. Psychol. **64**(4), 829–864 (2011)

20. Oelkers, G., Elsey, B.: The 'Strategic Magnifier' – a cognitive tools for strategic thinking. Probl. Perspect. Manag. **3**, 196–209 (2004)

21. Mintzberg, H.: Managers Not MBAs. A Hard Look at the Soft Practice of Managing and Management Development. Berrett-Koehler Publishers, San Francisco (2004)

22. O'Shannassy, T.: Modern strategic management: balancing strategic thinking and strategic planning for internal and external stakeholders. Singap. Manag. Rev. **25**(1), 53–67 (2003)

23. Gollwitzer, P.M., Fujita, K., Oettingen, G.: Planning and the implementation of goals. In: Baumeister, R., Vohs, K. (eds.) Handbook of Self-regulation: Research, Theory and Applications, pp. 211–228. Guilford Press, New York (2004)

24. Benito-Ostolaza, J.M., Sanchis-Llopis, J.A.: Training strategic thinking: experimental evidence. J. Bus. Res. **67**(5), 785–789 (2014)

25. Jacobs, T.O.: A Guide to the Strategic Leader Development Inventory. National Defense University Industrial College of the Armed Forces, Washington (1994)

26. Jelenc, L., Swiercz, P.M.: Strategic thinking capability: conceptualization and measurement. In: Conference Paper on the 56th Annual ICSB World Conference, Stockholm (2011)

27. Meyer, R.: Mapping the Mind of the Strategist. A Quantitative Methodology for Measuring the Strategic Beliefs of Executives. Series Research in Management, no. 106 (2007)

28. Reece, S.: Strategic thinking. Leadersh. Excel. **25**(1), 19 (2008)

29. Pisapia, J., Reyes-Guerra, D., Coukos-Semmel, E.: Developing the leader's strategic mindset: establishing the measures. Leadersh. Rev. **5**, 41–68 (2005)

30. Lawshe, C.H.: A quantitative approach to content validity. Pers. Psychol. **28**, 563–575 (1975)

31. Brown, T.A.: Confirmatory Factor Analysis for Applied Research. Guilford, New York (2006)

32. Jöreskog, K., Sörbom, D.: Recent developments in structural in equation modelling. J. Mark. Res. **19**, 404–416 (1982)

33. Cronbach, L.J.: Coefficient alpha and the internal structure of tests. Psychometrika **16**, 297–334 (1951)

34. Peterson, R.A.: A meta-analysis of Cronbach's coefficient alpha. J. Consum. Res. **21**(2), 381–391 (1994)

35. Whitney, D.R., Sabers, D.L.: Two generalizations of the item discrimination index to multi-score items. J. Exp. Educ. **39**(3), 88–92 (1971)

36. Myers, I.B.: Introduction to Type. Consulting Psychologists Press, Palo Alto (1987)

37. Steinley, D.: K-means clustering: a half-century synthesis. Br. J. Math. Stat. Psychol. **59**(1), 1–34 (2006)

38. Fairholm, M.R.: Leadership and organizational strategy. Innov. J.: Public Sect. Innov. J. **14**(1), 1–16 (2009)

39. Friedman, L., Gyr, H.: Creating the Dynamic Enterprise: Strategic Thinking Tools for HR Practitioners. The Enterprise Development Group, Jossey-Bass/Pfeiffer (1998)

40. Goldman, E., Casey, A.: Building a culture that encourages strategic thinking. J. Leadersh. Organ. Stud. **17**, 119–128 (2010)

# Author Index

© Springer Nature Switzerland AG 2020
Z. Wilimowska et al. (Eds.): ISAT 2019, AISC 1052, pp. 311–312, 2020.
https://doi.org/10.1007/978-3-030-30443-0

Printed in the United States
By Bookmasters